BREAKDOWN IN HUMAN ADAPTATION TO 'STRESS' VOLUME II

BREAKDOWN IN HUMAN ADAPTATION TO 'STRESS'
Towards a multidisciplinary approach

VOLUME I

Part 1: Psychological and sociological parameters for studies of breakdown in human adaptation
J. Cullen and J. Siegrist (editors)

Part 2: Human performance and breakdown in adaptation
H.M. Wegmann (editor)

VOLUME II

Part 3: Psychoneuroimmunology and breakdown in adaptation: interactions within the central nervous system, the immune and endocrine systems
R.E. Ballieux (editor)

Part 4: Breakdown in human adaptation and gastrointestinal dysfunction: clinical, biochemical and psychobiological aspects
J.F. Fielding (editor)

Part 5: Acute effect of psychological stress on the cardiovascular system: Models and clinical assessment
A. L'Abbate (editor)

Compendium of papers presented in workshops sponsored by the Commission of the European Communities as advised by the Committee on Medical and Public Health Research. (Dublin, Ireland (Dec. 1982), Utrecht, the Netherlands (Dec. 1982), Köln, FRG (Jan. 1983), Bad Homburg, FRG (Feb. 1983), Dublin, Ireland (March 1983) and Pisa, Italy (April, 1983))

BREAKDOWN IN HUMAN ADAPTATION TO 'STRESS'

Towards a multidisciplinary approach

VOLUME II

Part 3: Psychoneuroimmunology and breakdown in adaptation: interactions within the central nervous system, the immune and endocrine systems

R.E. Ballieux (editor)

Dept. of Clinical Immunology, University Hospital, Utrecht, The Netherlands

Part 4: Breakdown in human adaptation and gastrointestinal dysfunction: clinical, biochemical and psychobiological aspects

J.F. Fielding (editor)

Dept. of Gastroenterology, The Charitable Infirmary, Dublin, Ireland

Part 5: Acute effect of psychological stress on the cardiovascular system: Models and clinical assessment

A. L'Abbate (editor)

Dept. of Clinical Physiology, National Research Council, Pisa, Italy

1984 **Springer-Science+Business Media, B.V.**

Distributors

for the United States and Canada: Kluwer Boston, Inc., 190 Old Derby Street, Hingham, MA 02043, USA
for all other countries: Kluwer Academic Publishers Group, Distribution Center, P.O.Box 322, 3300 AH Dordrecht, The Netherlands

Library of Congress Cataloging in Publication Data

Main entry under title:

Breakdown in human adaptation to "stress."

"Compendium of papers presented in workshops
sponsored by the Commission of the European Communities
as advised by the Committee on Medical and Public Health
Research.(Dublin, Ireland (Dec. 1982), Utrecht, the
Netherlands (Dec. 1982), Köln, FRG (Jan. 1983), Bad
Homberg, FRG (Feb. 1983), Dublin, Ireland (March 1983),
and Pisa, Italy (April 1983))"
    Vol. 2.edited by R.E. Ballieux, J.F. Fielding, and
A. L'Abbate.
    Contents: v. 1. pt. 1. Psychological and sociological
parameters for studies of breakdown in human adaptation
J. Cullen and J. Siegrist, editors. pt. 2. Human performance and breakdown in adaptation / H.M. Wegmann, editor
-- v. 2. pt. 3. Psychoneuroimmunology and breakdown in
adaptation / R.E. Ballieux, editor -- [etc.]
    1. Medicine, Psychosomatic--Addresses, essays, lectures.
2. Adaptation (Physiology)--Addresses, essays, lectures.
3. Stress (Psychology)--Addresses, essays, lectures.
I. Cullen, John H. II. Ballieux, R.E. III. Commission
of the European Communities. IV. Commission of the
European Communities. Committee on Medical and Public
Health Research. [DNLM: 1. Stress, Psychological--
Complications--Congresses. 2. Adaptation, Psychological
--Congresses. 3. Disease--Etiology--Congresses. WM 172
B828 1982-82]
RC49.B727 1983   616.08   83-19333

EUR 8943 II EN

ISBN 978-94-010-7975-4   ISBN 978-94-009-3285-2 (eBook)
DOI 10.1007/978-94-009-3285-2

Book information

Publication arranged by: Commission of the European Communities, Direc-
torate-General Information Market and Innovation, Luxembourg

Copyright/legal notice

© 1984 by Springer Science+Business Media Dordrecht
   Originally published by ECSC, EEC, EAEC, Brussels-Luxembourg 1984
   Softcover reprint of the hardcover 1st edition 1984

PREFACE

The widespread interest in "stressful" aspects of contemporary society which contribute to its burden of illness and diseases (e.g. gastro intestinal, cardiovascular) has led to a large number of statements and reports which relate the manifestations to a maladaptation of the individual. Furthermore, recent research suggests that under some conditions stress may have a more generalized effect of decreasing the body's ability to combat destructive forces and expose it to a variety of diseases.

Breakdown in adaptation occurs when an individual cannot cope with demands inherent in his environment. These may be due to an excessive mental or physical load, including factors of a social or psychological nature and task performance requirements ranging from those which are monotonous, simple and repetitive to complex, fast, decision-taking ones. Experience shows however that not all people placed under the same conditions suffer similarly, and it follows that to the social and psychological environment should be added a genetic factor influencing, through the brain, the responses of individuals.

It is clear that, besides human suffering, this "breakdown in adaptation" causes massive losses of revenue to industry and national health authorities. Thus a reduction in "stress", before "breakdown" occurs, or an improvement in coping with it would be very valuable.

An area of the Commission of the European Communities multiannual coordination programme in Medical and Public Health research considers "breakdown in human adaptation" occurring when facing the industrial life style in countries of the European Community or a too rapid change of the urban/industrial environment. The problem is of great economic importance from the points of view of industrial productivity, movement of workers and health care of all the population.

The principal aim of the programme is to determine a series of measurable parameters which could indicate, first a-posteriori but then a-priori the tendency or initiation of breakdown in adaptation and the form it may take if no counter measure was introduced. Such a programme, it is hoped, could contribute to a wider integration of European know-how, promote a better understanding of biological parameters underlying organic correlates of stress, and lead eventually to applied routine procedures. Harmonized and standardized measurements would give values for parameters from various laboratories in the European Community; these will be compared, integrated in a wider context and, results fed back at regular intervals to scientists taking part in the exercise for discussion and interpretation.

The broad outline of this programme calls for studies to:

1.- Investigate, simultaneously if possible, the sociological, psychological and endocrinological factors which may indicate "high risk groups" among the population,

2.- Develop reliable but relatively simple indicators which will allow the identification of persons likely to suffer a breakdown in adaptation,

3.- Establish the mechanisms by which certain factors lead to a higher risk in adaptive breakdown and the manifestations of performance, cardiovascular and gastro intestinal dysfunction,

4.- Use the information gathered to establish situations which will avoid or minimize breakdown in adaptation.

It is evident that to be successful, the combined expertise of researchers in biology, biochemistry, physiology, sociology, psychology, epidemiology and clinical sciences would be required. Through coordinated studies, these scientists could adapt and compare methods, follow results and test hypotheses since standardized or at least harmonized protocols would have been used by the various disciplines. This coordination will, it is hoped, allow the identification of real gaps and make specific recommendations as research develops to achieve the aims stated.

In this context one should mention studies to determine whether or not different categories or causes of adaptive breakdown are selectively correlated with pathogenic tendencies at the level of specific organs. The answer to this question would greatly facilitate the assessment of the relevancy of specific versus non specific organic markers of potential diseases.

Finally, because the programme would be directed towards one single but crucial goal with a well coordinated multinational and interdisciplinary approach, it is hoped to establish firm, productive and long lasting links between scientists, in Europe and elsewhere.

The publications in this series stem from a number of workshops held in 1982-1983 under the auspice of the Commission of the European Communities in the framework of the above mentioned programme, as advised by the Committee for Medical and Public Health Research.

These are not proceedings, but a compendium of recent overviews, in some cases examples of methods, and reports on specific topics, or approaches used or envisaged in the various disciplines where there could be fruitful collaboration in the overall study of breakdown in human adaptation to some stressful conditions. The purpose is to help identify important problems, indicate interdisciplinary research possibilities offered and through the references quoted make available sources of further information on methods and aspects outside the normal specific field of the scientist concerned.

The separation into five parts each dealing more specifically with one aspect is only done for the ease of the reader and the publication should be considered as a whole.

The Commission of the European Communities and its Committee for Medical and Public Health Research wish to thank all those who contributed through their participation in the elaboration of this programme and the organization of the meetings.

TABLE OF CONTENTS

VOLUME I

PART 5 — ACUTE EFFECT OF PSYCHOLOGICAL STRESS ON CARDIOVASCULAR

SYSTEM: MODELS AND CLINICAL ASSESSMENT

I. SYSTEMS INTERPLAY IN STRESS RESPONSE

LIST OF CONTRIBUTORS

J.Aagard,
Institute of psychiatric demography
Aarhus psychiatric hospital,
DK-8420 Riskov, Denmark

A.L.'Abbate,
Fisiologia Clinica,Istituto del
Consiglio Nazionale delle Ricerche
presso l'Università degli Studi di
Pisa,
Via Savi 8, Pisa, Italia.

R. Ader,
University of Rochester,
300 Crittenden Bvard.,
Rochester, NY 14642, USA.

T. Akerstedt,
Laboratory for clinical stress
research,
National Institute for psychosocial
factors and health,
10401 Stockholm, Sweden.

C.P. Allott,
Bioscience Department, Imperial
Chemical Industries, Plc,
Pharmaceuticals Division,
Alderley Park, Macclesfield,
Cheshire, U.K.

D. Alpers,
Division of gastroentrology,
Washington University school of
Medecine,
660 S. Euclid Ave.
St. Louis, MO 63110, USA.

F. Amenta,
Centro internazionale radiomedico,
CIRM,
via dell'Archittetura 41,
I-00144 ROMA, Italy

A. Appels,
State University of Limburg,
School of Medecine, Dept. of medical
psychology,
MAASTRICHT, The Netherlands

J. Armour,
Dalhousie University, Department of
physiology and biophysics,
Halifax, Nuova Scotia B3H 4H7,
Canada

G.Athanassenas,
State mental hospital,Psychiatric
department III, DAPHNI-ATHENS,Greece

R.Ballieux,
Department of clinical immunology,
University hospital,
Catharijnesingel 101,
3511 GV Utrecht, The Netherlands

E. Bassenge,
Albert Ludwigs Universität,
Herman Herder Str. 7,
D-7800 Freiburg 1, FRG.

J. Bennett,
Hull Royal infirmary,
Anlaby Road,
Kingston upon Hull HU3 2JZ, U.K.

G. Bertinieri,
Università di Milano,
Istituto di clinica medica IV,
via F.Sforza 35,
20122 MILANO, Italy

G. Bertolotti,
Centro di riabilitazione di Veruno,
VERUNO (NO), Italy

A. Biaggini,
Institute of clinical physiology-CNR
Via Savi 8,
I-56100 PISA, Italy

D. Blom,
Netherlands institute for preventive
health care / TNO,
P.O.B. 124,
2300 AC Leiden, The Netherlands

H. Boehm,
Bremen polytechnic, Department of
nautical sciences,
Werderstr. 73,
2800 Bremen, FRG.

B. Bohus,
Department of animal physiology,
University of Groningen,
P.O.B. 14,
9750 AA Haren, The Netherlands.

N. Boon,
Cardiac department, John Radcliffe
Hospital,
Headington, Oxford, U.K.

P. Branton,
22 Kings Gardens,
London NW6 4PU, U.K.

D.Broadbent,Medical Research Council,
University of Oxford, Oxford, U.K.

F. Brooks,
University of Pennsylvania hospital,
Department of medecine .GI section,
574A Maloney Bdg.,3400 Spruce St.
Philadelphia, Pa 19104, USA

A. Burette
Université Libre de Bruxelles,
Hopital Brugmann, G.I. departement
Bruxelles, Belgium

R. Busse
Albert Ludwigs Universität,
Herman Herder Str. 7,
D-7800 Freiburg 1, FRG.

G. Cesana,
Istituto di medicina del lavoro,
Clinica del lavoro L.Devoto dell'
università,
Via S. Barnaba 8,
20122 Milano, Italy

J. Cloix,
INSERM, U 7, CNRS LA 318,
Faculté de médecine, departement
de pharmacologie,
Necker-Enfants malades,
156 rue de Vaugirard,
75015 Paris, France

N. Cohen,
Agricultural university, Department
of experimental animal morphology and
cell biology,
Wageningen, The Netherlands

W. Colquhon,
Medical Research Council percep-
tual and cognitive performance unit,
Laboratory of experimental psycho-
logy, University of Sussex,
Brighton, Sussex BN1 9QG, U.K.

R. Condon,
Medical Research Council perceptual
and cognitive performance unit,
Laboratory of experimental
psychology,
University of Sussex,
Brighton, Sussex BN1 9QG, U.K.

J. Conway,
Cardiac department, John Radcliffe
hospital,
Headington, Oxford, U.K.

T. Cox,University of Nottingham,
Department of psychology - stress
research, Nottingham NG7 2RD, U.K.

J. Cullen,
Health care and psychosomatic unit,
Garden Hill, E.H.B. Box 41A,
Dublin 8, Ireland

M. Deltenre,
Université Libre de Bruxelles,
Hopital Brugmann, GI Department
Bruxelles, Belgium

M. Devynck,
Inserm U7, CNRS LA 318, Faculté
de médecine Necker-Enfants malades
156 Rue de Vaugirard,
75015 Paris, France

N. van Dijkhuizen,
SWO/DPKM, Ministry of defence,
P.O.B. 20702,
2500 ES Den Haag, The Netherlands

K. Dittmann,
Institute of medical sociology,
Faculty of medecine,
University of Marburg,
Marburg, FRG.

L. Donato,
CNR Institute of clinical physio-
logy and institute of pathology,
University of Pisa,
Via Savi 8,
56100 Pisa, Italy

A. von Eiff,
Medizinische Universitätsklinik,
Venusberg, D-5300 Bonn 1, FRG.

J. Elghozi,
INSERM U7, CNRS LA 318, Faculté
de médecine Necker-Enfants malades
156 Rue de Vaugirard,
75015 Paris, France

P. Falger,
State University of Limburg school
of medecine, Department of medical
psychology,
Maastricht, The Netherlands

S. Farrow,
Department of epidemiology and
community medecine, Welsh National
School of medecine, Heath Park,
Cardiff, Wales, U.K.

J. Fielding,
Department of medecine and gastroentorology, The Charitable Infirmary, Jervis St.,Dublin 1, Ireland

H. Foushee,
Man-vehicle systems research division, NASA, Ames Research center,
Moffett Field, California 94035 USA

D. Ganten,
German Institute for high blood pressure research and Department of
Pharmacology, University of
Heidelberg, Im Neuenheimer Feld 366
D-6900 Heidelberg, FRG.

M. Giesler,
Albert Ludwigs Universität,
Herman Herder Str. 7,
D-7800 Freiburg, FRG.

A. Giordano,
Centro di riabilitazione di Veruno
Fondazione clinica del lavoro
Veruno (NO), Italy

H. Goethe,
Bernhard -Nocht-Institute for nautical and tropical diseases, Department of nautical medecine,
2000 Hamburg, FRG.

A. Goldstein,
Department of biochemistry,
The George Washington University
College of health sciences,
Washington DC 20037, USA.

R. Gorczynski,
Ontario cancer institute,
500 Sherbourne St.
Toronto, Ontario M4X 1K9, Canada

R. Graeber,
Man-vehicle systems research division, NASA, Ames research center,
Moffett Field, California 94035 USA.

D.T. Greenwood,
Bioscience Department, Imperial
Chemical Industries Plc, Pharmaceutical division, Alderley Park,
Macclesfield, Cheshire, U.K.

A. Grieco,
Istituto di medicina del lavoro,
Clinica del lavoro L. Devoto dell'
Università, via S. Barnaba 8,
20122 Milano, Italy

N.Hall,
Department of biochemistry, The
George Washington University, college of health sciences,Washington
DC 20037, USA.

F. Hawkins,
Schiphol airport,
P.O.B. 75577,
1118 ZP Amsterdam, The Netherlands

D. Healy,
Pregnancy research branch, National
Institute of child health and development,
Bethesda, Maryland 20205, USA.

H. Hermann,
ESSO A.G.
Kapstadring 2,
D-2000 Hamburg 60, FRG.

A. Holte,
Institute of behavioural sciences
in medecine, University of Oslo,
P.O.B. 1111, Blindern, Oslo 3,
Norway

J. Holtz,
Albert Ludwigs Universität,
Herman Herder Str. 7
D-7800 Freiburg 1, FRG.

G. Huse-Kleinstoll,
Department of Medical Psychology
Medical Clinic, University Hospital
Hamburg-Eppendorf, FRG.

E. Isaksen,
Institute of physiological psychology, University of Bergen,
Bergen, Norway

L. Jacomini,
INSERM U7, CNRS LA 318, Faculté de
médecine Necker-Enfants malades,
156 Rue de Vaugirard,
75015 Paris, France

C. Jenkins,
Department of preventive medecine
and community health, University of
Texas medical branch,
Galveston, Texas 77550, USA.

L. Kamal,
INSERM U7, CNRS LA 318, Faculté de
médecine Necker-Enfants malades,
156 Rue de Vaugirard,
75015 Paris, France

M. Kennedy,
Ontario cancer institute,
500 Sherbourne St.,
Toronto, Ontario M4X 1K9, Canada

Margit von Kerekjarto,
Universitäts-Krankenhaus Eppendorf
Medizinische Klinik,
Martinistr. 52,
2000 Hamburg, FRG.

F. Klimmer,
Institut für Arbeitsphysiologie an
der Universität Dortmund,
Ardeystr. 67,
D-4600 Dortmund 1, FRG.

P. Knauth
Institut für Arbeitsphysiologie an
der Universität Dortmund,
Ardeystr. 67,
D-4600 Dortmund 1, FRG

J. Koolhaas,
Department of animal physiology,
State University of Groningen,
P.O.B. 14,
9750 AA Haren, The Netherlands

K. Kraft,
German Institute for high blood pres-
sure research and Department of
Pharmacology, Universität of
Heidelberg,
Im Neuenheimer Feld 366,
D-6900 Heidelberg,FRG.

Th. Küchler,
Department of medical psychology,
University hospital,
Hamburg-Eppendorf, FRG

R.E. Lang,
German Institute for high blood pres-
sure research and Department of
Pharmacology, University of
Heidelberg,
Im Neuenheimer Feld 366,
D-6900 Heidelberg, FRG.

J. Lauber,
Man-vehicle systems research divi-
sion, NASA, Ames research center,
Moffett Field, California 94035 USA

M. Lemke,
Volkswagenwerk AG, Nutzfahrzeug-
Entwicklung,
D-3180 Wolfsburg, FRG.

A. Low,
Bernard Nocht Institute for nautical
and tropical diseases, Department of
nautical medecine,
2000 Hamburg, FRG.

R. Lulofs,
State University of Limburg school
of medecine, Department of medical
psychology,
Maastricht, The Netherlands

G. Lyketsos,
Dromokaiton mental hospital,
Athens, Greece

S. MacRae
Ontario cancer Institute,
500 Sherbourne St.,
Toronto, Ontario M4X 1K9 Canada

C. Mackay,
Medical Division, Health and safety
executive,
25 Chapel St.,
London NW1 5DT, U.K.

G. Mancia,
Università di Milano, ospedale
Maggiore, Istituto di clinica
medica IV, via F. Sforza 35,
20122 Milano, Italy

M. Marmot,
London School of Hygiene and tro-
pical medecine,
Keppel St.,
London WC1 7HT, U.K.

D.W. Marshall,
Bioscience Department, Imperial
Chemical Industries Plc, Pharma-
ceuticals Division, Alderley Park,
Macclesfield, Cheshire, U.K.

G. Mazzuero,
Centro di riabilitazione di Veruno,
Fondazione clinica del lavoro,
Veruno (NO), Italy

J. McGillis,
Department of biochemistry, The
George Washington University
college of health sciences,
Washington DC 20037, USA.

K. McIntyre,
Harvard University School of Public
health,
Boston, Massachusetts 02115, USA

P. Meyer,
INSERM U7, CNRS LA 318, Faculté de
Médecine Necker-Enfants malades,
156 Rue de Vaugirard,
75015 Paris, France

Aslaug Mikkelsen,
Institute of behavioural sciences in
medecine, University of Oslo,
P.O.B. 1111, Blindern,
Oslo, Norway

S. Mohler,
Wright State University School of
Medecine,
Dayton, Ohio 45401, USA.

J. Moraal,
Institute for perception TNO,
3769 ZG Soesterberg, The Netherlands

R. Murison,
Institute of physiological psycholo-
gy, University of Bergen,
Bergen, Norway

R. Mykletun,
Rogoland Research Institute,
Stavanger, Norway

H. Neus,
Medizinische Universitätsklinik,
Venusberg, D-5300 Bonn 1, FRG.

H. Nichamin,
Wright State University School of
Medecine, Dayton, Ohio 45401, USA.

A. Nicholson,
Royal Air Force Institute of
Aviation Medecine,
Farnborough, Hampshire, U.K.

G. Parati,
Universitá di Milano, Istituto di
clinica medica IV, via F. Sforza 35,
20122 Milano, Italy

Katherine Parkes,
Department of experimental psychology
University of Oxford, Oxford, U.K.

A. Pena,
University hospital, Department of
gastroentorology,
Leiden, The Netherlands

M. Pernollet,
INSERM U7, CNRS LA 318, Faculté de
Médecine Necker-Enfants malades,
156 Rue de Vaugirard,
75015 Paris, France

A. Perski,
National Institute for psychosocial
Factors in Health, P.O.B. 60210,
10401 Stockholm, Sweden

W. Pierpaoli,
Institute for integrative bio-
medical research,
Ebmatingen, Switzerland

M. Pokorny,
Netherlands Institute for preventive health care/TNO,
P.O.B. 124,
2300 AC Leiden, The Netherlands

A. Raedler,
Department of Medical psychology
University hospital,
Hamburg-Eppendorf, FRG.

A. Ramirez,
Università di Milano, Istituto di
clinica medica IV,
Via F. Sforza 35,
20122 Milano, Italy

M. de Reuck,
Université Libre de Bruxelles,
Hopital Brugmann, GI Department
Bruxelles, Belgium

J. Riemersma,
Institute for perception / TNO
3769 ZG Soesterberg,The Netherlands

N. Rizzo,
Centro Internazionale Radiomedico
C.I.R.M.,
via dell'Architettura 41,
00144 Roma, Italy

D. Russell,
Cardiovascular Research Unit,
University of Edinburgh
Edinburgh, U.K.

J. Rutenfranz,
Institut für Arbeitsphysiologie an
der Universität Dortmund,
Ardeystr. 67,
4600 Dortmund1, FRG.

E. Sand,
Université Libre de Bruxelles,
Campus Erasme, Laboratoire
d'épidemiologie et de Médecine
sociale,
880 Route de Lennik,
1070 Bruxelles, Belgium

P.Schwartz,
Institute of cardiovascular research
Giogio Sisini, University of Milano
via F.Sforza 35, 20122 Milano,Italy

XXII

W. Schulte,
Medizinische Universitätsklinik,
Venusberg, D-5300 Bonn, FRG.

K. Schulz,
Department of medical psychology,
University hospital
Hamburg-Eppendorf, FRG.

S. de Servi,
Division of Cardiology,
Policlic S. Matteo, University of
Pavia,
27100 Pavia, Italy

Patricia Shipley,
Department of occupational psycho-
logy, Birkbeck College,
University of London,
Malet Street,
London WC1 E7HX, U.K.

Karin Siegrist,
University of Marburg, Faculty of
Medecine, Department of Medical
Sociology,
Marburg, FRG.

J. Siegrist,
University of Marburg, Faculty of
Medecine, Department of Medical
Sociology,
Marburg, FRG.

P. Sleight,
Cardiac Department, John Radcliffe
Hospital,
Headington, Oxford, U.K.

G. Solomon,
Fresno County Department of Health,
P.O.B. 11867, Fresno, California
93775, USA.

B. Spangelo,
Department of Biochemistry, The
George Washington University,
College of Health Sciences,
Washington DC 20037, USA.

G. Specchia,
Division of Cardiology, Policlinic
S. Matteo, University of Pavia,
27100 Pavia, Italy.

A. Steptoe,
St. George's Hospital Medical
School, University of London, Depart
ment of psychology, Cranmer Terrace
London SW17 ORE, U.K.

L. Tavazzi,
Centro di riabilitazione di Veruno
Fondazione Clinica del Lavoro,
Veruno (NO), Italy

H. de The
INSERM U7, CNRS LA 318, Faculté de
Médecine Necker-Enfants malades
156 Rue de Vaugirard,
75015 Paris, France

T. Theorell,
National Institute for psychosocial factors in health,
P.O.B. 60210,
10401 Stockholm, Sweden.

O. Tønder,
The Gade Institute, Department of
Microbiology and Immunology,
University of Bergen,
Bergen, Norway

L. Torsvall,
Laboratory for Clinical Stress
Research, National Institute for
psychosocial factors and health,
1040 Stockholm, Sweden

Th. Urger,
German Institute for high blood
pressure Research and Department of
Pharmacology, University of Heidel-
berg, Im Neuenheimer Feld 366,
D-6900 Heidelberg, FRG.

U. Ursin,
Institute of physiological psycho-
logy, University of Bergen,
Bergen, Norway.

R. Vaernes,
Institute of physiological
psychology, University of Bergen,
Bergen, Norway

J. vann Jones
Cardiac Department, John
Radcliffe Hospital,
Headington, Oxford, U.K.

E. Vanoli,
Institute of cardiovascular
Research Giorgio Sisini, Università di Milano, via F. Sforza,
20122 Milano, Italy

Joan Vernikos,
Biomedical Research Division,NASA
Ames Research Center,Moffett Field,
California 94035, USA.

L. Vogt,
DFVLR, Institut für Flugmedizin,
P.O.B. 906058, 5000 Köln 90, FRG.

M. Wadsworth,
Medical Research Council, National
Survey of Health and Development,
University of Bristol, Department of
Community Health,
Whiteladies Road,
Bristol BS8 2PR, U.K.

M. Waltz,
Universität Oldenburg,
Westerstr. 2
Oldenburg, FRG.

I. Weber,
University of Marburg, Faculty of
Medecine, Department of Medical
Sociology,
Marburg, FRG.

H. Wegmann,
DFVLR, Institut für Flugmedizin,
P.O.B. 906058,
5000 Köln 90, FRG.

H. Weiner,
Neuropsychiatric Institute, UCLA,
760 Westwood Plaza,
Los Angeles, California 90024 USA.

A. Zanchetti,
Università di Milano, Istituto di
Clinica Medica IV,
via F. Sforza 35,
20122 Milano, Italy.

A. Zaza,
Institute of Cardiovascular Research
Giorgio Sisini, Università di Milano
via F. Sforza 35,
20122 Milano, Italy.

A. Zotti,
Centro di riabilitazione di Veruno,
Veruno (NO), Italy.

PART 3

Psychoneuroimmunology and breakdown in adaptation:
interactions within the central nervous system,
the immune and endocrine system

edited by

R.E. Ballieux

IMMUNOLOGY FOR NONIMMUNOLOGISTS: SOME GUIDELINES

FOR INCIPIENT PSYCHONEUROIMMUNOLOGISTS

Nicholas Cohen[1]

Department of Experimental Animal Morphology and Cell Biology,

Agricultural University, Wageningen, The Netherlands

One critical message of this workshop is that the immune system in-
teracts with other physiological systems to produce an immune response.
Indeed, the term "psychoneuroimmunology" was coined a few years ago by
Ader (1981) primarily to highlight the interactive nature of the immune
and the central nervous systems. Although by definition, psychoneuroimmunology is an integrative approach to the study of adaptation, those who
might like to actively contribute towards understanding these interactions are often confronted by a significant logistic obstacle, namely
the incredible and bewildering array of factual and conceptual information (often presented in an incomprehensible jargon) that constitutes the
state of the art in the subdisciplines of immunology, neuroendocrinology
and behavior. My purpose in writing these few pages and in presenting a
glossary of "immunologese" is to try to outline a few of the facts, concepts and paradigms in immunology that may prove useful for the nonimmunologist who wishes to enter the arena of psychoneuroimmunology. My selection of concepts has been limited to those that, in my opinion, may
prove relevant to understanding the connections between the immune system
and the central nervous system. For example, although immunologists are
now beginning to understand, in precise molecular and genetic terms, how
an organism can produce antibodies of so many different combining specificities, the major conceptual advances underlying the generation of
antibody diversity (Gearhart, 1982) are irrelevant (at least for the present) for promoting advances in psychoneuroimmunology. As psychoneuroimmunologists, we should be more interested in studying how the organism
regulates amounts of antibody produced rather than in how somatic diver-
sification comes about at the genomic level. With this rationale in mind,

1

Present address: Department of Microbiology, Division of Immunology,
University of Rochester, School of Medicine, Rochester (NY), USA.

I will consider the following topics: lymphocyte circulation, the major histocompatibility complex, lymphocyte heterogeneity, and cell interactions and interleukins.

1. THE LYMPHOID SYSTEM AND LYMPHOCYTE CIRCULATION

The lymphoid system is comprised of so-called primary and secondary lymphoid organs and the blood vessels and lymphatics that connect them. Primary lymphoid organs include the spleen, lymph nodes and gut-associated lymphoid tissues (e.g., tonsils, Peyer's patches, appendix). The main physiological function of the lymphoid system is, of course, the recognition of antigen and the dissemination of the immune response that is initiated by the recognition of antigen. These functions are facilitated by the continual recirculation of lymphocytes between and through the secondary lymphoid organs (de Sousa, 1981). In the adult mammal, the bone marrow is the primary site of hemopoiesis (Schrader, 1981). Lymphoid cells of varying stages of differentiation and committment to a certain function, migrate either directly to the secondary lymphoid organs or first to the thymus and then to the peripheral lymphoid tissues. Those lymphocytes that have sojourned in the thymus are called T cells; those that have not, are called B cells. Secondary lymphoid organs are compartmentalized (e.g. the mucosal IgA immune system of the gut; Strober et al., 1981). Of special interest to the psychoneuroimmunologist is the fact that lymphoid organs are innervated (Williams et al., 1981; Felton et al. 1981).

As mentioned, lymphocytes actively and continuously change their anatomical locations. The basic patterns of lymphocyte movement are known. The first is a migration or homing of lymphocytes from one site to another; the second is the continuous recirculation between and through the lymphoid tissues (de Sousa, 1981). Both blood vessels and lymphatics serve as conduits for this traffic. Circulation depends on the structure of the lymphocytes' cell membranes and on the surface structure of specialized cells of the postcapillary venules. Of interest to the psychoneuroimmunologist is that lymphocyte traffic is modifiable and that such modifications can affect immunity. For example, antigen exerts nonspecific effects on the flow of lymphocytes. In animals that have previously encountered antigen, a reexposure to the same antigen effects an immediate drop in the output of lymphocytes from a secondary lymphoid organ such as a lymph node. This phenomenon, known as cell shutdown, is followed by a marked increase in the output of lymphocytes from that node. During cell shutdown, there is a significant increase of the blood supply to the node, an increase of lymphoid cells in the node, but no reduction in the efferent flow of lymph (Cahill et al., 1976). In principle, shutdown may facilitate antigen-lymphocyte or lymphocyte-lymphocyte interactions. Recent studies suggest that cell shutdown is mediated by the local production of prostaglandin E_2 (Hopkins et al., 1981). Thus, mediators that alter prostaglandin E_2 production, in addition to factors that stimulate cyclic AMP production and ACTH (de Sousa, 1981), can affect lymphocyte circulation which, in turn, can alter immunity.

2. THE MAJOR HISTOCOMPATIBILITY COMPLEX (MHC)

Cellular components of the immune system must interact with each other as well as with antigen to produce an immune response. Important (some believe essential) for optimization of these cellular interactions in the secondary lymphoid tissues are products of a cluster of tightly linked genes that comprise the MHC. Every vertebrate studied has a homologous MHC and even at the phylogenetic level of anuran amphibians (frogs), the products of this genetic region appear to subserve the same functions as they do in mammals. Simply stated, antigen, be it particulate or soluble, is recognized by receptors on the surface of T lymphocytes, not by themselves, but in association with products of certain genes within this MHC. Such recognition of antigen (nonself) in association with self MHC, is involved in the production of antibody and in the guidance of effector (killer) lymphocytes to their appropriate cellular targets. This critical concept of associated recognition of antigen plus self-MHC (Bevan, 1981) is mentioned in the context of psychoneuroimmunology because MHC gene products are components of the membranes of cells of the immune system. As such, their synthesis and expression (e.g. density) can be modified and as a result, those cellular interactions that characterize the response to antigen can also be modified. It may prove noteworthy to the psychoneuroimmunologist that certain products of the MHC (those of the so-called class II region genes) which had been thought to be restricted to certain cells of the lymphoid and macrophage-monocyte series, have recently been found in the brain on glial cells (Nixon et al., 1982).

3. LYMPHOCYTE HETEROGENEITY

Cells of the lymphoid series that either produce antibody themselves or differentiate into antibody-producing cells are called bone marrow-derived (in chickens, bursa-derived) B cells (Nieuwenhuis, 1981). Immature B cells are characterized by the presence of immunoglobulins of the isotype (class) IgM in the cytoplasm but not on the cell surface. During the maturation of B cells, small lymphocytes differentiate from surface IgM negative to surface IgM positive cells. These cell surface immunoglobulins serve as the receptors for antigen. In addition, B cells express MHC products, receptors for a certain component of complement, receptors for the Fc portion of the IgG molecules (Kung & Paul, 1983), and unique differentiation antigens that can be recognized by monoclonal antibodies (McKenzie & Zola, 1983).

The lineage of T lymphocytes that directly mediate functions of the immune system that do not involve antibody is quite distinct from that of B cells. These T cells spend an early period of their life in the thymus where they are thought to become restricted with respect to recognizing self-MHC products. The use of monoclonal antibodies and certain lectins (peanut agglutinin) together with immunofluorescence, flow cytometry, complement-dependent antibody lysis, and functional studies, have revealed differences between those thymocytes that are found in the cortex and those in the medulla of the thymus. Once thymocytes leave the periphery (and become bonafide T cells), they differentiate further and can be recognized by the acquisition and loss of certain cell surface differentiation antigens and by their abilities to perform certain functions. Thus, peripheralized T cells can be separated into functionally and structurally distinguishable effector T cells and regulatory T cells. Effector T cells (Schwartz, 1982) include those that are capable of directly killing target cells (e.g. virus-infected autolous cells) or causing delayed-type hypersensitivity reactions by producing a variety of factors (lymphokines) that affect the behavior of other cell types. Regulatory T cells are those that interact with other populations of lymphocytes (T and/or B cells) and with "macrophages" to either amplify or dampen immune responses (Reinherz & Schlossman, 1981). An example of the latter type of regulatory T cells are suppressor T cells that affect B cell function by an exceedingly complex circuitry of cell-cell-factor interactions. T cells also function as helper cells in the mediation of either T cell effector function or in the production of antibody by B cells (Howie & McBride, 1982). Essential for the successful interactions of these diverse populations of lymphocytes are accessory cells (macrophages, interdigitating cells, dendritic cells, Langerhans cells; Nussenzweig & Steinman, 1982) that serve to present antigen, in association with products of their MHC, to the appropriate T cells.

4. INTERLEUKINS AND CELL INTERACTIONS

Immunologists focus on two major questions in this arena of cell interactions. What cells interact? How do they interact? Psychoneuroimmunologists might also ask where do these interactions occur in vivo and how can they be directly or indirectly modified by the CNS? It seems important to emphasize the obvious point that any significant perturbation in the network of cell interactions could have significant repercussions on immunity. It is equally important to point out that we really do not know how components of other physiological systems interact with the network of immunocytes and their products. With this in mind, it should be appreciated that interactions between macrophages and T cell subsets not only involve antigen, self MHC molecules and their specific receptors (Reinherz et al., 1983) but soluble mediators that have been called interleukins. Interleukin-1 (Il-1) is synthesized by macrophages that have been activated by a variety of compounds such as adjuvants (lipopolysaccharide, muramyl dipeptide) and factors produced by lymphocytes, themselves. Immunologists are interested in Il-1 (Oppenheim & Gery, 1982), since it induces helper T cells to differentiate into lymphocytes that produce another interleukin known as Il-2 (Smith et al., 1980). When T cells are activated by antigen (or lectin), they express a receptor for Il-2 and can proliferate so long as Il-2 is available. Thus, any factors that affect the production of Il-1 will also affect production of Il-2, and hence, the proliferation of cells that will function in a helper or cytotoxic capacity. Pharmacological concentrations of hydrocortisone (Smith et al., 1980) inhibit both the release and the mitogenic effects of Il-1. Prostaglandin E_2 also inhibits production of Il-1 (Oppenheim & Gery, 1982), whereas inhibitors of prostaglandin synthetase such as indomethacin can enhance Il-1 production by activated macrophages.

It now appears that production of Il-1 is not restricted to macrophages. Keratinocytes (Luger et al., 1981), astrocytes, and glioma cells (Fontana et al., 1981) all produce a factor with the same biochemical and biological characteristics of Il-1. Furthermore, as reviewed by Oppenheim and Gery (1981), T cells are not the only targets of macrophage-derived interleukins. For example, cells of the hypothalamic fever center bind Il-1-like endogenous pyrogen. The production of acute phase proteins by hepatocytes, the production of prostaglandins by isolated rheumatoid synovial cells, and the growth of fibroblasts are also effected by Il-1. Whether interleukins (clearly a misnomer) may provide molecular bridges between the immune and the central nervous systems remains to be seen.

5. GLOSSARY[*] OF "IMMUNOLOGESE"

Any immunologist worth his/her salt will recognize that the incomplete information in this chapter has been grossly simplified and superficially treated. (But then, this chapter was not written for immunologists.) Therefore, the nonimmunologist who wishes to pursue immunology in greater depth during the course of a research program in psychoneuroimmunology will be well advised to read a basic text in the field, peruse all past issues of that marvellously topical monthly journal known as Immunology Today, and seek the sympathetic and sagacious consultation of the card-carrying immunologist. The glossary that follows is offered to facilitate this communication.

ADCMC: Antibody-dependent, cell-mediated cytotoxicity. See CMC.

Acquired immunity: Immunological resistance to disease developed after birth.

Adherent cell: A nonlymphoid cell which adheres to plastic or glass.

Adjuvant: As used in immunology, a substance mixed with antigen in order to elicit a stronger or more sustained response; e.g. Freund's complete adjuvant (a mixture of mineral oil, lanolin and killed mycobacteria).

Adoptive transfer: The transfer of previously sensitized immune elements (e.g. lymphocytes, antibody) to a nonimmune recipient.

Affinity: A measure of the strength of binding, e.g. a single antibody combining site to a single antigenic determinant.

Agglutination: The clumping together of cells (including bacteria) via a multivalent molecule or agglutinin.

Allergic: Refers to a state of altered reactivity. Commonly used to describe a state of hypersensitivity.

Allo: Between genetically different members of the same species.

Antibody: Molecule produced in response to antigen which has the property of combining specifically with the antigen.

---

[*] Selectively compiled from glossaries in the texts by McConnell et al. (1981), Golub (1981) and Clark (1980).

Antigen: Molecule or particle which induces the formation of and combines with antibody.

Ascites: Serous fluids accumulating in the peritoneal cavity. Tumor cells growing in the peritoneal cavity usually induce ascitic fluid.

Atopy: IgE-mediated allergic hypersensitivity.

Autoantibody: Antibody produced by an animal which reacts against its own antigens.

Autograft: A graft from one part of the body to another in the same individual.

Autoreactivity: The ability to mount an immune response against self-antigens.

Avidity: Net combining power of an antibody molecule with its antigen; related to both the affinity and the valencies of the antibody and the antigen.

B cell: SIg^+, Lyt^- lymphocyte of bone marrow (bursal) origin which are the precursor of the antibody forming plasma cell.

β_2 microglobulin: A polypeptide component (11,000 m.w.) of the MHC class I molecules.

B-mice: Mice depleted of T lymphocytes. A commonly used procedure is to thymectomize neonatal or adult mice. In the latter case, the thymectomized mice are irradiated to destroy mature T cells and then reconstituted with bone marrow.

Blast cell: A cell shortly before, during, or shortly after cell division. Usually has a higher cytoplasminucleus ratio than in the resting state.

Blastogenesis: The production of blast cells; used in association with cell activation by antigen or mitogen.

Blocking antibody: Antibody that blocks access by other cells or molecules with specific receptors to a particular antigen.

Bursa of Fabricius: Lymphoepithelial organ unique in birds, located at the junction of the hind gut and cloaca.

Capping: Process of redistribution of cell surface determinants to one small part of the cell surface. Usually accomplished by antibody which must be at least divalent.

Carrier: Immunogenic molecule to which a hapten is coupled.

Chemotaxis: Attraction of a cell across a chemical gradient.

Chimerism: A situation in which cells from two genetically different individuals coexist in one body.

Clone: A family of cells (or organisms) of genetically identical constitution derived asexually from a single cell by repeated division.

CMC: Cell-mediated cytotoxicity; the process of destruction of a cell by another cell. The common types of CMC in immunity are direct, cytotoxic T lymphocyte-mediated CMC, in which target cell recognition is through a membrane-bound T cell receptor; ADCMC, mediated by any leukocyte with an Fc receptor against an antibody-coated target cell, and LDCMC, mediated by any leukocyte against a target cell coated with an appropriate lectin.

Complement: Series of interacting serum proteins which combine with antigen-antibody complexes and cause lysis of cells.

Con A: Concanavalin A, a lectin extracted from jack beans that binds to the surface of most cells. Depending on the species of animals involved, and the form in which it is utilized, Con A may completely activate T cells or B cells or both.

Congenic: Identical in genetic composition except at a defined chromosomal segment, where alternate allelic gene forms are present.

Cytophylic: Having affinity for cells (i.e. cytophylic antibody).

Differentiation antigens: Antigens expressed by cells at certain times during differentiation and not at other times.

DNP: Dinitrophenol. A commonly used hapten, as is TNP or trinitrophenol.

DTH: Delayed-type hypersensitivity. Cell-mediated raction measured as a skin reaction.

Effector cell: Effector cells, the end products of maturation, are those cells which actually produce the observed effect.

Endotoxin: Lipopolysaccharides localized in the cell walls of gram-negative bacteria.

Enhancement: The prolongation of graft survival by antibody to the graft.

Epitope: Single antigenic determinant. The portion of a molecule which will combine with a particular antibody combining site.

E-rosettes: Cluster of sheep erythrocytes around a human T cell.

Fc receptor: Receptor found on a wide range of cells in the immune system that is specific for the Fc portion of certain immunoglobulin molecules.

Germinal center: A histologically discernible region of lymph nodes and spleens populated mostly by B lymphocytes. During immune reactions leading to antibody production, cells within the germinal centers undergo extensive proliferation.

GVH: Graft-versus-host reaction. Cell-mediated reaction in which T cells in a grafted tissue react with antigens of the host.

HLA: The major histocompatibility gene complex of humans.

H-2: The major histocompatibility gene complex of the mouse.

Haplotype: When used in reference to the MHC, a haplotype is defined as the particular collection of alleles of MHC genes present on a given chromosomal segment.

Hapten: A low molecular weight compound that cannot, by itself, induce an immune response although it can combine with antibody.

Hapten-carrier: A hapten chemically bounded to a protein or other carrier. Hapten-carrier conjugates are used to evoke an immune response to a hapten.

H-chain: Heavy chain of immunoglobulins.

Helper cell: Subpopulation of T cells which cooperates with precursors of effector cells.

Hemagglutinin: Any molecule capable of agglutinating red blood cells (antibodies, lectins, etc.).

Hemolysis: The disintegration (lysis) of an erythrocyte.

Heterologous: Not the same as the original (antigen, cell, antibody).

Histocompatibility antigen: Any cell surface antigen capable of provoking graft rejection.

Humoral: In solution or suspension. Refers particularly to the antibody and complement immune mechanism, in contrast to cell-mediated immunity.

Hybridoma: The name given to cell lines created by fusing B lymphocytes with a plasmacytoma. Hybridomas produce monoclonal antibodies.

Hypervariable regions: Regions of the Ig molecule which have great variation of amino acid sequence when compared to other Ig molecules and serve as antigen binding sites.

Ia antigens: I region-associated antigens. Serologically defined class II antigens coded in the I region of the mouse MHC (or the D/DR region of HLA).

Idiotype: Antigenic marker of the antigen combining site on an immunoglobulin molecule.

Ig: Immunoglobulin.

Immune complex: A term used to denote the products of antibody-antigen reactions and often refer to the small soluble complexes containing two or three antibody molecules associated with antigen.

Immunofluorescence: Method involving the use of fluorochrome-labeled antibody.

Immunogen: A substance capable of provoking an immune response. All immunogens are antigens but not all antigens need be immunogens (e.g. most haptens).

Immunoglobulin: A globular serum glycoprotein with antibody activity.

Indirect plaque assay: Method for detecting IgG antibody-producing cells.

Interferon: Endogenous glycoproteins which have a nonspecific anti-viral activity.

Iso: Same, of identical genetic constitution - isologous, isogeneic (syn. for syngeneic).

K cells: Also called null cells. Nonphagocytic cells of unknown lineage, similar to lymphocytes but with neither T nor B cell surface markers. K (for "killer") cells can destroy antibody-coated target cells by ADCMC.

Lectins: A group of plant-derived proteins capable of binding to the surfaces of animal cells (see PHA, Con A).

Ligand: A substance which links two molecules together.

Lymphokine: Generic term for molecules other than antibodies produced by or through the aid of lymphocytes. Lymphokines have a variety of biological activities. At least some lymphokines are now called interleukins.

Local immunity: Immunity that develops in a local site, independent of the network of lymph glands and ducts throughout the body. Production of IgA antibodies to bacterial antigens in the gut is an example.

LPS: Lipopolysaccharide. A bacterial endotoxin, usually obtained from E. coli, capable of activating B cells in an antigen-independent fashion.

Memory: An altered immunological response to a given antigen resulting from a prior exposure to that antigen (syn. anamnesis).

MHC: Major histocompatibility complex. Originally defined as the genetic locus coding for the cell surface antigens presenting the major barrier to transplantation between allogeneic individuals. Now known to encompass a wide range of genes coding for or controlling immunological functions.

MLC: Mixed leukocyte (or lymphocyte) culture. Generally interpreted to include both the proliferation and generation of cytotoxic effectors against allogeneic lymphocytes in vitro.

MLR: Mixed leukocyte (or lymphocyte) reaction. Often used in a more restricted sense than MLC to refer to just the proliferative phase of the in vitro reaction of T lymphocytes to allogeneic cells.

Monoclonal: Derived from a single clone.

Network hypothesis: Hypothesis which states that the immune response regulates itself through a series of idiotype-antiidiotype responses.

NK cells: Natural killer cells. Cells that can kill target cells, particularly tumor cells, without previous sensitization, and by an unknown mechanism. NK cells have not yet been assigned to any of the defined leukocyte classes.

Nude mouse: A genetically athymic mouse which also carries closely linked defects in hair production.

Null cells: See K cells.

Opsonin: Any substance that enhances phagocytosis of a cell or particle. Antibodies function as opsonins.

Opsonization: Usually involves coating a particle or cell with an antibody enabling it to adhere firmly to the surface of a phagocyte bearing Fc recpetor. The stable contact thus created enhances phagocytosis.

PBA: Polyclonal B cell activator. Most substances mitogenic for B cells such as LPS are effective PBA.

PBL: Peripheral blood lymphocytes (leukocytes).

PHA: Phytohemagglutinin, a plant lectin from kidney beans, which activates principally T lymphocytes.

Phagocytosis: The process of ingestion of material into a cell by closing off an invagination of the protoplasm.

Plasma cell: Terminally differentiated antibody forming cell with a short half life (2-3 days).

Plasmacytoma: Tumor of a plasma cell that almost always secretes a homogeneous immunoglobulin.

Prime: To give a first exposure to an antigen.

Primary lymphoid organs: Thymus and bursa bone marrow; sites of lymphopoiesis.

Reagin: Old term for IgE antibody.

Reticulendothelial system: A diffuse system of macrophages associated with the connective tissue framework of the liver, spleen, lymph nodes, serous cavities, lungs, etc.

Secondary lymphoid organs: Peripheral concentrations of lymphocytes in spleen, lymph nodes, gut-associated lymphoid tissues.

Splenomegaly: Increase in spleen size. Used as an assay for GVH reactions.

636

Syngeneic: Animals which have been produced by repeated brother-sister mating until homozygous at all measurable loci.

T cells: Mature lymphocytes that have been influenced in their differentiation by the thymus.

Thy-1: Alloantigen unique to thymus-derived lymphocytes from mice.

Thymocytes: Lymphoid cells present in or extracted from the thymus.

Titer: A term used to connote the relative strength of an antiserum. An antiserum is progressively diluted until some measurable property of the antiserum (agglutination, faciliation of complement-mediated lysis, etc.) is reduced by some predetermined amount. That dilution (i.e. 1:512) is then defined as the titer for that antiserum.

Tolerance: State of specific immunological unresponsiveness induced by exposure to the antigen. Can be actively (suppressor T cells) or passively (clonal elimination) induced/maintained.

Xenogeneic: Immunogenetic term defining the relationship between individuals belonging to different species.

Xenograft: Graft transplanted from one species to another.

6. ACKNOWLEDGEMENTS

This manuscript was written during a sabbatical year as Fulbright Senior Research Scholar at the Agricultural University, Wageningen, The Netherlands. The financial and cultural enrichment provided by the University, the E.D.C., The Netherlands-America Commission for Educational Exchange, and USPHS grant HD 07901 are gratefully appreciated.

7. REFERENCES

Ader, R., 1981.
    Psychoneuroimmunology. Academic Press, New York, 661 pp.
Bevan, M.J., 1981.
    Thymic education. Immunol. Today, 2:216-219.
Cahill, R.N.P., Frost, H. and Trnka, Z., 1976.
    The effects of antigen on the migration of recirculating lymphocytes through single lymph nodes. J. exp. Med., 143:870-888.
Clark, W.R., 1981.
    The experimental foundations of Modern Immunology. John Wiley and Sons Inc., New York, 372 pp.
de Sousa, M., 1981.
    Lymphocyte circulation. John Wiley and Sons Inc, Chichester, 259 pp.
Felton, D.L., Overhage, J.M., Felton, S.Y. and Schmedtje, J.F., 1981.
    Noradrenergic sympathetic innervation of lymphoid tissue in the rabbit appendix: Further evidence for a link between the nervous and immune systems. Brain Res. Bull., 7:595-612.
Fontana, A., Kristensen, F., Dubs, R., Gemsa, D. and Weber, E., 1982.
    Production of prostaglandin E and an interleukin-1-like factor by cultured astrocytes and C_6 glioma cells. J. Immunol., 129:3413-3419.
Gearhart, P.J., 1982.
    Generation of immunoglobulin variable gene diversity. Immunol. To-

day, 3:107-112.

Golub, E.S., 1981.
The cellular basis of the immune response. Sinauer, MA, 2nd edition, 330 pp.

Hopkins, J., McConnell, I. and Lachmann, P.J., 1981.
Lymphocyte traffic through lymph nodes. II. Role of prostaglandin E$_2$ as a mediator of cell shutdown. Immunology, 42:225-231.

Howie, S. and McBride, W.H., 1982.
Cellular interactions in thymus-dependent antibody responses. Immunol. Today, 3:273-278.

Kung, J.T. and Paul, W.E., 1983.
B-lymphocyte subpopulations. Immunol. Today, 4:37-41.

Luger, T.A., Stadler, B.M., Katz, W.I. and Oppenheim, J.J., 1981.
Epidermal cell (keratinocyte)-derived thymocyte-activating factor. (ETAF). J. Immunol., 127:1493-1498.

MacLennan, I.A.C., Gray, D., Kumararatne, D.S. and Bazin, H., 1982.
The lymphocytes of splenic marginal zones: A distinct B-cell lineage. Immunol. Today, 4:10-15.

McConnell, I., Munro, A. and Waldmann, H., 1981.
The immune system. Blackwell, Oxford, 2nd edition, 219 pp.

McKenzie, I.F.C. and Zola, H., 1983.
Monoclonal antibodies to B cells. Immunol. Today 4: 10-15.

Nieuwenhuis, P., 1981.
B cell differentiation in vivo. Immunol. Today, 2:104-110.

Nixon, D.F., Pan-Yung Ting, J. and Frelinger, J., 1982.
Ia antigens on nonlymphoid tissues. Immunol. Today, 3:65-68.

Nussenzweig, M.C. and Steinman, R.M., 1982.
The cell surface of mouse lymphoid dendritic cells. Immunol. Today, 3:65-68.

Oppenheim, J.J. and Gery, I., 1982.
Interleukin-1 is more than an interleukin. Immunol. Today, 3:113-119

Reinherz, E.L. and Schlossman, S.F., 1981.
The characterization and function of human immunoregulatory T lymphocyte subsets. Immunol. Today, 2:69-75.

Reinherz, E.L., Meuer, S.C. and Schlossman, S.F., 1983
The delineation of antigen receptors on human T lymphocytes. Immunol. Today, 4:8.

Schrader, J., 1981.
Stem cell differentiation in the bone-marrow. Immunol. Today, 2: 7-12.

Schwartz, R.H., 1982.
The cloning of T lymphocytes. Immunol. Today, 3:43-46.

Smith, K.A., Lachmann, L.B., Oppenheim, J.J. and Favata, M.F., 1980.
The functional relationship of the interleukins. J. exp. Med. 151: 1551-1556.

Strober, W., Richman, L.K. and Elson, C.O., 1981.
The regulation of gastrointestinal immune responses. Immunol. Today, 2:156-162.

Williams, J.M., Peterson, R.G., Shea, P.A., Schmedtje, J.F. Bauer, D.C. and Felton, D.L., 1981.
Sympathetic innervation of murine thymus and spleen: Evidence for a functional link between the nervous and immune systems. Brain Res. Bull. 6:83-94.

NEUROENDOCRINE INTERACTIONS WITH BRAIN AND BEHAVIOR: A MODEL FOR PSYCHO-
NEUROIMMUNOLOGY?

B. Bohus
Department of Animal Physiology, University
of Groningen, P.O. Box 14, 9750 AA Haren, The Netherlands.

ABSTRACT

Observations on animal and man suggest that psychosocial stimuli are among the most adverse stressor that elicit the release of stress hormones. The organization of neuroendocrine responses occurs at four levels: in the limbic system, the hypothalamus, the pituitary gland and in the target organs including the brain. By actions on the brain hormonal states induced by psychosocial stimuli modulate adaptive behavioral processes such as learning, memory, extinction. The possibility that the brain-immune system interactions are similarly organized and that the neuroendocrine system plays a role herein is considered.

INTRODUCTION

Adaptation to environmental alterations that impose physical or psychosocial demands on animal and man requires a chain of behavioral, neuroendocrine, autonomic and metabolic responses in order to maintain homeostasis. Cannon (1915) was the first who suggested that such homeostatic function may be the result of interactions between brain, behavior and the endocrine system, particularly of the adrenomedullary one. That the pituitary-adrenal system also subserves the physiological adaptation to environmental demands as emerged from Selye's (1950) stress theory seemed to be a continuation of Cannon's psychoneuroendocrine efforts. Although Selye (1950) himself commented that "even mere emotional stress" such as immobilization activates the pituitary-adrenal system, it has taken a long time to recognize that psychological or social stimuli are among the most adverse ones that activate the pituitary-adrenal system (e.g., Mason, 1968; Levine et al., 1972). Furthermore, the importance of extrahypothalamic brain structures that were oftenly related to emotion (e.g., Papez, 1937) have been recognized in the control of pituitary-adrenal function (e.g., Lissák and Endröczi, 1960; Bohus, 1975a, Smelik and Vermes, 1980). It appeared too that the release of other pituitary hormones such as prolactin (PRL), vasopressin, α-MSH, endorphins are also activated by psychosocial stimuli (e.g. Brown et al., 1974; Thompson and De Wied, 1973; Rossier et al., 1977; etc.) Finally, it has been one of the most important discoveries in neurobiology that the brain is not only the "controler" of the endocrine systems but it also serves as a major target and production organ for peptide and other hormones (e.g., De Wied, 1969; Bohus, 1975a, O'Donohue and Dorsa, 1982). The actions of hormones on the brain serve the organization of specific and unspecific behavioral and other physiological responses (e.g., De Wied, 1969; Bohus, 1981, 1982; McGaugh, 1983; Bohus and Koolhaas, 1983; etc.).

This paper is dealing with the question how neuroendocrine responses are organized to environmental challenges, what are the major variables in psychosocial environment that determine neuroendocrine responses, and how hormonal states that mimick the ones which are induced by psychosocial stress affect behavioral adaptation. Finally, an attempt is made to incorporate the knowledge on the immune system into the organization schemes of the brain-neuroendocrine interactions in order to establish a more uniform view on the physiology (and pathology) of stress.

PSYCHOSOCIAL STRESSORS AND NEUROENDOCRINE RESPONSES: IMPORTANCE OF ENVIRONMENTAL VARIABLES

As it has been mentioned in the Introduction the notion that psychosocial stressors represent very strong stimuli for the pituitary-adrenocortical system is well documented. The importance of factors such as expectancy, certainty or uncertainty, establishment of coping strategy in determining the pattern of acute stress responses has been emphasized (Levine et al., 1972; Bassett et al., 1973; Ursin, 1980; etc.). In addition, the status of an animal in the social hierarchy is of importance both for the resting levels and the responsiveness to stress of the pituitary-adrenal system (Sassenrath, 1970; Louch and Higginbotham, 1967; Ely and Henry, 1978; etc.).

Additional environmental variables that play a role in the temporal aspects of the pituitary-adrenal responsiveness are the possibility or absence of possibility for behavioral coping (Bohus, 1975b). It was observed that if rats were trained in an inhibitory (passive) avoidance situation their plasma corticosterone level gradually increased by reaching a maximum 15 min after the onset of the test provided that the rats were allowed to determine their own behavior - i.e. avoiding the formerly adverse environment. However, if the rats were placed directly in the adverse environment - i.e. no avoidance or escape possibility is provided - the maximal increase in plasma corticosterone level occurred already after 5 min.

The importance of environmental and/or social factors in pituitary-adrenal responsiveness in social-aggressive interactions has been suggested

by the studies of Schuurman (1981) in this department. He found that increases in plasma corticosterone during social-aggressive interactions in
male rats of an aggressive strain are always higher in unfamiliar than in
familiar environment. In addition, the duration of the elevation of the
plasma corticosterone level is longer and the maximal increase is also
higher in the looser of the fight as compared to the winners. As the consequence of loosing a fight, a marked increase in corticosterone levels
occurs in the looser even when a fightless meeting occurs with the former
winner. Furthermore, the environment in which a fight was lost is highly
stressful for some of the loosers even without the presence of the winner.

These data suggest that diverse parameters of the pituitary-adrenal
system response to psychosocial challenges is regulated in a very subtle
way. The knowledge is rather insufficient as to whether the same kind of
patterns hold true for other hormones that are released following environmental stress stimuli. Psychological stress accompanying inhibitory avoidance behavior results in a generalized release of pituitary hormones such
as of ACTH, α-MSH and vasopressin (van Wimersma Greidanus et al., 1977;
De Rotte et al., 1982; Laczi et al., 1983b). In contrast, Henry (1980)
suggested that the kind of hormonal response to social situations may depend on the coping style of the animal. If the animal is controlling the
situation (aggressive) than noradrenaline and testosterone levels are high;
at the case of avoidance adrenaline level is high and when an animal is a
victim of the situation the corticosterone level is high and testosterone
level is low. This suggestion is in agreement with the findings that plasma
testosterone level is increased in the winner and it decreases at least
temporarily in the loosers following an aggressive interaction, and the
latter is accompanied by an increase of defensive behavior (Schuurman,
1981).

ENDOCRINE RESPONSES TO ENVIRONMENTAL STIMULI: ORGANIZATION AT FOUR LEVELS

Little is known about the overall organization of endocrine responses
to diverse stressful conditions in animal and man. The main body of knowledge stems again from experiments concerning the pituitary-adrenocortical
system. This may not be surprizing because of the classical coupling
between stress and the adrenal cortex. The view of the organization of
neuroendocrine responses as presented here presupposes a comparable organization of pituitary-adrenal and other systems. In this view the neuroendocrine responses in conjunction with other physiological alterations are

organized at four levels (see also Bohus and Koolhaas, 1983).

The limbic-midbrain system, eventually in conjunction with cerebral cortical areas is considered to be the first level of organization. This level is composed of the amygdaloid complex, the septal nuclei and the hippocampus (limbic structures) and the ascending reticular activation system (RAS). The RAS may be envisaged in accordance with classical views as a general activational system that induces more specific arousal in limbic structures (Moruzzi and Magoun, 1949). One should, however, include here the ascending projections of the aminergic systems which originate from noradrenergic, dopaminergic and serotonergic nuclei of the brainstem (Lindvall and Björklund, 1974). The major function of the limbic-midbrain system can be summarized as follows: this system integrates sensory information (inputs from the actual environment), and visceral and endocrine signals (inputs from the milieu interieur) on one hand. On the other hand, organization of adaptive behavioral processes such as learning, memory, motivation, etc. occurs in the limbic system. In addition, this system plays a role in the emotional background of behavior (see Bohus, 1975a; Gray, 1982; Isaacson, 1974; Routtenberg, 1972). Studies in relation to the limbic regulation of the pituitary-adrenal system showed that the role of amygdaloid structures is primarily an activating one, while the septal complex and the hippocampus are more of an inhibitory nature. The reticular activating system in general seems to be activatory while the ascending serotonergic fibers may carry inhibitory informations (see Bohus, 1975a; Smelik and Vermes, 1980).

The second level of organization is at the hypothalamus. The hypothalamus was long considered as the integrative centre of neuroendocrine regulation. While this view may still be valid for physical stressors, such an integrative function in the case of psychosocial stimuli is rather questionable. Extensive anatomical connections between the limbic-midbrain system and the hypothalamus may provide a morphological basis of the communication between the first and second level of organization. The function of the hypothalamus in the organization of neuroendocrine response to stressors has, however, more than one aspect. First of all, the informations as provided by the limbic-midbrain system in the form of neural message by means of neurotransmission are "translated" into hormonal messages: the synthesis and/or release of releasing or inhibiting factors/hormones are activated or inhibited depending on the kind of information provided by the limbic-

midbrain system e.g. for the corticotrophin releasing hormone via various
neurotransmitter mechanisms (see Smelik and Vermes, 1980). It should be
emphasized that the releasing factor producing cells are the subject of
circadian variation and of the feedback effect of peripheral hormones. The
hypothalamus plays a role in integrating these functions. The releasing/
inhibiting factors reach the pituitary gland and thereby regulate the
release of stress-related hormones. Another function of releasing hormones
is to affect directly brain functions. It has been shown recently that corticotrophin releasing hormone may cause behavioral effects that mimic the
action of stressful environmental stimuli (Sutton et al., 1982). Besides
the releasing/inhibiting hormones the hypothalamus is the site of synthesis
of neurosecretory peptides that are transported to the posterior pituitary.
In addition, the release of adrenomedullary catecholamines is directly
under the control of the hypothalamus by neural communication through the
spinal cord.

That the hypothalamus contains cell bodies of extensive peptidergic
neuronal systems that terminate in the limbic-midbrain structures, but also
in hindbrain and the spinal cord represents another aspect of its neuroendocrine function. The arcuate nucleus contains the cell bodies of the multiple peptidergic opiomelanocortin system: β-endorphin, α-MSH/ACTH and
γ-MSH may originate from the large proopionelanocortin precursor by proteo-
lytic degradation (Bloom and McGinty, 1981; Swaab et al., 1981; Chrétien
et al., 1979). In addition, synaptosomal plasma membrane-bound enzymes
assure the formation of β-endorphin fragments related to γ- and α-endorphin
(Burbach et al., 1980). Vasopressinergic and oxytocinergic cell bodies are
located in the supraoptic and paraventricular nuclei while the parvocellu-
lar neurosecretory cells of the suprachiasmatic nuclei produce only vasopressin and these neurons terminate exclusively in the brain (see Sofroniew
and Weindl, 1981). It is not clear yet how stressful stimuli affect the
central release of brain-born peptides and whether their release serves
transmitter or hormone-like modulator function or both. The few data that
are available suggest indirectly the release of the stored peptide. Electric
footshock (Rossier et al., 1977), active avoidance training, pseudocondit-
ioning, habituation training (Izquierdo et al., 1980) result in a decrease
of β-endorphin immunoreactivity in the hypothalamus.

Similarly, changes in immunoreactive arginine-vasopressin content of
several hypothalamic and extrahypothalamic brain areas in rats immediately

after the retention test of an inhibitory avoidance behavior suggest central release of this peptide (Laczi et al., 1983a).

The third level of organization occurs at the level of the pituitary gland. The release of stress-related hormones from the hypophysis is controled by hypothalamic (releasing or inhibiting) factors and by the circulating peripheral hormones. One may distinguish between two generations of stress hormones. The classically known stress hormones such as ACTH and adrenal cortical and medullary hormones belong to the first generation. The second generation is represented by vasopressin, prolactin and the endorphins. Prolactin is released by stressful stimuli such as handling, novelty (Brown and Martin, 1974; Euker et al., 1975) while vasopressin (antidiuretic hormone) is secreted after footshock or emotional stimuli related to avoidance behavior (Thompson and De Wied, 1973; Laczi et al., 1983b). That these hormones function as stress hormones has been suggested by behavioral studies (see later in this paper). Stress-related function of the recently discovered endorphins has been suggested by the co-release of ACTH and β-endorphin in response to stressful stimuli (Guillemin et al., 1977). In addition, the release of β-LPH, α- and γ-MSH in conjunction with stress is keeping with the coexistence of these peptides in the opiocortin molecule both in the anterior and intermediate lobe of the pituitary gland (see for details O'Donohue and Dorsa, 1982). Recent observations suggest, however, that the release of opiomelanocortins from the two lobes of the pituitary is regulated by separate mechanisms during stress. Emotional stressors probably via circulating catecholamines result in the release of β-endorphin and α-MSH from the intermediate lobe while the corelease of ACTH from the anterior lobe is not mediated by adrenergic mechanisms (Berkenbosch et al., 1983). That the release of the various stress-hormones is interrelated is suggested by a number of observations. Stress-induced release of prolactin can be blocked by pretreatment with dexamethasone (Euker et al., 1975; Rossier et al., 1980) and by the opiate antagonist naloxone (Rossier et al., 1980). Since dexamethasone blocks the release of both ACTH and β-endorphin in response to stress (Guillemin et al., 1977), β-endorphin mediation of prolactin release to stress has been suggested (Rossier et al., 1980). In addition, the involvement of a catecholaminergic mechanism in the release of prolactin is not unlikely. β-adrenergic mechanisms stimulate the release of prolactin at least in vitro (Denef et al., 1982).

The fourth level of organization is at the level of the target organs. The classical theories on stress considered the peripheral organs or organ systems such as the cardiovascular, neuromuscular and immune systems, the liver, the kidney, etc. (Target Organs I) as the only sites at which stress hormones affect bodily functions. The discovery that the brain is a major target site for stress hormones (Target Organ II) has opened new vistas in the understanding and interpreting the adaptive functions of the neuroendocrine systems. It appears that stress hormones modulate physiological functions of the brain related to basic adaptive processes such as learning, memory, motivation, etc. (see later in this paper). In addition, neuroendocrine influences on stress-related physiological responses such as cardiovascular changes in relation to emotional behavior are also related to action at Target Organ II level (Bohus et al., 1983a).

What a major importance of neuroendocrine actions on the brain is, is that this link closes the circuit of the brain-neuroendocrine interactions. The limbic-midbrain system may be considered as the major site of action of stress hormones on adaptive behavioural functions (e.g. Bohus et al., 1982a, b) or central cardiovascular regulation (Bohus et al., 1983b).

NEUROENDOCRINE INFLUENCES ON BEHAVIORAL ADAPTATION

The stress concept as formulated by Selye (1950) emphasized the adaptive aspects of nonspecific reactions to environmental challenges. Adaptation should be considered as a chain of events including behavioral and neuroendocrine responses that serve to preserve homeostasis. Behavioral adaptation is viewed here as the most effective form of proper responding to both psychosocial and physical environmental events. Behavioral stress responses may be of specific and nonspecific nature. Learning, retention (memory) and extinction are considered as the specific one. Learning means the acquisition of new behavioral patterns that cope with a given environment (physical or psychosocial). Retention (memory) processes ensure the storage (consolidation) of new experience that takes place shortly after learning (McGaugh, 1961). Furthermore, retrieval or recall of learned behavior is a major adaptive response. The retrieval processes care about the organization of the proper behavioral response whenever the same environmental stimuli or stimuli that signal those events reoccur that led to learning. Extinction processes serve the elimination of the performance (but not forgetting) of a behavioral response that is no more relevant. The specificity of these responses lies on the assumption that learning of one particular behavioral response means adaptation to a particular environment. Even minor changes in the environment may require new strategies in which older behavioral patterns may be incorporated, but those may also negatively interfere with the acquisition of new patterns.

The nonspecific behavioral stress responses may occur in different environments with or without interfering with specific responses. Exploration, displacement behaviors such as grooming, wet shaking, teeth chattering, defecation, urination, and analgesia and reflex immobility are considered to belong to this groups of adaptive behavioral responses. These behavioral responses do occur as unconditioned responses in novel environment. Many of these responses are subject of learning (conditioning) by reoccurrence of the same stimuli. Such a way, exploratory activities may diminish, analgesia may be conditioned (see Watkins and Mayer, 1982) and the duration of immobility reflexes such as the dorsal immobility response (Webster et al., 1981) is also decreased following repeated induction (Meyer and Bohus, 1983).

Considerable evidence suggests that hormonal responses to environmental challenges   result in an alteration in the hormonal state both in the periphery and the brain and thereby modulate specific and nonspecific behavioral stress responses. The evidences have been obtained mostly by a classical endocrine approach: the removal of the organ of the production of the hormones and subsequent replacement with the missing hormones in order to restore physiological hormonal state.

As far as the specific behavioral responses are concerned, the pituitary gland plays an important modulatory role in learning, retention and extinction processes. Hypophysectomy results in impairments of these processes. The deficits can be amanded by peripheral administration of ACTH and related peptides (α-MSH, ACTH fragments) or of vasopressin (see De Wied, 1969; Bohus and De Wied, 1980; Bohus, 1982).

The importance of pituitary peptides in specific adaptive behaviors may be, however, difficult to reconcile with the existence of peptidergic systems in the brain. It should be emphasized that the behavioral deficits that follow hypophysectomy are not absolute. Increasing the intensity of punishment may also correct for the deficits (Lissák and Bohus, 1972; Fekete et al., 1983). Brain peptides related to ACTH may be involved in the correction of behavior. Avoidance learning in hypophysectomized rats is not improved by higher shock punishment when the centrally available peptides

are neutralized by antiserum against ACTH 1-16 (Fekete et al., 1983).
Accordingly, brain peptidergic system may be activated by stronger demands
and   take over the role of the pituitary gland in the behavioral adaptation. The contribution of the central systems in the intact rat particularly at lighter demands seems to be negligible.

Observations in intact rats suggest that beside ACTH/α-MSH⁻and vasopressin,prolactin and β-endorphin also improve adaptive behavior. While
the action of prolactin is restricted to learning (Drago et al., 1983), the
action of β-endorphin is on retention and extinction rather than on learning (Bohus, 1982; Bohus et al., 1983c).

Interestingly, the classical stress hormones of the adrenal medulla
and the cortex do not play a primary role in learning. However, retention
behavior is dependent upon circulating catecholamines (Borrell et al.,
1983) while corticosteroids through specific receptors in the brain serve
extinction processes (Bohus et al., 1982b).

Although nonspecific behavioral responses are less dependent on the
integrity of the neuroendocrine system (see Bohus and De Wied, 1980), endogenous opioid mechanisms (most probably a propiomelanocortin system) seem
to be involved in their organization and/or modulation. For example, opiate
antagonist reduces exploratory activity in stressful environments (Katz,
1979). β-endorphin and ACTH-related peptides induce some forms of displacement behaviors such as grooming (Gispen and Isaacson, 1981). Stress-induced
analgesia is in part dependent on endogenous opiates (Watkins ane Mayer,
1982). Interestingly, prolactin may also induce excessive grooming and
analgesia in which opioid mechanisms are also involved (Drago et al., 1983).
Finally, the importance of the opioid system in specific behavioral stress
responses is of interest. Action of ACTH and prolactin on acquisition behavior is blocked by opiate antagonists and these antagonists have their own
inhibitory action⁻on avoidance acquisition (Bohus et al., 1983c; Drago et
et., 1983).

These observations suggest that hormonal state as the consequence of
the neuroendocrine component of the stress response modifies and/or contributes to the organization of specific and nonspecific behavioral responses
to novel or repeated environmental challange. The divers stress hormones
are differentially involved in the various phases of behavioral adaptation.

There are at least two possible ways how stress hormones exert their
behavioral functions. It may be that various stages and forms of behavioral

adaptation are accompanied by a specific pattern of release of stress hormones and thereby alteration in the hormonal states may have specific function in the organization of the most adequate behavioral response. Although such a suggestion is consonant with Henry's (1970) hypothesis, evidences concerning specific release of pituitary hormones are missing.

Alternatively, the effectiveness of nonspecific hormonal states to modulate behavioral stress responses may depend on the variable hormonal sensitivity of those limbic-midbrain mechanisms that control the organization of adaptive behavior. This alternative thus emphasizes the importance of the target organ. Behavioral evidences concerning the localization of actions of both steroid and peptide hormones in the brain are in favour of this alternative (Bohus, 1975b; Bohus et al., 1982a; Van Wimersma Greidanus et al., 1983).

NEUROENDOCRINE AND IMMUNE SYSTEMS AND STRESS: INTEGRATED RESPONSES WITH SIMILAR ORGANIZATIONAL PRINCIPLES

Stressful stimuli cause the organisms to respond behaviorally, autonomically, endocrinologically, etc. in order to maintain homeostasis. The endocrine and cardiovascular responses are similarly organized at four levels and interaction does take place between the two systems. While such organizational principles can be considered as physiological ones, alterations at each of the four levels or dichotomy between the various responses may contribute to stress pathology (Bohus and Koolhaas, 1983; Bohus et al., 1983a).

There are a number of facts that support the concept that the immune and neuroendocrine responses are similarly organized and the neuroendocrine responses may serve an important link in the course of host defense (see for details Ader and Hall et al., in this volume). Briefly, the immune system is conditionable. The central nervous system is able to modulate immunogenesis. Furthermore, products of the immune system such as thymosins may regulate the activity of the pituitary-adrenal system, but the release of other stress hormones such as β-endorphin is also affected. Regulation of immune activities by these hormones closes the circuit between brain and the immune system involving a neuroendocrine link. That other pituitary hormones such as prolactin, a stress hormone of second generation play a role in the regulation of humoral immunity (Nagy and Bérczi, 1978; Bérczi et al., 1983) further reinforce the importance of the neuroendocrine link. It is of interest that vasopressin (and oxytocin) possess lymphokine-like activity

648

(Johnson et al., 1982) thereby affecting γ-interferon production. Additional capacities of the immune system should also be considered. Lymphoid cells may be the source of ACTH and β-endorphin (Smith and Blalock, 1981) and thymosin α_1-like immunoreactivity has been found in the arcuate nucleus and medial eminence of the hypothalamus and in the pituitary (Palaszynski, 1981; Hall et al., 1982). The former system may form a micro-environmental circuit in which lymphatic β-endorphin receptors of nonopioid nature (Hazun et al., 1979) may be of importance. The latter may serve a regulation of pituitary-adrenal axis which is different from the corticotropin releasing factor controlled ACTG release.

CONCLUDING REMARKS AND PERSPECTIVES

The complexity of organization of the physiological stress responses opens the possibility of integration via the neuroendocrine system, but malfunctioning in the environment-body interaction may easily occur when the function in one of the four levels is disturbed. A psychoneural view on the immune system allows the recognition of the integrated organization of behavior, host defense, neuroendocrine and autonomic response. The organization of the neuroendocrine system and its function as a major bidirectional link between brain and body may serve as a model for the brain-immune system interactions. It should, however, be realized that a large number of questions are still open. Little is known about the nature of environmental and/or behavioral variables that may affect the course of host defense. Furthermore, it remains to be shown whether thymosins and lymphokines affect such complex brain functions such as the specific or nonspecific behavioral stress responses. The exact nature of the neuroendocrine macro- and micro-circuits in immune regulation remains to be solved too. Multidisciplinary approach to these problems is not only promising but also imperative and it is expected to contribute to the understanding of stress pathology and breakdown in adaptation.

ACKNOWLEDGEMENTS

Extensive discussions on the theses of psycho-neuro-immunology with R.Ader and R.E.Ballieux greatly improved my understanding the functioning of the immunosystem and contributed to forming an integrated concept. The secretarial assistance by Mrs. Joke Poelstra-Hiddinga is greatly acknowledged.

REFERENCES

Bassett, J.R., Cairncross, K.D. and King, M.G. 1973.
Parameters of novelty, shock predictability and response contingency in corticosterone release in the rat. Physiol. Behav., 10: 901-907.

Berczi, I., Nagy, E., Kovács, K. and Horvath, E. 1983. Regulation of humoral immunity in rats by pituitary hormones. Acta Endocrinol., in press.

Berkenbosch, F., Vermes, I. and Tilders, F.J.H. 1983. βendorphin secretion during some stress conditions can be prevented by propanolon. Nat.,i.p.

Bloom, F.E. and McGinty, J.F. 1981. Cellular distribution and function of endorphins. In: Endogenous Peptides and Learning and Memory Processes. J.L.Martinez, R.A.Jensen, R.B.Messing, H.Rigter and J.M.McGaugh (Eds.) Academic Press, New York, pp. 199-230.

Bohus, B. 1975a. The hippocampus and the pituitary-adrenal system hormones. In: The Hippocampus, vol. 1: Structure and development. R.L.Isaacson and K.H.Pribram (Eds.) Plenum Press, New York and London, pp. 323-353.

Bohus, B. 1975b. Environmental influences on pituitary-adrenal system functions. In: Les Endocrines et la Milieu. H.-P.Klotz (Ed.) Problèmes Actuels D'Endocrinologie et de Nutrition. Série No. 19. Expansion Scientifique Francaise, Paris, pp. 55-62.

Bohus, B. 1981. Neuropeptides in brain functions and dysfunction. Int.J. Ment.Health, 9: 6-44.

Bohus, B. 1982. Neuropeptides and memory. In: Expression of Knowledge. R.L. Isaacson and N.E.Spear (Eds.) Plenum Press, New York, pp. 141-177.

Bohus, B. and De Wied, D. 1980. Pituitary-adrenal system hormones and adaptive behaviour. In: General, Comparative and Clinical Endocrinology of the Adrenal Cortex. I.Chester-Jones and I.W.Henderson (Eds.) Vol. 3, Academic Press, London, pp. 256-347.

Bohus, B. and Koolhaas, J.M. 1983. Psychosocial stress: endocrine and brain interactions and their relevance for cardiovascular processes. In: Acute Cardiovascular Responses to Stress. L'Abatte (Ed.) Martinus Nijhof, 's-Gravenhage, in press.

Bohus, B., Conti, L., Kovacs, G.L., and Versteeg D.H.G. 1982a. Modulation of memory processes by neuropeptides: interaction with neurotransmitter systems. In: Neuronal Plasticity and Memory Formation. C.Ajmone Marsan and H.Matthies (Eds.) Raven Press, New York, pp. 75-87.

Bohus, B., de Kloet, E.R., and Veldhuis, H.D. 1982b. Adrenal steroids and behavioral adaptation: relationship to brain corticoid receptors. In: Adrenal Action on Brain. D.W.Pfaff and D.Ganten (Eds.) Springer, Berlin, pp. 108-140.

Bohus, B., Versteeg, C.A.M., de Jong, W, Cransberg, K. and Kooy, J.G. 1983a. Neurohypophyseal hormones and central cardiovascular control in the neurohypophysis. B.A.Cross and S.Leng (Eds.) Progr.Brain Res. vol. 60, Elsevier Amsterdam, pp. 463-475.

Bohus, B., de Jong, W., Hagan, J.J., de Loos, W., Maas, C.M. and Versteeg, C.A.M. 1983b. Neuropeptides and steroid hormones in adaptive autonomic processes: implications for psychosomatic disorders. In: Integrative Neurohumoral Mechanisms. E.Endröczi,D.de Wied, L.Angelucci and U.Scapagnini (Eds.) Elsevier Biomedical Press, Amsterdam, pp. 35-49.

Bohus, B., de Boer, S., Zanotti, A. and Drago, F. 1983c. Opiocortin peptides, opiate agonists and antagonists, and avoidance behavior. In: Integrative Neurohumoral Mechanisms. E.Endröczi, D.de Wied, L.Angelucci and U.Scapagnini (Eds.) Elsevier Biomedical Press, Amsterdam, pp. 107-113.

Borrell, J., de Kloet, E.R. Versteeg, D.H.G. and Bohus, B. 1983. Inhibitory avoidance deficit following short term adrenalectomy in the rat: the role of adrenal catecholamines. Behav.Neurol.Biol. in press.

Brown, G.M. and Martin, J.B.M. 1974. Corticosterone, prolactin, and growth Hormone responses to handling and new environment in the rat. Psychosom.Mev., 36: 241-247.

Brown, G.M., Uhlir, I.V., Seggie, J., Schally, A.V. and Kastin, A.J. 1974. Effect of septal lesions on plasma levels of MSH, corticosterone, GH and prolactin before and after exposure to novel environment: role of MSH in septal syndrome. Endocrinology 94: 583-587.

Burbach, J.P.H., Loeber, J.G., Verhoef, J., Wiegant, V.M., de Kloet, E.R. and de Wied, D. 1980. Selective conversion of β-endorphin into peptides related to γ- and α-endorphin. Nature 283: 96-97.

Cannon, W.B. 1915. Bodily changes in pain, hunger, fear and rage. Appleton New York.

Chrétien, M., Benjannet, S., Gossard, F., Gianoulakis, C., Crine, P., Lis, M. and Seidah, N.G. 1979. From β-lipotropin to β-endorphin and 'pro-opio-melanocortin'. Canad.J.Biochem. 57: 1111-1121.

Denef, C. and Baes, M. 1982. β-adrenergic stimulation of prolactin release from superfused pituitary cell aggregates. Endocrinology 111: 356-358.

De Rotte, A.A., Egmond, M.A.H. and Wimersma Greidanus, Tj.B.van. 1982. α-MSH levels in cerebrospinal fluid and blood of rats during manipulations. Physiol.Behav. 28: 765-768.

De Wied, D. 1969. Effects of peptide hormones on behavior. In: Frontiers in Neuroendocrinology. W.F.Ganong and L.Martine (Eds.) Oxford Press, New York, pp. 97-140.

Drago, F., Bohus, B., Gispen, W.H., Ree, J.M.van, Scapagnini, U. and De Wied, D. 1983. Behavioral changes in short-term and long-term hyper-prolactinaemic rats. In: Integrative Neurohumoral Mechanisms. E.Endröczi, D. de Wied, L.Angelucci and U.Scapagnini (Eds.) Developments in Neuroscience, Vol. 16. Elsevier, Amsterdam, pp. 417-427.

Ely, D.L. and Henry, J.P. 1978. Neuroendocrine response patterns in dominant and subordinate mice. Horm.Behav. 10: 156-169.

Euker, J.E., Meites, J. and Riegle, G.D. 1975. Effects of acute stress on serum LH and prolactin in intact, castrate and dexomethosone-treated male rats. Endocrinology 96: 85-95.

Fekete, M., Bohus, B. and De Wied, D. 1983. Comparative effects of ACTH-related peptides on acquisition of shuttle-box avoidance behavior of hypophysectomized rats. Neuroendocrinology 36: 112-118.

Gispen, W.H. and Isaacson, R.L. 1981. ACTH-induced excessive grooming in the rat. Pharmacol.Ther. 12: 209-246.

Gray, J.A. 1982. The Neuropsychology of Anxiety: An Enquiry into the Functions of the Septo-Hippocampal System. Oxford Univ. Press, New York.

Guillemin, R., Vargo, T., Rossier, J., Minick, S., Ling, N., Rivier, J., Vale, W. and Bloom, F. 1977. Beta-endorphin and adrenocorticotropin are sevreted concomitantly by the pituitary. Science 197: 1367-1369.

Hall, N.R., McGillis, J.P., Spangelo, B., Palaszynski, E., Moody, T. and Goldstein, A.L. 1982. Avidence for a neuroendocrine-thymis axis mediated by thymosin peptides. In: Current Concepts in Human immunology and cancer Immunomodulation. B.Serrou (Ed.) Elsevier, Amsterdam, pp. 653-660.

Hazum, E., Chang, K.J. and Cautrecasas, P. 1979. Specific nonopiate receptors for β-endorphins. Science 205: 1033-1035.

Henry, J.P. 1980. Present concepts of stress theory. In: Catecholamines and stress: Recent Advances. E.Usdin, R.Kvetnansky and I.J.Kopin (Eds.) Elsevier/North Holland, Amsterdam, pp. 557-571.

Isaacson, R.L. 1964. The Limbic System. Plenum Press, New York.

Izquierdo, I., Souza, D.O., Carrasco, M.A., Dias, R.D., Perry, M.L., Eisinger, S., Elisabetsky, E. and Vendite, D.A. 1980. Beta-endorphin causes retrograde amnesia and is released from the rat brain by various forms of training and stimulation. Psychofarmacologia 70: 173-177.

Johnson, H.M., Farrar, W.L. and Torres, B.A. 1982. Vasopressin replacement of interleukin 2 requirement in gamma interferon production: lymphokine activity of a neuroendocrine hormone. J.Immunol. 129: 983-986.

Katz, R.J. 1979. Naltrexone antagonism of exploration in the rat. Int.J. Neurosci., 9: 49-51.

Laczi, F., Gaffori, O., De Kloet, E.R. and De Wied, D. 1983a. Differential responses in immunoreactive arginine-vasopressin content of micro-dissected brain regions during passive avoidance behavior. Brain Res., 260: 342-346.

Laczi, F., Fekete, M. and De Wied, D. 1983b. Antidiuretic and immunoreactive arginine-vasopressin levels in eye plexus blood during passive avoidance behavior in rats. Life Sci., 32: 577-589.

Levine, S., Goldman, L. and Coover, G.D. 1972. Expentancy and the pituitary-adrenal system. In: Physiology, Emotion and Psychosomatic Illness. CIBA Foundation Symposium 8 (new series). Elsevier, Excerpta Medica, North Holland, Amsterdam, pp- 281-291.

Lindvall, O. and Björklund, A. 1984. The organization of the ascending catecholamine neuron system in the rat brain as revealed by the glyoxylic acid fluorescence method. Acta physiol.Scand. Suppl.,412: 1-48.

Lissák, K. and Endröczi, E. 1960. Die neuroendokrine Steuerung der Adaptationstätigkeit. Akadémiai Kiado, Budapest.

Lissák, K. and Bohus, B. 1972. Pituitary hormones and avoidance behavior in the rat. Int.J.Psychobiol., 2: 103-115.

Louch, C.D. and Higginbotham, M. 1967. The relation between social rank and plasma corticosterone levels in mice. Gen.comp.Endocrinol., 8: 441-444.

Mason, J.W. 1968. A review of psychoneuroendocrine research on the pituitary-adrenal cortical system. Psychosom.Med., 30: 576-607.

McGaugh, J.L. 1961. Facilitative and disruptive effects of strychnine sulfate on maze learning. Psychol.Rep., 9: 99-104.

McGaugh, J.L. 1983. Hormonal influences on memory. Ann.Rev.Psychol., 34: 297-323.

Meyer, M.E. and Bohus, B. 1983. Modulation of dorsal immobility response in the adult male Wistar rat: the opposite effects of ACTH 4-10 and [D-Phe[7]] ACTH 4-10. Behav.Neurol.Biol. In press.

Moruzzi, G. and Magoun, H.W. 1949. Brain-stem reticular formation and activation of the EEG. Electroenceph.clin.Neurophysiol., 1: 455-473.

Nagy, E. and Berczi, I. 1978. Immunodeficiency in hypophysectomized rats. Acta Endocrinol., 89: 530-537.

O'Donohue, T.L. and Dorsa, D.M. 1982. The opiomelanotropinergic neuronal and endocrine systems. Peptides 3: 353-395.

Palaszynki, E. 1982. Doctoral Dissertation. George Washington University.

Papez, J.W. 1937. A proposed mechanism of emotion. Arch.Neurol.Psychiat., 38: 725-744.

Rossier, J., French, E.D., Rivier, C., Long, N., Guillemin, R. and Bloom, F.E. Footshock induces stress and increases β-endorphin levels in blood but not in brain. Nature 270: 618-620.

Rossier, J., French, E.D., Rivier, C., Shibasaki, T., Cuillemin, R. and Bloom, F.E. 1980. Stress-induced release of prolactin: blockade by dexamethasone and naloxone may indicate β-endorphin mediation. Proc. Natl.Acad.Sci., 77: 666-670.

Routtenberg, A. 1972. Memory as input-outpot reciprocity: an integrative neurobiological theory. Ann.N.Y.Acad.Sci., 193: 159-174.

Sassenrath, E.N. 1970. Increased adrenal responsiveness related to social stress in rhesus monkeys. Horm.Behav., 1: 283-298.

Schuurman, T. 1981. Endocrine processes underlying victory and defeat in the male rat. Doctoral dissertation. University of Groningen.

Selye, H. 1950. Stress. The Physiology and Pathology of Exposure to Stress. Acta Medica. Publ. Montreal.

Smelik, P.G. and Vermes, I. 1980. The regulation of the pituitary-adrenal system in mammals. In: General, Comparative and Clinical Endocrinology of the Adrenal Cortex. I.Chester Jones and I.W.Henderson (Eds.) Vol. 3. Academic Press, New York, pp. 1-55.

Smith, E.M. and Blalock, J.E. 1981. Humane lymphocyte production of corticotropin and endorphin-like substances: association with leukocyte interferon. Proc.Natl.Acad.Sci., 78: 7530-7540.

Sofroniew, M.V. and Weindl., A. 1981. Central nervous system distribution of vasopressin, oxytocin, and neurophysin. In: Endogenous Peptides and Learning and Memory Processes. J.L.Martinez Jr., R.A.Jensen, R.B. Messing, H.Rigter and J.L.McGaugh (Eds.) Academic Press, New York, pp. 327-369.

Sutton, R.E., Koob, G.F., Le Moal, J., Rivier, J. and Vale, W. 1982. Corticotropin releasing factor produces behavioral activation in rats. Nature 297: 331-333.

Swaab, D.F., Achterberg, P.W., Boer, G.J., Dogterom, J. and van Leeuwen, F.W. 1981. The distribution of MSH and ACTH in the rat and human brain and its relation to pituitary stores. In: Endogenous Peptides and Learning and Memory Processes. J.L.Martinez Jr., R.A.Jensen, R.B. Messing, H.Rigter and J.L.McGaugh (Eds.) Academic Press, New York, pp. 7-36.

Thompson, E.A. and De Wied, 1973. The relationship between the antidiuretic activity of rat eye plexus, blood and passive avoidance behavior. Physiol.Behav., 11: 377-380.

Ursin, H. 1980. Affective and instrumental aspects of fear and aggression. In: Functional states of the brain: their determinants. M.Koukkou and D.Lehmann (Eds.) Elsevier, Amsterdam, pp. 119-130.

Watkins, L.R. and Mayer, D.J. 1982. Organization of endogenous opiate and nonopiate pain control systems. Science 216: 1185-1192.

Webster, D.C., Lanthorn, T.H., Dewsbury, D.A. and Meyer, M.E. 1981. Tonic immobility and dorsal immobility response in twelve species of muroid rodents. Behav.Neurol.Biol. 31: 31-42.

Wimersma Greidanus, Tj.B. van, Rees, L.H., Schott, A.P., Lowry, P.J. and De Wied, D. 1977. ACTH release during passive avoidance behavior. Brain Res.Bull., 2: 101-104.

Wimersma Greidanus, Tj.B. van, Bohus, B., Kovács, G.L., Versteeg, D.H.G., Burbach, J.P.H. and De Wied, D. 1983. Sites of behavioral and biochemical action of ACTH-like peptides and neurohypophyseal hormones. Neurosci.Biobehav.Rev., in press.

PSYCHONEUROIMMUNOLOGY

R. Ader

Division of Behavioral and Psychosocial Medicine
Department of Psychiatry
University of Rochester School of Medicine and Dentistry
Rochester, New York 14642, USA

It is significant that a workshop on "psychoneuroimmunology" is being held within the context of discussions of "breakdowns in adaptation". It is recognition of the rapid growth of a new integrative field of study involving interdisciplinary research on the neuroendocrine mediation of the effects of behavior in modifying immune function. Because immune responses also influence neuroendocrine function and may thus influence behavior, psychoneuroimmunology may be defined, more generally and simply, as the study of the interactions between the central nervous system (CNS) and the immune system. The term "interaction" should be taken literally to mean that psychoneuroimmunology is concerned with understanding how the CNS and the immune system influence each other - or, as Ernest Sorkin described it, the "dynamic flow of information" between the CNS and the immune system. By definition, psychoneuroimmunology is an integrative approach to the study of adaptation.

Processes of adaptation include the integrated pattern of biologic, psychologic, and social changes that occur in response to environmental circumstances and that reflect the organism's attempt to maintain homeostatis. Integration of these processes is ultimately regulated by the brain. "Breakdowns" in these adaptational processes due, primarily, to more extreme environmental demands (i.e., stressful" circumstances) have behavioral, social and biologic consequences, including alterations in the predisposition to and/or the precipitation or perpetuation of a variety of disease processes. Understanding and intervening in these processes, however, is predicated on understanding the mechanisms underlying the normal operation and integration of adaptational processes.

To discuss the immune system within this context implies acceptance of the proposition that the CNS is capable of exerting some influence on immune processes. The assumption that the immune system is an autonomous agency of defense is no longer tenable. Like any other system operating in the interests of homeostasis, the immune system is integrated with other physiological processes and is subject to regulation or modulation by the brain.

For purposes of discussion, "stress" can be referred to as any natural or experimental perturbation of the organism or its environment. One can add to this the individual's perception that environmental circumstances pose some threat to the psychological or physiological integrity of the individual and the personal or social resources that are available to the individual for coping with such events. In the absence of independent criteria, however, this remains a circular and inadequate definition; but it characterizes the available literature. Most authors have accepted (if not used) Selye's (1950) distinction between "stress" and "stressor", including the (implicit) assumption of adrenal activation. This is more than a conceptual weakness since it dictates how many experiments are designed and conducted. For the present purpose, I have excluded extraordinary traumatic stimuli (burns, surgery) and concentrated on events which, processed by the CNS, are, by general consensus if not by definition, referred to as being stressful. A more elaborate discussion of the heuristic value of the concept of "stress" is given elsewhere (Ader, 1980a; Ader, 1980b).

First, what is the evidence relating immune function to breakdowns in adaptation?

STRESS AND DISEASE

To the extent that some cancers may have a viral etiology or involve immune surveillance, reference should be made to the extensive review of studies in humans (Fox, 1981) and animals (Riley & Spackman, 1981). Animal experiments have shown that the manipulation of early life experiences (Ader & Friedman, 1965a; Ader & Friedman, 1965b; LaBarba et al., 1970a; LaBarba et al., 1970b; Levine & Cohen, 1959; Newton et al., 1962; Otis & Scholler, 1967; Winokur et al., 1958), social factors (Ader & Friedman, 1964; Andervont, 1944; DeChambre & Gosse, 1973; Henry et al., 1975; Kaliss & Fuller, 1968; LaBarba et al., 1972; LeMonde, 1959; Muhlbock, 1951), or other noxious stimuli (Amkraut & Solomon, 1972; Gottfried & Molomut, 1963; Kavetsky et al., 1966; Khaletskaia, 1954; Kozhevnikova, 1953; Marsh et al., 1959; Molomut et al., 1963; Newberry et al., 1972; Plaut et al., 1980; Rashkis, 1952; Vinogradova, 1960) or, conversely, the minimization of environmental disturbances (Riley, 1975) can influence the development of and/or the response to spontaneously developing or experimentally-induced neoplastic disease. Also, the capacity to cope with stressful environmental circumstances can attenuate tumor growth and mortality (Sklar & Anisman, 1979). Actually, stress has been reported to increase and/or decrease the incidence of disease, depending upon the response measure chosen for analysis. The variety of manipulations and the variety of pathologic processes studied yield a seemingly contradictory pattern of results that preclude definitive generalizations at this time.

Abundant clinical, mostly retrospective evidence documents a relationship between "life change" or stress, if you will, and a variety of disease processes including those that are (or may be) immunologically mediated (Gunderson & Rahe, 1974; Weiner, 1977). Both susceptibility to and recovery from infectious, allergic, and autoimmune disease have been related to life stresses in humans (Engels & Wittkower, 1975; Greenfield et al., 1959; Imboden et al., 1961; Jacobs et al., 1970; Kasl et al., 1979; Meyer & Haggerty, 1962; Solomon, 1981). In animals, too, a variety of environmental circumstances (e.g., avoidance conditioning and/or noise, painful stimulation, or manipulation of social interactions) have been shown to influence susceptibility or response to experimentally-induced infectious, parasitic, allergic, and autoimmune diseases (Amkraut et al., 1971; Chang & Rasmussen, 1965; Davis & Read, 1958; Friedman et al., 1965; Friedman et al. 1973; Friedman et al., 1969; Gross & Siegel, 1965; Hamilton, 1974; Jensen & Rasmussen, 1963; Johsnon et al., 1959; Levine et al., 1962; Marsh et al., 1963; Plaut et al., 1969; Rasmussen et al., 1957; Rogers et al., 1980a; Rogers et al., 1980b; Soave, 1964; Weinmann & Rithman, 1967; Yamada et al., 1964). As in the case of neoplastic disease, the observed increases and decreases in susceptibility depend upon the nature of the environmental circumstances and the nature of the pathophysiologic process.

STRESS AND IMMUNOLOGIC REACTIVITY

A variety of genetically and experimentially determined host factors e.g., age (Makinodan & Yunis, 1977; Yunis et al., 1976), nutritional state (Suskind, 1977), sleep deprivation (Palmblad, 1981), and circadian rhythms (Fernandes et al., 1977) influence immunologic reactivity as measured by a variety of assays that reflect different specific and nonspecific immune defense mechanisms. The death of a spouse reflects a common experience that can be extremely stressful and Bartrop et al. (1977) and Schleifer et al. (1980) have noted a depression in lymphocyte function associated with bereavement which was, incidentally, independent of the hormonal responses that were also measured. Immunologic effects of separation experiences in monkeys (Laudenslager et al., 1982; Reite et al., 1981) and rats (Michaut et al., 1981) have also been reported.

Increases as well as decreases in immunologic reactivity have been observed following a variety of stressful experiences in animals. Repeated sampling procedures have uncovered a biphasic response: first, a depression and then an increase in the lymphocyte response to mitogenic stimulation (Folch & Waksman, 1974; Monjan & Collector, 1976). A suppressed activity of splenic lymphocytes has also been observed in mice subjected to acceleration or anesthesia, the effect being a function of strain (Gisler et al., 1971). While these in vitro effects could be mimicked by ACTH (Gisler & Schenkel-Hulliger, 1971), the in vivo effects of these stimuli could not be reproduced by ACTH (Solomon & Amkraut, 1981).

In response to a topically applied irritant, stressful stimulation has been reported to increase immunologic reactivity in guinea pigs (Guy, 1952; Mettrop & Visser, 1969). Using mice, others (Christian & Williamson, 1958; Funk & Jensen, 1967; Smith et al., 1960) have reported that stress reduces inflammatory responses. Delayed hypersensitivity was reduced (Pitkin, 1965), and survival of a skin allograft was prolonged (Wistar & Hildemann, 1960) in stressed mice. Also, a graft-vs-host response was depressed in rats subjected to a limited feeding schedule and the effect could not be attributed to elevated adrenal activity (Amkraut et al., 1973).

Avoidance conditioning results in a less severe anaphylactic shock response (Rasmussen et al., 1959) which is generally attributed to elevated steroid levels. Individually-housed mice, however, are more resistant to anaphylaxis than group-housed animals (Treadwell & Rasmussen, 1961) depending upon challenge dose. Keller et al. (1981) reported a reduction in lymphocyte activity following stressful stimulation that was related to the degree of environmental stimulation. These same effects were also obtained in adrenalectomized animals.

A variety of noxious stimuli and, again, social factors have been found to modify nonspecific defense reactions and increase or decrease in vitro and primary and secondary humoral responses to different antigens in several species (Edwards & Dean, 1977; Edwards et al., 1980; Glenn & Becker, 1969; Hill et al., 1967; Joasoo & McKenzie, 1976; Pavlidas & Chirigos, 1980; Perlmutter et al., 1973; Solomon, 1969). Stimulation during early life has also been shown to influence the subsequent response to a bacterial antigen (Solomon et al., 1968). Reviews of these data, the hormonal mediation of stress-induced alterations in immunologic reactivity, and the applicability of a "stress" conceptualization have recently been prepared (Ader, 1980b; Solomon & Amkraut, 1981; Vessey, 1964; Solomon et al., 1968; Monjan, 1981).

SUMMARY OF STRESS EFFECTS

Stress can influence the predispostion to and/or the precipitation and perpetuation of immunologically mediated disease. Stress is neither a necessary nor sufficient condition for the development of disease, but, rather, a risk factor - a factor which, in combination with pathogenic stimuli and with a variety of other host factors, can be of etiologic significance in disease.

Furthermore, the effects of stress are neither uniformly detrimental nor beneficial to the organism. A given experimental situation defined by an observer as "stressful" can exert different effects on different pathologic processes and, with respect to a given disease process, it may differentially affect the initiation of and the response to that disease.

Stress can also directly affect humoral and cell-mediated immune responses in vitro and in vivo (although the two are not necessarily related). Again, however, there is no uniformity to the results obtained. In general, the impact of stress on immune function is determined by several major factors: (1) The quality and quantity of stressful stimulation; (2) The quality and quantity of immunogenic stimulation; (3) The temporal relationship between stress and immunologic stimulation; (4) The myriad host factors upon which stress and immunogenic stimulation are superimposed; (5) Procedural factors such as the nature of the dependent variable and sampling parameters; and (6) The interaction among any or all of the above.

Considering the complexity of CNS-endocrine-immune interactions, it is not surprising that so little definitive data is available. Relatively speaking, little research has yet been undertaken and it is only recently that the autonomy of the immune system has been questioned and the effects of CNS (and environmental events operating through the CNS) have been experimentally examined. The extent to which the CNS can modulate immune responses, then, has yet to be determined - and there are formidable obstacles to such research. Some of these are methodological.

Two events of concern are the stimulus and response characteristics of stress and the stimulus and response characteristics of pathogenic or antigenic stimulation. Little can be said about the former, except that the methodological problems are compounded by conceptual ones. There is no uniformity in the data because there is no uniformity in the stressful stimulation that has been used; few attempts have been made to define the psychophysiologic sequelae of the stimulation imposed or, for that matter, the difference in the pattern of neuroendocrine responses induced by different stimuli; and, beyond a reference to adrenocortical steroids, few investigators even allude to the possible involvement of other neuroendocrine changes and their influence on immune responses. Consequently, much of the seeming confusion in the data stems from the premature generalization implied by a reference to "stress" (implicitly, at least) as a unitary and unifying concept.

In contrast to stressful stimulation, immunogenic stimulation can be relatively precisely administered, titrated, and reported. Problems do arise, however, with respect to the measurement of disease susceptibility or of immunologic competence. For example, one problem is the sensitivity of in vivo immunologic assays. When one addresses the issue of a CNS involvement in immune responses, the questions are not so much a matter of "whether or not", but of "how much", and many current techniques may not be sufficiently sensitive to the latter question. There is more precision in in vitro analyses, but in vitro effects are not always reproducible in vivo - and it is, after all, the response of the immune system operating within its natural neuroendocrine milieu that is of primary concern in evaluating the effects of "breakdowns in adaptation".

Many of the problems are not inherent in the technology. They relate more to the need for parametric analyses necessitated by the study of interrelated systems about which little is yet known. The problem exists at several levels, ranging from the need for serial sampling, through the analysis of differences between individuals (or strains of animals) and within individuals (e.g., as a function of circadian rhythms), to the analysis of the effects of defined, quantifiable forms of stimulation on different aspects of immune function that are, in turn, adequately defined and uniformly reported.

CNS-IMMUNE SYSTEM INTERACTIONS

The fact that "stress" has immunologic consequences implies that CNS function can influence the immune system. More direct evidence of CNS-immune system interactions, however, can be cited. At a behavioral level, for example, learning is a primary means by which organisms adapt to their environment and is a primary function of brain activity - the most complex, perhaps, of all CNS functions. Recently, it has been demonstrated that learning processes, as studied by classical (Pavlovian) conditioning techniques, are capable of modifying humoral and cell-mediated immune responses. Several studies by Ader and his colleagues (Ader & Cohen, 1975; Ader & Cohen, 1981; Ader et al., 1979) have shown that a single pairing of consumption of a novel, distinctively flavored drinking solution, the conditioned stimulus (CS), with an immunosuppressive drug, the unconditioned stimulus (US), could result in an attenuated antibody response to sheep erythrocytes when previously conditioned animals were reexposed to the CS at the time of antigenic stimulation. Such conditioned immunopharmacologic effects have been independently verified (Rogers et al., 1976; Wayner et al., 1978) and extended to graft-vs-host (Bovbjerg et al., 1982) and delayed type hypersensitivity (Bovbjerg et al., 1982) responses. Conditioned enhancement of a delayed type hypersensitivity response has also been demonstrated (Bovbjerg, 1982) and, in other recent research (Gorczynski et al., 1982) enhanced responses to allografts in mice were demonstrated in a paradigm in which the immunogenic stimulus itself, served as the unconditioned stimulus. In these latter studies, the imposition of unreinforced presentations of the CS and resulting extinction of the conditioned response support the notion that associative (learning) processes were involved in these alterations in immunologic reactivity. A conditioning paradigm has also been used to reduce the amount of cyclophosphamide required to delay the onset of autoimmune disease in the NZBxNZW mice (Ader & Cohen, 1982; Ader & Cohen, 1983).

That there are behavioral influences on immune responses should not be surprising in light of the neuroanatomic and neurochemical links between the CNS and the immune system or the effects of neuroendocrine manipulations and pharmacologic agents (including psychotropic drugs) on immune function. For example, lesions and electrical stimulation of the hypothalamus result in alterations in humoral and cell-mediated immunity (e.g., Cross et al., 1980; Roszman et al., 1982; Jankovic & Isakovic,

1973; Spector & Korneva, 1981; Stein et al., 1981). Not only do hypothalamic interventions influence immunologic reactivity, but elicitation of
an immune response influences hypothalamic activity (Besedovsky et al.,
1977).

The central role of the hypothalamus in regulating neuroendocrine
function and autonomic nervous system activity implies a role for hormones and neurotransmitters in the modulation of immune responses - a role
that has since been reinforced by reports (Abraham & Bug, 1976; Arrenbrecht, 1974; Cake & Litwak, 1975; Gillette & Gillette; 1979; Hadden et
al., 1970; Helderman & Strom, 1978; Hollenberg & Cuatrecasas, 1974; Rosz-
kowski et al., 1977; Singh et al., 1979; Strom et al., 1974) that lymphocytes bear receptors for a variety of hormones and neurotransmitter substances. Not only do variations in hormonal state and neurotransmitter
substances (as well as pharmacologic agonists and antagonists of neurotransmitter activity) influence immunocompetence, but antigenic stimulation (or the immune response to antigenic stimulation) elicits neuroendocrine changes (e.g., Besedovsky & Sorkin, 1981).

Insofar as the endocrine system is concerned, there is actually a
voluminous literature on hormonal modulation of immune responses and several recent reviews are available (e.g., Ahlqvist, 1981; Comsa et al.,
1982). The vast majority of this research has concentrated on the thymus.
Indeed, because of its critical function in the ontogeny of immune competence, the endocrine functions of the thymus and its prominent role in
the development of neuroendocrine function has, with few exceptions (e.g.
Pierpaoli, 1981), not received the attention it deserves. Conversely,
hormones other than the thymus (with the possible exception of the adrenal) have not received sufficient attention as modulators of immune responses. Even in the case of the corticosteroids, though, parametric data
on the effects of endogenous steroids or physiological doses of steroids
are limited. The parallel development of neuroendocrine and immune function (Pierpaoli et al., 1970; Pierpaoli et al., 1977) and the long-term
effects of behavioral, neuroendocrine or immunologic interventions during
early life on subsequent behavioral, neuroendocrine, or immune function
(Ader, 1982) represent other fruitful areas of research into CNS-immune
system interactions.

It is not unlikely that breakdowns in adaptation which can be reflected in altered susceptibility or response to autoimmune or other
immunologically mediated disease are mediated or influenced by endocrine

and neurotransmitter changes that are ultimately regulated by the brain. Recent experiments have also implicated the endorphins in (stress-induced) immunologic reactivity (Gilman et al., 1982; Miller et al., 1982; Shavit et al., 1982). Some of these changes, at least, have been hypothesized to exert their influence on the immune system via their action on cyclic nucleotides (e.g., Bourne et al., 1974). It is also of interest, then, that behaviorally active neuropeptides have been found to influence immunologic reactivity (Johnson et al., 1982) and that still other data suggest that neuropeptides could alter immunologic reactivity through their effects on thymic hormone secretion (MacLean & Reichlin, 1981) or their effects on levels of cyclic nucleotides via alterations in neurotransmitter metabolism (Iuvone et al., 1978; Lichtensteiger & Monnet, 1979; Versteeg & Wurtman, 1975; Weigant et al., 1979).

These, and other recent data providing neuroanatomic and neurochemical evidence of CNS innervation of lymphoid tissue (Besedovsky & Sorkin, 1981; Bulloch & Moore, 1980; Hall & Goldstein, 1981; Felten et al., 1981; Reilly et al., 1979; Williams et al., 1981), document the multiple connections between the brain and the immune system and provide the basis for psychoneuroimmunologic interactions. Hormones, neurotransmitters, and neuropeptides are endogenous substances regulated by the CNS and directly or indirectly involved in processes of adaptation. Data already available indicate that they are also implicated in the modulation of immune responses. Although the self-regulatory capacities of the immune system are immense and can be experimentally dissected, the adaptive functions of the immune system that are of ultimate concern are those that take place within a constantly changing neuroendocrine environment that is demonstrably sensitive to external events, including behavioral and social processes of adaptation. Exploration of these relationships is likely to contribute much to our understanding of the immune system - and the brain.

REFERENCES

Abraham, A.D. and Bug, G., 1976.
[3]H-testosterone distribution and binding in rat thymus cells in vivo. Mol. Cell. Biochem., 13: 157-163.
Ader, R. and Friedman, S.B., 1964.
Social factors affecting emotionality and resistance to disease in animals: IV. Differential housing, emotionality, and Walker 256 carcinosarcoma in the rat. Psychol. Rep., 15: 535-541.
Ader, R. and Friedman, S.B., 1965a.

Differential early experiences and susceptibility to transplanted tumor in the rat. J. comp. physiol. Psychol., 59: 361-364.

Ader, R. and Friedman, S.B., 1965b.
Social factors affecting emotionality and resistance to disease in animals: V. Early separation from the mother and response to a transplanted tumor in the rat. Psychosom. Med., 27: 119-122.

Ader, R. and Cohen, N., 1975.
Behaviorally conditioned immunosuppression. Psychosom. Med., 37: 333-340.

Ader, R., Cohen, N. and Grota, L.J., 1979.
Adrenal involvement in conditioned immunosuppression. Int. J. Immunopharmac., 1: 141-145.

Ader, R., 1980a.
Animal models in the study of brain, behavior, and bodily disease. Res. Publ. nerv. ment. Dis., 59: 11-26.

Ader, R., 1980b.
Psychosomatic and psychoimmunologic research. Psychosom. Med., 42: 307-322.

Ader, R. and Cohen, N., 1981.
Conditioned immunopharmacologic responses. In: R. Ader (ed),Psychoneuroimmunology, Academic Press, New York, pp. 281-320.

Ader, R., 1982.
Developmental psychoneuroimmunology. Presented at the meetings of the International Society for Developmental Psychobiology, Minneapolis, MN.

Ader, R. and Cohen, N., 1982
Behaviorally conditioned immunosuppression and murine systemic lupus erythematosus. Science, 215: 1534-1536.

Ader, R. and Cohen, N., 1983.
Behaviorally conditioned immunosuppression: Effects on the course of autoimmune disease in New Zealand hybrid mice. Adv. Immunopharmacol. (in press).

Ahlqvist, J., 1981.
Hormonal influences on immunologic and related phenomena. In: R. Ader (ed), Psychoneuroimmunology. Academic Press, New York, pp 355-403.

Amkraut, A.A., Solomon, G.F. and Kraemer, H.C., 1971.
Stress, early experience and adjuvant-induced arthritis in the rat. Psychosom. Med., 33: 203-214.

Amkraut, A and Solomon, G.F., 1972.
Stress and murine sarcoma virus (Moloney)-induced tumors. Cancer Res., 32: 1428-1433.

Amkraut, A.A., Solomon, G.F., Kasper, P. and Purdue, P., 1973.
Stress and hormonal intervention in the graft-versus-host response. In: B.D. Jankovic and K. Isakovic (eds), Microenvironmental aspects of immunity. Plenum, New York.

Andervont, H.B., 1944.
Influence of environment on mammary cancer in mice. J. Nat. Cancer Inst., 4: 579-581.

Arrenbrecht, S., 1974.
Specific binding of growth hormone to thymocytes. Nature, 252: 255-257.

Bartrop, R.W., Lazarus, L., Luckhurst, E., et al., 1977.
Depressed lymphocyte function after bereavement. Lancet, i, 834-836.

Besedovsky, H.O., Sorkin, E., Felix, D. and Haas, H., 1977.
Hypothalamic changes during the immune response. Eur. J. Immunol.,

7: 325-328.

Besedovsky, H.O. and Sorkin, E., 1981.
Immunologic-neuroendorine circuits: Physiological approaches. In: R. Ader (ed), Psychoneuroimmunology. Academic Press, New York, pp 545-574.

Bourne, H.R., Lichtenstein, L.M., Melmon, K.L., Henney, C.S., Weinstein, Y. and Shearer, G.M., 1974.
Modulation of inflammation and immunity by cyclic AMP. Science, 184: 19-28.

Bovbjerg, D., 1982.
Behaviorally conditioned alterations in cell-mediated immune response. Ph.D. dissertation, University of Rochester.

Bovbjerg, D., Ader, R. and Cohen, N., 1982.
Behaviorally conditioned suppression of a graft-vs-host response. Proc. Nat. Acad. Sci. (USA), 79: 583-585.

Bulloch, K. and Moore, R.Y., 1980.
Nucleus ambiguous projections to the thymus gland: Possible pathway for regulation of the immune response and the neuroendocrine network. Anat. Rec., 196: 25A.

Cake, M.H. and Litwak, G, 1975.
The glucocorticoid receptors. In: G. Litwak (ed), Biochemical action of hormones. Academic Press, New York, pp 317-390.

Chang, S. and Rasmussen, A.F. jr., 1965.
Stress-induced suppression of interferon in virus-infected mice. Nature, 205: 623-624.

Christian, J.J. and Williamson, H.O., 1958.
Effect of crowding on experimental granuloma formation in mice. Proc. Soc. exp. Biol. Med., 99: 385-387.

Comsa, J., Leonhardt, H. and Wekerle, H., 1982.
Hormonal coordination of the immune response. Rev. Physiol. Biochem. Pharmacol., 92: 115-189.

Cross, R.J., Markesberry, W.R., Brooks, W.H. and Roszman, T.L., 1980.
Hypothalamic-immune interactions: I. The acute effect of anterior hypothalamic lesions on the immune response. Brain Res., 196: 79-87.

Davis, D.E. and Read, C.P., 1958.
Effect of behavior on development of resistance in trichinosis. Proc. Soc. exp. Biol. Med., 99: 269-272.

DeChambre, R.P. and Gosse, C., 1973.
Individual versus group caging of mice with grafted tumors. Cancer Res., 33: 140-144.

Edwards, E.A. and Dean, L.M., 1977.
Effects on crowding of mice on humoral antibody formation and protection to lethal antigenic challenge. Psychsom. Med., 39: 19-24.

Edwards, E.A., Rahe, R.H., Stephens, P.M. and Henry, J.P., 1980.
Antibody response to bovine serum albumin in mice: The effects of psychosocial environmental change. Proc. Soc. exp. Biol. Med., 164: 478-481.

Engels, W.D. and Wittkower, E.D., 1975.
Psychophysiological allergic and skin disorders. In: A.M. Freedman, H.J. Kaplan and B.J. Sadock (eds), Comprehensive textbook of psychiatry, II. Williams and Wilkins, Baltimore, pp 1685-1694.

Felten, D.L., Overhage, J.M., Felten, S.Y. and Schmedtje, J.F., 1981.
Noradrenergic sympathetic innervation of lymphoid tissue in the rabbit appendix: Further evidence for a link between the nervous and immune systems. Brain Res. Bull., 7: 595-612.

Fernendes, G., Yunis, E.J. and Halberg, F., 1977.

Circadian aspects of immune responses in the mouse. In: J.P. McGarvern, M.H. Smolensky and G. Reinberg (eds), Chronobiology in allergy and immunology. Charles C. Thomas, Springfield, pp 233-249.

Folch, H. and Waksman, B.H., 1974.
The splenic suppressor cell: Activity of thymus dependent adherent cells: Changes with age and stress. J. Immunol., 113: 127-139.

Fox, B.H., 1981.
Psychosocial factors and the immune system in human neoplasia. In: R. Ader (ed), Psychoneuroimmunology. Academic Press, New York, pp 103-157.

Friedman, S.B., Ader, R. and Glasgow, L.A., 1965.
Effects of psychological stress in adult mice inoculated with Coxsackie B viruses. Psychosom. Med., 27: 361-368.

Friedman, S.B., Glasgow, L.A. and Ader, R., 1969.
Psychosocial factors modifying host resistance to experimental infections. Ann. NY. Acad. Sci., 164: 381-392.

Friedman, S.B., Ader, R. and Grota, L.J., 1973.
Protective effect of noxious stimulation in mice infected with rodent malaria. Psychosom. Med., 35: 535-537.

Funk, G.A. and Jensen, M.M., 1967.
Influence of stress on granuloma formation. Proc. Soc. exp. Biol. Med., 124: 653-655.

Gillette, S. and Gillette, R.W., 1979.
Changes in thymic estrogen receptor expression following orchidectomy. Cell. Immunol., 42: 194-196.

Gilman, S.C., Schwartz, J.M., Milner, R.J., Bloom, F.E. and Feldman, J.D. 1982.
B-endorphin enhances lymphocyte proliferative responses. Proc. Nat. Acad. Sci. (USA), 79: 4226-4230.

Gisler, R.H., Bussard, A.E., Mazie, J.C. and Hess, R., 1971.
Hormonal regulation of the immune response: I. Induction of an immune response in vitro with lymphoid cells from mice exposed to acute systemic stress. Cell. Immunol., 2: 634-645.

Gisler, R.H. and Schenkel-Hulliger, L., 1971.
Hormonal regulation of the immune response: II. Influence of pituitary and adrenal activity on immune responsiveness in vitro. Cell Immunol., 2: 646-657.

Glenn, W.G. and Becker, R.E., 1969.
Individual versus group housing in mice: Immunological response to time-phased injections. Physiol. Zool., 42: 411-416.

Gorczynski, R., MacRae, S. and Kennedy, M., 1982.
Conditioned immune response associated with allogenic skin grafts in mice. J. Immunol., 129: 704-709.

Gottfried, B. and Molomut, N., 1963.
Effect of surgical trauma and other external stress agents on tumor growth and healing of cancer. Proc. 8th Anti-Cancer Congr., 3: 1617-1620.

Greenfield, N.S., Roessler, R. and Crosley, A.P., 1959.
Ego strength and length of recovery from infectious mononucleosis. J. nerv. ment. Dis., 128: 125-128.

Gross, W.B. and Siegel, H.S., 1965.
The effect of social stress on resistance to infection with Escherichia coli or Mycoplasma gallisepticum. Poult. Sci., 44: 98-1001.

Gunderson, E.K. and Rahe, R.H. (eds), 1974.
Life stress and illness. Thomas, Springfield, Illinois.

Guy, W.B., 1952.

Neurogenic factors in contact dermatitis. Arch. Dermatol. Syphil., 66: 1-8.

Hadden, J.W., Hadden, E.M. and Middleton, E., 1970.
Lymphocyte blast transformation: I. Demonstration of adrenergic receptors in human peripheral lymphocytes. Cell. Immunol., 1: 583-595.

Hall, N.R. and Goldstein, A.L., 1981.
Neurotransmitters and the immune system. In: R. Ader (ed), Psychoneuroimmunology. Academic Press, New York, pp 521-543.

Hamilton, D.R., 1974.
Immunosuppressive effects of predator induced stress in mice with acquired immunity to Hymenolepsis nana. J. Psychosom. Res., 18: 143-153.

Helderman, J.H. and Strom, T.B., 1978.
Specific insulin binding site on T and B lymphocytes as a marker of cell activation. Nature, 274: 62-63.

Henry, J.P., Stephens, P.M. and Watson, F.M.C., 1975.
Force breeding, social disorder and mammary tumor formation in CBA/USC mouse colonie: A pilot study. Psychosom. Med., 37: 277-283.

Hill, C.W., Greer, W.E. and Felsenfeld, O., 1967.
Psychological stress, early response to foreign protein, and blood cortisol in vervets. Psychosom. Med., 29: 279-283.

Hollenberg, M.D. and Cuatrecasas, P., 1974.
Hormone receptors and membrane glycoproteins during in vitro transformation of lymphocytes. In: B. Clarkson and R. Baserga (eds), Control and proliferation of animal cells. Cold Spring Harbour Laboratory, New York, pp 423-434.

Imboden, J.B., Canter, A. and Cluff, L.E., 1961.
Convalescence from influenza: A study of the psychological and clinical determinants. Arch. Int. Med., 108: 393-399.

Iuvone, P.M., Morasco, J., Delaney, R.L. and Dunn, A.J., 1978.
Peptides and the conversion of ^{3}H-tyrosine to catecholamines: Effect of ACTH analogs, melanocyte stimulating hormones, and lysine vasopressin. Brain Res., 139: 131-139.

Jacobs, M.H., Spelker, A., Norman, M.M. et al., 1970.
Life stress and respiratory illness. Psychosom. Med., 32: 233-242.

Jankovic, B.D. and Isakovic, K., 1973.
Neuroendocrine correlates of immune response: II. Effects of brain lesions on antibody production, arthus reactivity and delayed hypersensitivity in the rat. Int. Arch. Allergy, 45: 360-372.

Jensen, M.M. and Rasmussen, A.F. jr., 1963.
Stress and susceptibility to viral infections: II. Sound stress and susceptibility to vesicular stomatitis virus. J. Immunol., 90:21-23

Joasoo, A. and McKenzie, J.M., 1976.
Stress and the immune response in rats. Int. Arch. Allergy, 50: 659-663.

Johnson, H.M., Farrar, W.L. and Torres, B.A., 1982.
Vasopressin in replacement of interleukin 2 requirement in gamma interferon production: Lymphokine activity of a neuroendocrine hormone. J. Immunol., 129: 983-986.

Johsnon, R., Lavender, J.F. and Marsh, J.T., 1959.
The influence of avoidance learning stress on resistance to Coxsackie virus in mice. Fed. Proc., 18: 575.

Kaliss, N. and Fuller, J.L., 1968.
Incidence of lymphatic leukemia and methylcholanthrene-induced cancer in laboratory mice subjected to stress. J. Nat. Cancer Inst., 41: 967-983.

Kasl, S.V., Evans, A.S. and Neiderman, J.C., 1979.
  Psychosocial risk factors in the development of infectious mononu-
  cleosis. Psychosom. Med., 41: 445-466.
Kavetsky, R.E., Turkevitch, N.M. & Baltisky, K.P., 1966.
  On the psychophysiological mechanism of the organism's resistance to
  tumor growth. Ann. NY. Acad. Sci., 125: 933-945.
Keller, S.E., Weiss, J., Schleifer, S.J., Miller, N.E. and Stein, M.,
1981.
  Suppression of immunity by stress: Effects of a graded series of
  stressors on lymphocyte stimulation in the rat. Science, 213: 1397-
  1400.
Khaletskaia, F.M., 1954.
  Effect of overloading of the nervous system on the development of
  induced neoplasma in mice. Zh. vyssh. deiat. Pavlov, 4: 869-876.
Kozhevnikova, E.P., 1953.
  Effect of the higher nervous function on the development of experi-
  mental neoplasms. Arkh. Pat., 15: 22-27.
LaBarba, R.C., Klein, M.L., White, J.L. and Lazar, J., 1970a.
  Effects of early cold stress and handling on the growth of Ehrlich
  carcinoma in BALB/C mice. Develop. Psychol., 2: 312-313.
LaBarba, R.C., White, J.L., Lazar, J. and Klein, M., 1970b.
  Early maternal separation and the response to Ehrlich carcinoma in
  BALB/C mice. Develop. Psychol., 3: 78-80.
LaBarba, R.C., Lazar, J.M. and White, J.L., 1972.
  The effects of maternal separation, isolation and sex on the res-
  ponse to Ehrlich carcinoma in BALB/C mice. Psychosom. Med., 34: 557-
  559.
Laudenslager, M., Reite, M. and Harbeck, R.J., 1982.
  Immune status during mother-infant separation. Psychsom. Med., 44:
  303.
LeMonde, P., 1959.
  Influence of fighting on leukemia in mice. Proc. Soc. exp. Biol.
  Med., 102: 292-295.
Levine, S. and Cohen, C., 1959.
  Differential survival to leukemia as a function of infantile stimu-
  lation in DBA/2 mice. Proc. Soc. exp. Biol. Med., 102: 53-54.
Levine, S., Strebel, R., Wenk, E.J. and Harman, P.J., 1962.
  Suppression of experimental allergic encephalomyelitis by stress.
  Proc. Soc. exp. Biol. Med., 109: 294-298.
Lichtensteiger, W. and Monnet, F., 1979.
  Differential response of dopamine neurons to oc-melantropin and ana-
  logs in relation to their endocrine and behavioral potency. Life
  Sci., 25: 2079-2087.
MacLean, D. and Reichlin, S., 1981.
  Neuroendocrinology and the immune response. In: R. Ader (ed), Psy-
  choneuroimmunology. Academic Press, New York, pp 475-519.
Makinodan, R. and Yunis, E. (eds), 1977.
  Immunology and aging. Plenum, New York.
Marsh, J.T., Miller, B.E. and Lamson, B.G., 1959.
  Effect of repeated brief stress on growth of Ehrlich carcinoma in
  the mouse. J. Nat. Cancer Inst., 22: 961-977.
Marsh, J.T., Lavender, J.F., Chang, S. and Rasmussen, A.F.jr., 1963.
  Poliomyelitis in monkeys: Decreased susceptibility after avoidance
  stress. Science, 140: 1415-1416.
Mettrop, P.J.G. and Visser, P., 1969.
  Exteroceptive stimulation as a contingent factor in the induction

and elicitation of delayed-type hypersensitivity reactions to 1-chloro-, 2-4, dinitrobenzene in guinea pigs. Psychophysiology, 5: 385-388.

Meyer, R.J. and Haggerty, R.J., 1962.
Streptococcal infections in families: Factors altering individual susceptibility. J. Pediatr., 29: 339-349.

Michaut, R.J., DeChambre, R.P., Doumerc, S., Lesourd, B, Devillechabrolle, A. and Moulias, R., 1981.
Influence of early maternal deprivation on adult humoral immune response in mice. Physiol. Behav., 26: 189-191.

Miller, G.C., Murgo, A.J. and Plotnikoff, N.P., 1982.
The influence of leucine and methionine on immune mechanisms. Int. J. Immunopharmac., 4: 366-367.

Molomut, N., Lazere, F. and Smith, L.W., 1963.
Effect of audiogenic stress upon methylcholanthrene-induced carcinogenesis in mice. Cancer Res., 23: 1097.

Monjan, A.A. and Collector, M.I., 1976.
Stress-induced modulation of the immune response. Science, 196: 307-308.

Monjan, A.A., 1981.
Stress and immunologic competence: Studies in animals. In: R. Ader (ed), Psychoneuroimmunology. Academic Press, New York, pp 185-228.

Muhlbock, O., 1951.
Influence of environment on the incidence of mammary tumors in mice. Acta Un Int. Cancer, 7: 351-353.

Newberry, B.H., Frankie, G., Beatty, P.A., Maloney, B.D. and Gilchrist, J.C., 1972.
Shock stress and DMBA-induced mammary tumors. Psychosom. Med., 34: 295-303.

Newton, G., Bly, C.G. and McCrary, C., 1962.
Effects of early experience on the response to transplanted tumor. J. nerv. ment. Dis., 134: 522-527.

Otis, L.S. and Scholler, J., 1967.
Effects of stress during infancy on tumor development and tumor growth. Psychol. Rep., 20: 167-173.

Palmblad, J., 1981.
Stress and immunologic competence. In: R. Ader (ed), Psychoneuroimmunology. Academic Press, New York, pp 229-257-

Pavlidas, N. and Chirigos, M., 1980.
Stress-induced impairment of macrophage tumoricidal function. Psychosom. Med., 42: 47-54.

Perlmutter, A., Sarot, D.A., Yu, M.L., Filazzola, R.J. and Seeley, R.J., 1973.
The effect of crowding on the immune response of the blue gourami, Trichogaster trichopterus, to infectious pancreatic necrosis (IPN) virus. Life Sci., 13: 363-375.

Pierpaoli, W., Fabris, N. and Sorkin, E., 1970.
Developmental hormones and immunological maturation. In: G.E.W. Wolstenholme (ed), Hormones and the immune response. Churchill, London, pp 126-153.

Pierpaoli, W, Kopp, H.G., Muller, J. and Keller, M., 1977.
Interdependence between neuroendocrine programming and the generation of immune recognition in ontogeny. Cell. Immunol., 29: 16-27.

Pierpaoli, W., 1981.
Integrated phylogenetic and ontogenetic evolution of neuroendocrine and identity-defense, immune functions. In: R. Ader (ed), Psycho-

neuroimmunology. Academic Press, New York, pp 575-606.

Pitkin, D.H., 1965.
Effect of physiological stress on the delayed hypersensitivity reaction. Proc. Soc. exp. Biol. Med., 120: 350-351.

Plaut, S.M., Ader, R., Friedman, S.B. and Ritterson, A.L., 1969.
Social factors and resistance to malaria in the mouse: Effects of group vs. individual housing on resistance to Plasmodium berghei infection. Psychosom. Med., 31: 536-552.

Plaut, S.M., Esterhay, R.J., Sutherland, J.C. et al., 1980.
Psychosocial effects on resistance to spontaneous AKR leukemia in mice. Paper presented at the annual meetings of the American Psychosomatic Society, New York, NY.

Rashkis, H.A., 1952.
Systemic stress as an inhibitor of experimental tumors in Swiss mice. Science, 116: 169-171.

Rasmussen, A.F.jr., Marsh, J.T. and Brill, N.C., 1957.
Increased susceptibility to herpes simplex in mice subjected to avoidance-learning stress or restraint. Proc. Soc. exp. Biol. Med., 96: 183-189.

Rasmussen, A.F.jr., Spencer, E.S. and Marsch, J.T., 1959.
Decrease in susceptibility of mice to passive anaphylaxis following avoidance-learning stress. Proc. Soc. exp. Biol. Med., 100: 878-879.

Reilly, F.D., McCuskey, P.A., Miller, M.L., McCuskey, R.S. and Meineke, H.A., 1979.
Innervation of the periarteriolar lymphatic sheath of the spleen. Tissue & Cell, 11: 121-126.

Reite, M., Harbeck, R. and Hoffman, A., 1981.
Altered cellular immune response following peer separation. Life Sci., 29: 113-1136.

Riley, V., 1975.
Mouse mammary tumors: Alteration of incidence as apparent function of stress. Science, 189: 465-467.

Riley, V. and Spackman, D., 1981.
Psychoneuroimmunologic factors in neoplasia: Studies in animals. In: R. Ader (ed), Psychoneuroimmunology. Academic Press, New York, pp 31-101.

Rogers, M.P., Reich, P., Strom, T.B. and Carpenter, C.B., 1976.
Behaviorally conditioned immunosuppression: Replication of a recent study. Psychosom. med., 38: 447-451.

Rogers, M.P., Trentham, D., McCune, J., Ginsberg, B.I., Rennke, H.G., Reich, P. and David, J.R., 1980a.
Abrogation of Type II collagen-induced arthritis in rats by psychological stress. Arthr. Rheum., 23: 1337-1342.

Rogers, M.P., Trentham, D.E. and Reich, P., 1980b.
Modulation of collagen-induced arthritis by different stress protocols. Paper presented at the annual meetings of the American Psychosomatic Society, New York, NY.

Roszkwoski, W., Plaut, M. and Lichtenstein, L.M., 1977.
Selective display of histamine receptors on lymphocytes. Science, 195: 683-685.

Roszman, T.L., Cross, R.J., Brooks, W.H. and Markesberry, W.R., 1982.
Hypothalamic-immune interactions: II. The effect of hypothalamic lesions on the ability of adherent spleen cells to limit blastogenesis. Immunology, 45: 737-742.

Schleiffer, S.J., Keller, S.E., McKegney, F.P. and Stein, M., 1980.
Bereavement and lymphocyte function. Presented at the meeting of the

American Psychiatric Association, San Francisco, CA.

Selye, H., 1950.
Stress. Acta, Montreal.

Shavit, Y, Lewis, J.W., Terman, G.W., Gale, R.P. and Liebskind, J.C., 1982.
Opiod peptides may mediate the immunosuppressive effect of stress. Presented at the meetings of the Society for Neuroscience, Minneapolis, MN.

Singh, U., Millson, D.S., Smith, P.A. and Owen, J.J.T., 1979.
Identification of B adrenoceptors during thymocyte ontogeny in mice. Eur. J. Immunol., 9: 31-35.

Sklar, L.S. and Anisman, H., 1979.
Stress and coping factors influence tumor growth. Science, 205: 513-515.

Smith, L.W., Molomut, N. and Gottfried, B., 1960.
Effect of subconvulsive audiogenic stress in mice on turpentine induced inflammation. Proc. Soc. exp. Biol. Med., 103: 370-372.

Soave, O.A., 1964.
Reactivation of rabies virus infection in the guinea pig due to the stress of crowding. Amer. J. vet. Res., 25: 268-269.

Solomon, G.F., Levine, S. and Kraft, J.K., 1968.
Early experience and immunity. Nature, 220: 821-822.

Solomon, G.F., 1969.
Stress and antibody response in rats. Int. Arch. Allergy, 35: 97-104

Solomon, G.F., 1981.
Emotional and personality factors in the onset and course of autoimmune disease, particularly rheumatoid arthritis. In: R. Ader (ed), Psychoneuroimmunology. Academic Press, New York, pp 159-183.

Solomon, G.F. and Amkraut, A.A., 1981.
Psychoneuroendocrinological effects on the immune response. Ann. Rev. Microbiol., 35: 155-184.

Spector, N.H. and Korneva, E.A., 1981.
Neurophysiology, immunophysiology, and neuroimmunomodulation. In: R. Ader (ed), Psychoneuroimmunology. Academic Press, New York, pp 449-473.

Stein, M., Schleifer, S.J. and Keller, S.E., 1981.
Hypothalamic influences on immune responses. In: R. Ader (ed), Psychoneuroimmunology. Academic Press, New York, pp 429-447.

Strom, T.B., Sytkowsky, A.J., Carpenter, C.B. and Merrill, J.P., 1974.
Cholinergic augmentation of lymphocyte-mediated cytotoxicity. A study of the cholinergic receptor of cytotoxic T-lymphocytes. Proc. Nat. Acad. Sci. (USA), 69: 1330-1333.

Suskind, R., 1977.
Malnutrition and the immune response. Raven Press, New York.

Teshima, H., Kubo, C., Nagata, S., Imada, Y. and Ago, A., 1981.
Influence of stress on phagocytic function of macrophage. Shinshin-Igaku, 21: 99-103.

Treadwell, P.E. and Rasmussen, A.F.jr., 1961.
Role of the adrenals in stress induced resistance to anaphylactic shock. J. Immunol., 87: 492-497.

Versteeg, D.H.G. and Wurtman, R.J., 1975.
Effect of $ACTH_{4-10}$ on the rate of synthesis of (^3H)catecholamines in the brains of hypophysectomized and adrenalectomized rats. Brain Res., 93: 552-557.

Vessey, S.H., 1964.
Effects of grouping on levels of circulating antibodies in mice.

670

Proc. Soc. exp. Biol. Med., 115: 252-255.

Vinogradova, V.D., 1960.
The influence of conditioned reflex activity on blastomatous growth in animals. Vopr. Onkol., 6: 73-75.

Wayner, E.A., Flannery, G.R. and Singer, G., 1978.
Effects of taste aversion conditioning on the primary antibody response to sheep red blood cells and Brucella abortus in the albino rat. Physiol. Behav., 21: 995-1000.

Weigant, V.M., Dunn, A.J., Schotman, P. and Gispen, W.H., 1979.
ACTH-like neurotropic peptides: Possible regulators of brain cyclic AMP. Brain Res., 168: 565-584.

Weiner, H., 1977.
Psychobiology and human disease. Elsevier, New York.

Weinmann, C.J. and Rithman, A.H., 1967.
Effects of stress upon acquired immunity to the dwarf tapeworm Hymenolepsis nana. Exp. Parasitol., 21: 61-67.

Williams, J.M., Peterson, R.G., Shea, P.A., Schmedtje, J.F., Bauer, D.C. and Felten, D.L., 1981.
Sympathetic innervation of murine thymus and spleen: Evidence for a functional link between the nervous and immune systems. Brain Res. Bull., 6: 8-94.

Winokur, G., Stern, J. and Graham, D.T., 1958.
Effect of stress and gentling on hair loss due to epidermal carcinogen painting. J. Psychosom. Res., 2: 266-270.

Wistar, R.jr. and Hildemann, W.H., 1960.
Effect of stress on skin transplantation immunity in mice. Science, 131: 159-160.

Yamada, A., Jensen, M.M. and Rasmussen, A.F.jr., 1964.
Stress and susceptibility to viral infections: III. Antibody response and viral retention during avoidance learning stress. Proc. Soc. exp. Biol. Med., 116: 677-680.

Yunis, E.J., Fernandes, G. and Greenberg, L.J., 1976.
Tumor immunology, autoimmunity and aging. J. Amer. Geriat. Soc., 24: 253-263.

EMOTIONS, IMMUNITY AND DISEASE:
AN HISTORICAL AND PHILOSOPHICAL PERSPECTIVE

G. Solomon

Department of Psychiatry

University of California, San Francisco

Really to be historical about mind-body interactions let us go back to some very ancient concepts of medicine. When in India learning about traditional Hindu medicine, I was surprised to find out that Ayurveda, which dates back over two thousand years, had concepts of natural and acquired immunity, of psychophysiological response specificity, and believed that certain types of people, based on personality and somatotype, had greater resistance to disease (Shukla et al., 1979). Galen said that melancholy women are more prone to cancer than sanguine women, and Osler is reputed to have said that it is as important to know what is going on in a man's head as in his chest in order to predict the outcome of pulmonary tuberculosis. Jonas Salk pointed out in the early 1960's that all disease really relates to genetic, behavioral, nervous and immune interrelationships (and I think we should add endocrine)(Salk, 1962). We need to think of all disease in multifactoral ways. For example, in Mirsky's work, dependency and/or stress measures were lower in patients with peptic ulcer who had very high pepsinogen levels (Mirsky, 1957).

In 1963 I put a sign in front of my laboratory titled "Psychoimmunology Laboratory". Some of my colleagues thought I needed help by other of my colleagues. My experimental work in this field emerged as a result of prior clinical research on rheumatoid arthritis, actually beginning in medical school with a psychoanalytic bent (symbolism of motility and libidinization of the musculoskeletal system in rheumatoid patients) suggested by my father, Joseph C. Solomon, M.D. As a resident in psychiatry, I began working with the Rheumatic Disease Group of the University of California, San Francisco. Thus, I became further interested in emotional and stress factors in the onset and course of rheumatoid arthritis and other autoimmune diseases and began a fruitful research collaboration in this area with Rudolf H. Moos, Ph.D. (Solomon, 1981a). Becoming familiar with the work of Fudenburg and others, I became aware (long before an

understanding of the role of suppressor T cells), that autoimmunity somehow was related to relative immunologic incompetance. A mental "light-
bulb" went off, since patients with these diseases seemed to have failure
of psychological defenses and emotional distress, since such emotional
states had been shown to be associated with elevation of adrenal corticosteroids, and since these hormones appeared to be immunosuppressive
(Solomon & Moos, 1964). I then felt it would be necessary to prove experimentally that stress could be immunosuppressive; therefore, I went off
to the laboratory. Those initial ideas I feel now were most naive, but
they were critical in getting experimental work going in this area. I was
able to show stress and other experiential effects on humoral immunity
(Solomon, 1969; Solomon et al., 1968), and later with the able immunologist, Alfred A. Amkraut, on cellular immunity as well (Amkraut et al.,
1972). To go to the "other side of the coin", so to speak, as a consultant to the Rheumatic Disease Group, at first the only patients I was
asked to see were patients with systemic lupus erythematosis who had
psychosis in conjunction with that disease. I became struck by the similarity of the psychosis associated with that autoimmune disease and schizophrenia and wondered whether or not there might be autoimmune phenomena
in conjuction with schizophrenia and whether that disease itself might
not be an autoimmune disorder. That idea, subsequently taken up by Heath
(Heath & Krupp, 1967), was an attractive hypothesis (but one for which
evidence proved to be weak) because one could integrate stress, personality and genetic factors, all of which seem to play a role in schizophrenia. Luckily, I did the test for antinuclear antibodies inaccurately
and found a very high incidence of them in schizophrenic patients, encouraging me and Jeffrey Fessel to further research. What turns out to be
the case is that when one looks for any autoantibody, including antibrain
antibodies, there is a higher incidence among schizophrenic patients
(Solomon, 1981b), as will be mentioned later. I remain very dubious
whether they play a direct role in the illenss. Another extremely interesting phenomenon, which I think if ever explained will help understand
both diseases, is that in spite of the high incidence of the autoantibody
rheumatoid factor (an IgM anti-IgG) in schizophrenia, the two active diseases do not appear to co-exist (although non-schizophrenic psychoses
are found in rheumatoid arthritic patients) (Rothermich & Phillips,
1963).

If the central nervous system (CNS) and the immune systems are linked, there are several hypotheses that should be able to be proven. But philosophically why should we think of linking these two systems? Both systems serve functions of adaptation and defense and relate the organism to the outside world. In both systems defenses gone awry produce disease. In the immune system inappropriate defenses are called allergies, and in the central nervous system or the emotional system they are called neuroses. When the systems, if you will, turn against the self, in the psychiatric sphere, as Freud pointed out in terms of "retroflexed" hostility, one has depression, and in the immune system one has autoimmunity. Both systems have the function of memory and learn by experience. Immunization or prior experience in both systems can lead either to tolerance or to hypersensitivity. A person who had similar stresses before may be more resistant to subsequent traumatic events and have higher "ego strength". On the other hand, he or she may be hypersensitive and react excessively because of similar prior traumatic experience, such as loss for example. So, at least philosophically, there are "reasons" to postulate links. Now, if these systems be linked, the following are the hypotheses, it seems to me, that should be able to be proved. First, emotional upset and distress should alter the incidence or severity of those diseases to which there is immunologic resistance (infectious and neo-.plastic) and those associated with immunologic deficiency states or aberrant immunity (autoimmune and allergic). Second, severe emotional and mental dysfunction should be accompanied by immunologic abnormalities. Third, hormones regulated by the central nervous system, the neuroendocrines, should influence immune mechanisms. Fourth, experimental manipulation of appropriate portions of the central nervous system should have immunologic consequences. Fifth, experimental behavioral manipulation, such as stress, conditioning or differential early experience, should have immunologic consequences. Sixth, immunologically competent cells should have receptor sites for neuroendocrines, neurotransmitters, or substances regulated by them. Seventh, activation of the immune system should be accompanied by CNS phenomena, suggesting that feedback mechanisms in immune regulation act via the central nervous system. Obviously, I cannot give you all the data to support these various hypotheses, but I shall very superficially and briefly allude to evidence supporting each.

Personality and stress factors have been related to a variety of allergic diseases clinically. Experimentally, in guinea pigs the probability of delayed skin hypersensitivity to DNZB in sub-threshold concentration is increased in a stressed group (Mettrop & Visser, 1971). Asthmatic patients have been described in general as having varied personalities but as generally having unconscious dependency conflicts as a result of specific childhood experiences and parental attitudes. The asthmatic attack may occur when there is a frustration of dependency wishes (Weiner, 1977). In the autoimmune diseases (rheumatoid arthritis, systemic lupus erythematosus, multiple sclerosis, myasthenia gravis, Graves' diseases, and a number of others), people who are prone to such diseases are described generally as quiet, introverted, reliable, conscientious, restricted in the expression of emotion (particularly anger), conforming, self-sacrifying, tending to allow themselves to be imposed upon, sensitive to criticism, distant, overactive and busy, pseudo-independent to deny dependency, stubborn, rigid and controlling (Solomon, 1981a). The onset of disease occurs either after a period of psychological stress, such as loss of a significant person, or after the interruption of the ability to maintain previously successful patterns of adaptation and defense. Rapidity of progression of disease (Moos & Solomon, 1964), degree of incapacity by disease (Moos & Solomon, 1965), and lack of response to treatment in autoimmune disease (Solomon & Moos, 1965a), in our work have all been shown to be related to failure of psychological defenses and concomitant anxiety, depression, and alienation. One study of ours is of particular interest because of its correlation of psychological variables with a physiological variable in the absence of overt disease. We compared physically healthy individuals, as determined by history, x-ray and physical examination, differing between groups by whether or not they had rheumatoid factor (an autoantibody) in their sera (Solomon & Moos, 1965b). We found that the healthy individuals who did have rheumatoid factor were very emotionally healthy; whereas, the people negative for rheumatoid factor were like a cross section of a general population - from "together" to "crazy". All of the rheumatoid factor positive people were not significantly anxious, depressed or alienated, meaning to us that the combination of failure of psychological defenses and emotional distress in the presence of rheumatoid factor would have led to overt disease. Thus, in order to remain healthy if one has this autoantibody, which probably reflects genetic predisposition to rheumatoid disease, one has to be in good psychological condition as well. However, among the rheumatoid factor positive relatives there were more similarities similar in the types of adaptations as the case in arthritics, posing a question which is hard to answer. Perhaps genetic or constitutional factors might be integrated with a psychological theory by the assumptions that the biological capacity for production of rheumatoid factor might be triggered by physiological consequences of certain types of psychological traits and accompanied by disease in the presence of other more intense psychological factors. In a suggestive pilot study, I found that football players (who might possibly be expected to have a high incidence of "autoimmune-prone personalities" because of their interest in sports) who showed elevating on "neurotic" scales on the MMPI before the football season, developed low titers of rheumatoid factor after the stress of close losses to teams with which there was keen rivalry. In a similar vein, Goodman found that emotional stress heightens the tendency to produce thyroid autoantibodies (Goodman et al., 1963).

Emotional factors and personality factors have been tied to the presence and the rapidity of progression of cancer in a large number of studies, rapid progression of cancer being associated with unsuccessful psychological defenses and psychic distress, quite analogous to the case in autoimmune diseases. Again, in infectious diseases the clinical data are quite similar. There are two infectious diseases that interest me particularly. One is acute necrotizing ulcerative gingivitis or trench-mouth, which tends to occur after acute stress, such as in college students before finals (Cohen-Cole et al., 1981). Of special note is that the invading bacteria are the normal flora of the mouth; non-"pathogens" become invasive. This phenomenon means to me that there must be some change in immune balance, probably changes in secretory IgA, in such individuals. Another such disease is herpes. In afflicted individuals, the virus continues to live in the tissues and is suppressed by T cells. Under conditions of physical or emotional stress, it is activated. What are the psychophysiological factors that can lead to this failure of suppression or to activation?

Another line of evidence to show that the central nervous system is related to immunity is that in major dysfunction of the central nervous system, as occurs in mental illness, there are immunologic abnormalities. There have been descriptions of a variety of such abnormalities, particularly in schizophrenia, including a relatively high incidence of a variety of autoantibodies including, as said, antibrain antibodies, dysglobulinemias, purported "brain allergies", and alterations in immune responsivity (Solomon, 1981b).

Evidence that appears to hold up is that there are morphological differences of some lymphocytes in many schizophrenia patients, as well as higher incidence of such abnormalities in their relatives (Hirata-Hibi, 1982). There is a suggestion in as yet unpublished work at Loma Linda University that there is a deficiency of suppressor T cells in schizophrenic patients, which observation might account for the ubiquitous autoimmune phenomena in that illness.

Certainly, there is much more direct evidence for central nervous system involvement in immune function. Pioneered by Soviet ablation studies in the hypothalamus (as confirmed in our own laboratory), it appears that destruction of certain areas of the hypothalamus can lead to suppression of both cellular and humoral immune function (Korneva & Khai, 1963). Soviet workers claim enhancement of immunity by stimulation of these same areas (Korneva, 1967). The hypothalamus is rich in neuroendocrines, regulates pituitary function and has elaborate connections to the limbic system (the emotional system of the brain). We know there is pituitary-thymic interaction (Pierpaoli et al., 1971). It has been found that administered thymic hormone localizes in periventricular areas of the hypothalamic region. We know that thymic hormones affect neuroendocrines, and neuroendocrines affect thymic hormones. Serotonergic structures in the hypothalamus appear to participate in regulation of antibody production, and drugs altering biogenic amines can affect immune response, with such effect blocked by hypophysectomy or lesions of the pituitary stalk (Devoino et al., 1970).

There have been a variety of stress studies in animals - too extensive to be described here - showing effects on humoral and cellular immunity, including tumor immunity. We were able to show that early experience affects adult immune response (Solomon et al., 1968). Handling or gentling in infancy enhanced the later adult immune response of the animal. (Immunologists may come to find early life experiences as critical as psychiatrists have.). Chronic stress may well have quite different immunologic effects from acute stress; Monjan found a biphasic response (Monjan & Collector, 1977). The nature of the stress, the timing of the stress and the specific component of the immune response that is affected are all very relevant. It was interesting to us that one of the most immunosuppressive sorts of stresses that we could find was overcrowding, which may have evolutionary significance in the sense that when populations of animals become very dense their immune function decreases, which might be a means of population regulation in nature. Important work of Dr. Robert Ader has shown that conditioning can alter immune response, essentially proving a role of the central nervous system in immune regulation (Ader & Cohen, 1975). He has paired the immunosuppressive drug cyclophosphamide with saccharin and has been able to show humoral immunosuppression when saccharin alone is given. Recently, cellular immune response (graft rejection) has also been shown capable of conditioning (Gorczynski et al., 1982). Ader has been able significantly to reduce the amount of immunosuppressive drug when paired with a conditioned placebo and maintain life prolongation in a mouse model of autoimmune disease, of possible major significance to pharmacology (Ader & Cohen, 1982).

We now know that a variety of hormones regulated by the central nervous system and influenced by experience, as well as neurotransmitters themselves, impact on specific areas of the immune function including, of course, cortisol and also insulin, testosterone, estrogens, growth hormone, β-adrenergic agents, histamine, and acetylcholine. Receptor sites for these substances are being identified on lymphocytes or thymocytes. We must now also think about immunologic effects mediated by experiential influences on the variety of lymphokines (substances elaborated by immunologically competent cells and influencing functions of others), endorphins, and prostaglandins. Serum from stressed animals can alter (enhance or depress depending on duration of stress) the in vitro response of lymphocytes from control animals (Monjan et al., to be published). An additional direct link between the CNS and the immune system is suggested by the rich innervation of bone marrow and of thymus, the significance of which remains to be understood. We must think of multiplicities of hormone response and not the effect of a single substance in isolation. Hormonal response likely acts via impact on "second messengers", cyclic AMP and cyclic GMP, known to have immunologic influences. As mentioned before, there is likely a complex pituitary-thymic interaction that is important. I have wondered for some time whether thymic hormones themselves are neuroendocrines. We may yet find releasing or inhibitory factors from the central nervous system that affect thymic functions. CNS-immune interrelationships, again, are illustrated by the finding that pathogen-free animals reared in germ-free environments have lower levels of central nervous system neurotransmitters. Left handedness, more common in females, appears associated with an increased risk of autoimmune disease, also more frequent among females (Marx, 1982). "Personality", even among animals, may affect immune responsiveness. We found that those females among an inbred strain of mice that spontaneously developed fighting behavior were more resistant to an immunologically-resisted tumor (Amkraut & Solomon, 1972).

Human studies of immune function and experience are now being more regularly reported. Bereaved persons show immunosuppression, now confirmed in several studies (Bartrop et al., 1977), as do melancholic patients (Kronbol et al., 1982) (and infant monkeys separated from their mothers (Laudenslager et al., 1982). Bereavement-induced immunosuppression, which may, in part, account for higher mortality rates among the bereaved, seems to recover in conjunction with working through of the loss. Coping and effective psychological defenses seem to operate on physiological levels as well as psychic. Locke has shown that stress in people who are poor copers, can result in diminution of natural killer cell activity (Locke et al., 1978).

Finally, let me mention aging. A theory of aging relates the process to a complex variety of centrally "programmed", hypothalamically-mediated metabolic changes with resultant immunosuppression and cancer proneness, and psychological depression may be an involved variable, both in cause and effect roles (Dilman, 1981). Could the decrease in reactive T cells in older people, the increased incidence of autoimmune phenomena and of cancer in the elderly be the cumulative result of stress or influenced by the relative commonality of depressive affect in the older age group?

To conclude, important developments in the prophylaxis and treatment of disease should be expected from the emerging new field of psychoneuroimmunology. The possible role of psychological interventions in prevention and amelioration of distress-induced immunosuppression and for immune enhancement should be studied. Are euphoria or sense of well-being and the "relaxation response" with its vagotonia, as observed during grooming in animals and meditation, biofeedback, and progressive relaxation in man, the converse of stress and of depression and accompanied by immune enhancement? If the central nervous system and immune system are closely linked, not only may understanding psychophysiology enhance understanding of immune mechanisms, but cellular and molecular immunology may enhance the understanding of mechanisms within the central nervous system. Similar mechanisms may have evolved in the functions and regulation of both adaptive-defensive systems. The central nervous and immune systems surely appear to be mutually interacting.

REFERENCES

Ader, R. and Cohen, N., 1975.
  Behaviorally conditioned immunosuppression. Psychosom. Med., 37: 333-340.
Ader, R. and Cohen, N., 1982.
  Behaviorally conditioned immunosuppression and murine systemic lupus erythematosus. Science, 215: 1534-1536.
Amkraut, A.A. and Solomon, G.F., 1972.
  Stress and murine sarcoma virus (Maloney)-induced tumors. Cancer Res., 32: 1428-1433.
Amkraut, A.A., Solomon, G.F., Kaspar, P. and Purdue, A., 1972.
  Effects of stress and hormonal intervention on the graft versus host response. Adv. Exper. Med. Biol., 29: 667-674.
Bartrop, R.W., Luckhurst, E., Lozarus, L. and Kiloh, L.G., 1977.
  Depressed lymphocyte function after bereavement. Lancet, April 17, pp. 834-835.
Cohen-Cole, S., Cogen, R., Stevens, A., Kirk, K., Gaitan, E. and Hair, J., 1981.
  Psychosocial, endocrine and immune factors in acute necrotizing ulcerative gingivitis. Psychosom. Med., 43: 91 (Abstract).
Devoino, L.V., Eremina, O.F. and Yu Ilyutchenok, R., 1970.
  The role of the hypothalamic-pituitary system in the mechanic factor of reserpine and 5-hydroxytryptophan on antibody production. Neuropharm., 9: 67-72, 1970.
Dilman, V.M., 1981.
  The law of deviation of homeostasis and diseases in aging. John Wright P.S.G., Boston.
Goodman, M., Rosenblatt, M., Gottlieb, J.S., Miller, J. and Chen, C.H., 1963.
  Effect of age, sex and schizophrenia on autoantibody production. Arch. Gen. Psych., 8: 518-526.
Gorczynski, R.M., MacRae, S. and Kennedy, M., 1982.
  Conditioned immune response associated with allogenic skin grafts in mice. J. Immunol., 129: 704-709.
Heath, R.G. and Krupp, I.M., 1967.
  Schizophrenia as an immunologic disorder. Arch. Gen. Psych., 16: 1-33.
Hirata-Hibi, M., 1982.
  Stimulated lymphocytes in schizophrenia. Arch. Gen. Psych., 39:82-87.
Korneva, E.A. and Khai, L.M., 1963.
  The effect of destruction of hypothalamic areas on immunogenesis. Fiz. Zh. SSSR Imeni, 49: 42-46.
Korneva, E.A., 1967.
  The effect of stimulating different mesencephalic structures on protective immune response patterns. Sechenov Physiol. J. USSR, 53: 42-47.
Kronbol, Z., Silva, J., Greden, J., Dembinski, B.S. and Carroll, M.D., 1982.
  Cell-mediated immunity in melancholia. Psychsom. Med., 44: 304 (Abstract).
Laudenslager, M., Recte, M. and Harbeck, R., 1982.
  Immune states during mother-infant separation. Psychsom. Med., 44: 303 (Abstract).
Locke, S.E., Hurst, M.W. and Heisel, J.S., 1978.
  Paper delivered to American Psychosomatic Society, April 1.

Marx, J.L., 1982.
Autoimmunity in left-handers. Science, 217:141-144.

Mettrop, P.J.G. and Visser, P., 1971.
Influence on the induction and elicitation of contact dermatitis in guinea pigs. Psychophysiol., 8: 45-53.

Mirsky, I.A., 1957.
The psychosomatic approach to the etiology of clinical disorders. Psychosom. Med., 19: 424-430.

Monjan, A.A. and Collector, M.I., 1977.
Stress-induced modulation of the immune response. Science, 196: 307-308.

Monjan, A.A., Collector, M.I. and Guchhait, R.B., to be published.
Stress-induced modulation of lymphocyte reactivity: role of humoral factors.

Moos, R.H. and Solomon, G.F., 1964.
Personality correlates of rapidity of progression of rheumatoid arthritis. Ann. Rheum. Dis., 23: 145-151.

Moos, R.H. and Solomon, G.F., 1965.
Personality correlates of the degree of functional incapacity in patients with physical disease. J. Chron. Dis., 18: 1019-1038.

Pierpaoli, W., Fabris, N. and Sorkin, E., 1971.
In: Cohen, S., Cudhowicz, G. and McCluskey, R.T. (eds), Cellular interaction in the immune response. S. Karger, Basel, pp 25-30.

Rothermich, N.O. and Phillips, V.K., 1963.
Rheumatoid arthritis in criminal and mentally ill populations. Arthr. Rheum., 6: 639-640.

Salk, J., 1962.
Biological basis of disease and behavior. Perspect. Biol. Med., 5: 198-206.

Shukla, H.C., Solomon, G.F. and Doshi, R.P., 1979.
The relevance of some Ayurvedic (traditional Indian medical) cocncepts to modern holistic health. J. Holistic Health, 4: 125-131.

Solomon, G.F. and Moos, R.H., 1964.
Emotions, immunity and disease. A speculative theoretical integration. Arch. Gen. Psychiat., 11: 657-674.

Solomon, G.F. and Moos, R.H., 1965a.
Psychologic aspects of response to treatment in rheumatoid arthritis. GP, 32: 113-119.

Solomon, G.F. and Moos, R.H., 1965b.
The relationship of personality to the presence of rheumatoid factor in asymptomatic relatives of patients with rheumatoid arthritis. Psychosom. Med., 27: 350-360.

Solomon, G.F., Levine, S. and Kraft, J.K., 1968.
Early experience and immunity. Nature, 220: 821-822.

Solomon, G.F., 1969.
Stress and antibody response in rats. Int. Arch. Allergy, 35: 97-104

Solomon, G.F., 1981a.
Emotional and personality factors in the onset and course of autoimmune disease, particularly rheumatoid arthritis. In: Ader, R.A. (ed), Psychoneuroimmunology, Academic Press, New York, pp 159-182.

Solomon, G.F., 1981b.
Immunologic abnormalities in mental illness. In: Ader, R.A. (ed), Psychoneuroimmunology. Academic Press, New York, pp. 259-278.

Weiner, H., 1977.
Psychobiology and human disease. Elsevier, New York, pp. 223-317.

IMMUNOGLOBULINS AS STRESS MARKERS?

H. Ursin[+], R. Mykletun[++], E.Isaksen[+],
R. Murison[+], R. Værnes[+], and O. Tønder[+++]

+) Institute of Physiological Psychology
   University of Bergen, Norway
++) Rogaland Research Institute, Stavanger, Norway
+++) The Gade Institute, Department of Microbiology and Immunology,
     University of Bergen, Norway

ABSTRACT
    Immunoglobulins may represent a valid stress marker since they
react slowly and over a long time. They may, therefore, represent an
integrated function of sustained activation over prolonged time. Low
levels of immunoglobulins have been found to relate to personality and
subjectively experienced stress levels in a population subjected to
long lasting work load with feelings of incompetence and insufficiency.
No such relationships were found in a population which was subjected
only to brief exposures to intense fear. The psychological factors in
our own study explained up to 30% of the variance in the
immunoglobulins and complements, even if all levels were within the
normal range. The relationship between these findings and pathology
has not been proven, but is a tempting opening for psychosomatic
research and for identification of valid stress markers.

The available methods for measurement of stress are partly systems
for scaling subjective reports, partly recordings of somatic changes,
in particular of the autonomic nervous system and endocrine system. In
the following we will discuss whether immune processes may function as
such markers, with particular emphasis on the immunoglobulins. Are
there any links between the immune system and the central nervous
system, and to what extent are such links related to pathology and
useful as stress markers? This requires a critical review of what is
already available as stress markers. What is known about the
reliability and validity of the available measurements? Particular
emphasis will be placed on the search for possible markers that may be
useful in general epidemiological and preventive health work. This
requires a meaningful and generally acceptable definition of stress.

Recent Scandinavian legislation on work environment legislation
aims at eliminating unecessary "stress" from the work environment.
There seems to be a general agreement between trade unions, employers
and politicians that stress is related to disease. It would then be

reasonable if the the medical and psychological expertise had a clear and generally accepted stress definition, and that this stress could be measured with reliable and valid stress markers biochemically or physiologically. However, this is not the case. There is, therefore, talk about a "crisis" in stress research (Wolf et al., 1979), and a growing impatience with the present state of vagueness and contradiction in an area so important for issues of health and life quality.

A main problem with the term "stress" is the attribution power it has acquired in popular language. A variety of diseases are "explained" as due to "stress" and the self-evident remedy is to abolish stress. On the other hand, stress is also supposed to be good for you: it helps performance (Frankenhaeuser, 1975), it is actively sought by sensation seekers, and may have a training effect. Further confusion arises from the fact that the term is partly used about the stimulus situation, partly about the response.

The response is a general, non-specific alteration of the vegetative nervous system, the endocrine system, the brain level of activity, the muscle tonus and general behavior. All changes that have been reported in the literature are identical to the general, non-specific activation response initially described by the neurophysiologists (Moruzzi and Magoun, 1949, Lindsley, 1951). Recent criticisms of the activation concept concentrate either on to what extent specific ascending pathways contribute or play an essential role for the activation response rather than the reticular formation. Also, there has been concern as to why there are poor correlations between the various vegetative processes used as activation indicators. This is easily explained by learning factors and the dual innervation principle for these processes (see Ursin 1978 for further discussion and references). Even if there probably is more specificity to the system, at least for the cerebral mechanisms involved, the general activation concept is useful and valid. The use of the term activation rather than stress clarifies which somatic process we are dealing with. It also makes it clear when the response occurs, why it is so common, why people seek it, and why there are curvilinear relationships between it and performance, affective value and health consequences. The concept is reasonably simple to convey to the general public, and it has no erroneous attribution properties in the layman's language.

Since the "stress" response is identical to activation, which is a general, nonspecific response, many indicators can be used for measuring this process. The simplest operational definition of stress, therefore, is that it is the process which produces a change in your favorite physiological parameter. The decision of a stress "marker" depends on reliability, convenience and economy. But the choice must also consider whether the marker has any relevance from a pathogenic point of view.

The stimulus is much more difficult to define. There is a high number of stimuli and psychosocial situations that elicit the activation response. There is no particular stimulus or class of stimuli that will produce activation in all subjects at all times, except shortlasting activation without any pathogenic effects. The interesting aspects of the stimulus situation, or the "load" that the organism is subjected to, is whether or not it is related to pathology. The dominating pathogenic factor seems to be time course, longlasting activation from emotional pressure and unsolved conflicts produce disease. The empirical basis for this "sustained activation theory" (Ursin, 1980) derives partly from animal experiments, partly from studies of humans. Lack of control, poor or missing response feedback or lack of information on performance combined with high work load constitute pathogenic factors both in rat ulceration studies (see Miller, 1980) and in studies involving health and welfare issues for industrial workers (Gardell, 1977, Karasek, 1979) (Fig. 1).

The consistency in data across species dealing with complex psychological mechanisms suggest that we deal with essential brain mechanisms, and that these mechanisms dampen activation. The basic assumption is that when these mechanisms do not operate pathology occurs.

Strong activation has definite unpleasant aspects to it. This has adaptive value since it motivates or "drives" the individual into finding a solution to a dangerous situation. The activation also aids the individual in acquiring the solution (Frankenhaeuser, 1975) even if there is a danger of overactivation and overload. Since it is in our power to dampen this response chemically, and since the experience of this state has negative elements, it is important to acquire a proper understanding of this state to avoid misuse of drugs and

684

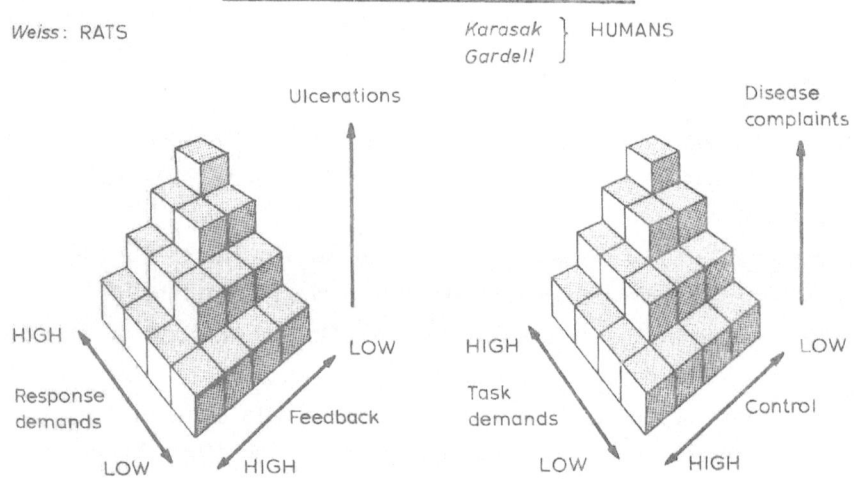

Fig. 1 - The rat model of Weiss (1972) and the human model of Karasek (1979)
    illustrating the relationship between task (response demand), con-
    trol (response feedback) and health variables. (Modified after
    Weiss 1972 and Karasek 1979).

misinterpretation of the activation ("stress") response. The hedonistic
attitude - or WHO health position - to avoid everything unpleasant may
be a dangerous position. It is normal and adaptive to be frightened by
dangers.

When a solution has been found to a threatening situation, and the
organism has learned that the solution is working, there is no need
for a continued high activation level. Accordingly, activation
decreases even if objectively the situation remains unchanged. This has
been demonstrated for rats in avoidance tasks (Coover, Ursin and Levine
1973) and humans in parachute training (Ursin, Baade and Levine 1978).
This may be referred to as "coping". Other words have been used, in
particular "adaptation". This seems an unfortunate choice since "adapt"
is used either for a passive process in sensory systems, or to express
an attribution idea that explains any biological phenomenon by its
existence. Coping implies that brains are able to find solutions to
problems, and, also, that brains learn that a solution works. More
formally, we assume that the subject develops a "positive response
outcome expectancy" (Bolles 1972, Seligman and Johnston 1973).

Humans may also meet threats by perceptual distortions of the threatening stimulus (Anna Freud 1937). These mechanisms may be measured by paper and pencil tests, by evaluation of clinical interviews, or by the tachistoscopic method described by Kragh (1960). The price for this activation dampening is that the situation is perceived less accurately and this may interfere with finding a proper solution. A high defense score on the Kragh test is correlated with poor performance in Swedish air force pilots and Danish attack divers (Kragh 1960), Norwegian parachutist trainees (Ursin, et. al. 1978) and Norwegian divers (Vaernes 1981). The defense mechanisms, therefore, differ from coping as defined here. It may be necessary for an individual to use defense, but coping is to be preferred. A low activation (stress) level at the first presentation of a threatening stimulus is not necessarily a good sign for coping capabilites. It could be a sign of high defense, and, therefore, low performance in dangerous situations.

To be interesting from a pathogenic or epidemiological point of view we must have a marker that relates to pathology. Personell with high levels of that marker must be identifiable as a risk, and there must be some remedy offered if the screening is to have any interest. There are reasons to question whether any of the traditionally available methods qualify this validity criterion. The validity for some of the very fast and shortlasting responses studied in traditional psychophysiology for the development of somatic disease is debatable. Heart rate changes lasting for a few seconds may be easy to record reliably, but are hard to accept as important for development of cardiovascular disease. On the other hand, some of these variables are very valid for psychosomatic research, as for instance blood pressure. Another advantage with these methods is that changes can be followed over long periods of time.

The emergence of reliable methods to measure endocrine levels in plasma has been a very promising development. Few if any of the handicaps above apply to endocrine activity. On the other hand, for most measurements blood samples are necessary. The endocrine changes are relatively slow, the rise time of the response to a stressor may vary from seconds to minutes, and the response may last for 15-60 min. Still, the response is usually shortlasting, and the

variance between stress and nonstress- groups declines over time. This
may be due to coping processes, or to physiological processes (Murison,
1980). According to the sustained activation hypothesis the hormonal
levels should be followed over time in order to give us an idea of who
is at risk. Any sample taken on any given day in a problem-ridden group
may not reveal the real risk-factor. High reactivity to stress
initially with high levels of tonic activation may not indicate a high
risk. On the contrary, it may be associated with high coping potentials
and a fast fall in activation.

The search for a physiological indicator, therefore, should
continue. The system should be slower, and have more "memory" aspects
to it, and, at the same time, have clear relationships to pathology.
In recent years, immunology has received increasing attention as a
possible mediator of psychosomatic disease. It is well established that
psychological stress influences the immune processes and the resistance
to infections and some neoplastic processes in animals (Friedman et
al., 1965; Riley, 1981; Ader, 1980). Sklar and Anisman (1979) have
shown that in animals the capacity to cope with environmental challenge
influences tumor growth and mortality in rats, in the same way as for
the endocrine system in general.

The mechanism through which the central nervous system (CNS)
exerts this influence could be the endocrine system, or via nerves to
the various lymphatic organs. The effect on the immune process could be
indirect, many hormones influence the immune process. But it is also
possible that the amount of immunoglobulin in itself could be affected.
If so, the plasma levels of the immunoglobulins could be used as a
marker, if there is enough variance within "normal" levels, and if this
variance is influenced by the activation process, in much the same way
as hormones are. There is, as far as I know, no direct evidence as to
the validity of any such stress marker, since there is no available
epidemiological data on the importance of fluctuations in the
immunoglobulins for somatic disease. Even so, the demonstration of the
importance of psyhological factors for the immune processes, and the
emergence of "psychoimmunology" makes this a very interesting and
promising area.

Recently we have found direct relationships between personality
factors and immunoglobulins in humans. In a study of 57 normal, healthy
primary school teachers, 40 females and 17 males, 26-65 years old

we have found significant correlations between immunoglobulins and personality traits (neuroticism on the Eysenck scale and defense measured with two different tests). The psychological factors explained 25-30% of the variance in the immunoglobulins and complement. Similar findings have been made by McClelland et al., (1980). They found low levels of immunoglobulin A (IGA) in saliva in college males with an inhibited and "stressed" "strong need for power". They also found that these men had high excretion of epinephrine in the urine, which correlated the low IGA-level, and, finally, the low levels of saliva IGA correlated significantly with reports of more frequent severe illnesses.

We found no relationships between personality and immunoglobulins in a population of young, healthy males that were not subjected to any chronic or sustained activation. However, their IgM rose sigificantly after one week of daily exposures to a frightening task, but this exposure only lasted for one hour or less each day.

Only in the group with sustained work stress was there a significant interaction between personality traits and the job-related problems and low levels of the relatively stable immunoglobulins IgG and IgA, and the complements. This was not evident in the group of healthy young men subjected to brief exposures to a frightening situation. Both groups showed relationships between situational factors and the fast-reacting IgM, but in opposite directions. The teachers had lower IgM if they reported high levels of psychological loads connected with their regular work. The Merchant Navy men increased their IgM levels during the days they were subjected to the brief exposures to the frightening experience. It should also be noted that their fear level decreased as a function of the repeated exposures. They learned to cope with the situation. In other similar situations this leads to a reduced physiological response in the variables related to pathology ("tonic activation"- norepinephrine, cortisol, free fatty acids), but a response persists in variables that may be related to training effects (epinephrine, testosterone increase) (Ursin et al., 1978, Ursin et al., 1982).

Personality traits are regarded as stable psychological patterns. High defense scores are generally believed to represent personality dimensions that may be disadvantageous, at least when the task is dangerous (Værnes, 1983). It may also lead to unfortunate, conflict-ridden human relationships which may even have health consequences (Vickers, 1982). Since the strong correlations were found only in the teachers, there does not seem to be any direct link between the defense mechanisms and the immunoglobulins. Only subjects with high defense mechanisms exposed to a psychological load for a long period of time seem to be affected. The high psychological defense may interfere with their coping capacity, and they may be subjected to "sustained activation", which may produce somatic pathology, at least in animals, and possibly also in humans (Ursin, 1980). Our data support the conclusion of McClelland et al. (1980) and their hypothesis that a chronic conflict between goals ("need for power") and the actual achievement may lead to a chronic vegetative and endocrine overactivty which, again, may lead to an immunosuppressive effect making individuals more susceptible to illness. The low values of immunoglobulins and complement are within the normal range, but high levels of saliva IgA have been found to be associated with better health (Yodfat and Silvian, 1977), and reports of less severe illnesses (McClelland et al., 1980).

The young men subjected to the repeated fear exposures did not show any changes in the IgA or IgG in the short time period over which they were observed. However, there was an increase in IgG which gives at least modest support to the position that the brief activation episodes may have a "training" effect rather than a "straining" effect (Ursin et al., 1982)

Low IgG and low IgA are most consistently related to high psychological defense. IgG is the dominating immunoglobulin in human serum. It has the longest half-life of the immunoglobulins (17-23 days) and appears in serum later than IgM and IgA. It is the dominating immunoglobulin in the secondary immunological responses. IgM, which has the shortest half-life (5-days), relates mainly to the situational factors. IgM is the first immunoglobulin class to be produced during an infection (the primary antibody response). It has clear importance under infections; a lack of this increases the risk for septicaemia.

All the immunological responses represent a much slower and longlasting physiological response to psychological stress than what has been studied traditionally. The vegetative changes are very shortlasting. Hormone responses are slower in occurrence, and last longer, but are still changes that only last for minutes, or hours, and which might be reset by a night's sleep. If the sustained activation theory is right, only changes that last long may produce pathology. "Stress" markers based on very short-lasting physiological changes are not adequate, or must be used with repeated sampling over prolonged time periods. They may tell more about the expectancies of the investigation than of the subject of investigation.

Epidemiological research has already pointed out the importance of psychosocial factors for organic disease (Weiner, 1977), but the rational model for the pathophysiological process must also be accounted for to have any impact on medicine. We must also be able to measure who is at risk, and when, in order to suggest screening for preventive interventions. We suggest that qualification of immunoglobulins should be a part of a stress battery. By repeated tests over longer timeperiods and individual's immunoglobulin profile may be of help in identifying individuals at risk. The combination of a particular psychological trait and sustained psychological loads may produce a lowered immunological defense in some individuals. This may produce a wide range of diseases, that all have more than one cause. Our postulate is that for all these diseases, including neoplastic processes and infections, there may be a psychosomatic factor as well. It may not account for much of the variance for each category of disease, and may therefore be overlooked in the classical approach to the analysis of pathology where each type of pathology is analyzed separately. But the total sum of pathological states where psychosomatic factors have a decisive influence may be much higher than we have realized.

REFERENCES

Ader, R., 1980. Psychosomatic and psychoimmunologic research. Psychosom. Med., 42: 307-321.

Bolles, R.C., 1972. Reinforcement, expectancy and learning. Psychol.Rev., 79: 394-409.

Coover, G., Ursin, H. and Levine, S., 1973. Plasma corticosterone levels during active avoidance learning in rats. J. Comp. Physiol. Psychol., 82:170-174.

Frankenhaeuser, M., 1975. Experimental approaches to the study of catecholamines and emotion. In: L.Levi (Ed), Emotions. Their parameters and measurement. New York, Raven Press.

Freud, A., 1937. The Ego and the Mechanisms of Defense, London, Hogarth Press.

Friedman, S.B., Ader, R. and Glasgow, L.A., 1965. Effects of psychological stress in adult mice inoculated with Coxsackie B viruses. Psychosom.Med., 27: 361-368.

Gardell, B., 1977. Psychological and social problems of industrial work in affluent societies. Int.J.Psychol., 12: 125-134.

Karasek, R.A. Jr., 1979. Job demands, job decision latitude, and mental strain: Implications for job redesign. Admin.Sci.Quart., 24: 285-308.

Kragh, U., 1960. The Defense Mechanism Test: A new method for diagnosis and personnel selection. J.Appl.Psychol., 44: 303-309.

Lindsley, D.B., 1951. Emotion. In: S.Stevens (Ed), Handbook of experimental psychology, New York, Wiley.

McClelland, D.C. Floor, E., Davidson,R.J. and Saron, C., 1980. Stressed power motivation, sympathetic activation, immune function, and illness. J.Human Stress, 6: 11-19.

Miller, N.E., 1980. A perspective on the effects of stress and coping on health and disease. In: S. Levine and H. Ursin (Eds), Coping and Health, New York: Plenum.

Moruzzi,G. and Magoun, H.W., 1949. Brain stem reticular formation and activation of the EEG. EEG. clin. Neurophysiol., 1: 455-473.

Murison,R., 1980. Experimentally induced gastric ulceration: a model disorder for psychosomatic disease. In: S.Levine and H.Ursin (Eds), Coping & Health, New York, Plenum.

Riley, V., 1981. Psychoneuroendocrine influences on immunocompetence and neoplasia. Science,212:1100-1109.

Seligman, M.E.P. and Johnston, S.C.A., 1973. A cognitive theory of avoidance learning. In: F.T. McGruigan & D.B. Lumsden (Eds), Contemporary approaches to conditioning and learning, Washington: V.H.Winston.

Sklar, L.S. and Anisman, H., 1980. Social stress influences tumor growth. Psychosom. Med.,42: 347-365.

Ursin, H., 1980. Personality, activation and somatic health. A new psychosomatic theory. In: S.Levine & H.Ursin (Eds), Coping and Health. New York: Plenum Press.

Ursin, H., Baade, E. and Levine, S. (Eds), 1978. Psychobiology of stress: A study of coping men. New York: Academic Press,

Ursin, H., Murison, R. and Knardahl, S., (1983). Sustained activation and disease. In: Ursin, H. and Murison, R. (Eds), Biological and psychological basis of psychosomatic disease, Pergamon Press, Oxford.

Vaernes, R.J., 1982. The Defense Mechanism Test predicts inadequate performance under stress. Scand.J.Psychol., 23: 37-43.

Vickers Jnr, R.R., (1983). Cardiovascular disease and psychological defense: Development of a working hypothesis. In: Ursin, H. and Murison, R. (Eds), Biological and psychological basis of psychosomatic disease, Pergamon Press, Oxford .

Weiner, H., (1977). Psychobiology and human disease. Elsevier, New York.

Weiss, J.M., (1972). Influence of psychological variables on stress-induced pathology. In: Physiology, Emotion and Psychosomatic Illness, CIBA Foundation Symposium, Elsevier, Amsterdam.

Wolf, S. Almy, T.D., Bachrach, W.H., Spiro, H.M., Sturdevant, R.A. and Weiner, H., 1979. The role of stress in peptic ulcer disease. J. Human Stress, 5:27-37.

Yodfat, Y. & Silvian, H., 1977. A prospective study of acute respiratory infections among children in a kibbutz. J.Infect.Dis.,136:26-30.

PROBLEMS OF CLINICAL INTERDISCIPLINARY RESEARCH. INVESTIGATION INTO BRONCHIAL ASTHMA AS A PARADIGM

Margit von Kerekjarto

Department of Medical Psychology, University of Hamburg (FRG)

Abstract: A report on a comparative clinical study on psychosomatic illnesses will allow to illustrate some problems facing collaborative interdisciplinary research.

Assessments were concentrated on bronchial asthma (a.br.) using the other disorders as controls to reveal similarities and differences between different psychosomatic illnesses. Data were gathered over a 7-year period from a group of 235 patients (Table 1).

During the data-gathering period there were a number of problems which resulted from our collaboration with several other clinics in the hospital. I'd like to take up these first. On the one hand they reflect the every-day research routine in a clinic and how it can come into conflict with the scientific goals stated at the outset of a project, as we learned during the course of our research. On the other hand, where the problems would have been avoidable, there are valuable lessons to be gained for the future.

In the collaboration between our department and the referring clinics we encountered problems of patient selection and patient motivation. Despite an agreement that patients would be referred to us on a random basis, we found in many cases that the doctors' subjective judgments were decisive in determining which patients would be referred. We got the impression that particularly those patients with problems of peculiarities in doctor-patient interaction were favored for referral to us.

Furthermore, we noted large differences in the readiness of patients to work with us. These differences might, among other things, have been due to the attitude of the referring physiciens to psychosomatics and psychosomatic research, which may have in turn influenced the individual patient's willing-

TABLE 1: Diagnostic-Groups and Research-Tools of comparative Clinical Investigations (1. Study 1974 - 1977; partial Cross-Validation Study 1978 - 1980)

	Asthma br. N =169	Neuro-dermatitis N = 46	Hives N = 22	Dyshidrosis N = 20	Funct.heart disorders N = 17	Surgical Control N = 42
Cutaneous Tests (in vivo-exposition)	n=53+44	n = 46	n = 10	n = 17	n = 7	n = 22
(P)RIST + RAST	n=34+20	n = 46	n = 10	n = 17	n = 0	n = 8
Lungfunction-Tests:Provocation-Test with Whole-Body-Pletismo-graphy Forced Oscillation Techn.	n=53+44	–	–	n = 20	n = 17	–
Polyphysiographic Parameters: ECG, EMG, PGR, Respiration	n=28+20	n = 30	n = 10	n = 20 (no PGR)	n = 17	n = 27
Standard psychometric Tests for 'Traits' and 'States' personality-characteristics	n=53+20	n = 44	n = 10	n = 20	n = 17	n = 40
Gottschalk-Gleser-Content Analysis for Anxiety and Aggression	n=33+11	n = 26	n = 22	n = 13	n = 8	n = 21
Audio-video-taped anamnestic interview (Ratings for verbal and non-verbal behavior)	n = 44	–	–	n = 20	n = 17	–

ness to cooperate. Aside from the significance of the general attitude of the medical staff toward psychosomatics, interest in the project often varied with changes in the medical personnel.

The technical difficulties were due fundamentally to problems of data collection, data processing and continuity of referrals. For our comparative studies it was necessary for patients to appear for testing several times. (Tab. 2) Not all patients kept their second or third appointments. Some of the patients were also reluctant to fill out all of the many tests which are necessary for a broad-scale study such as ours. Many questionnaires, even if they were filled out, were incorrectly filled out or left incomplete. A number of verbal samples for the Gottschalk-Gleser-test could not be evaluated because they fell under the neccessary minimum of 100 words. The data for the allergy investigation was not always complete because of technical problems connected with the laboratory. Blood samples sometimes deviated from the standard volume. In a few cases not all the required tests were made. And sometimes there were deviations from the standard series in allergy testing. Further difficulties were encurred by a change in methodology from RIST to PRIST procedure.

In the realm of psychophysiology, it was not possible to evaluate all the registered parameters. Sometimes the problems lay with errors in recording, sometimes the quality of the transcriptions of recorded values onto magnetic tape was inadequate.

In retrospect this seems like quite a catalog of problems and near disasters. If I have presented it in such lurid detail, it is only by way of feeling with those of you who have been involved in projects of a similar scope and as a word to the wise for those of you who may undertake such broad-scale studies in the future.

Bronchial asthma has been known to medical science for centuries. It is a condition caused by certain changes in the respiratory passages. Just what causes these changes to occur it still unknown. Usually bronchial asthma is seen as being

694

TABLE 2: The time-schedule of an out-patient entering
the study

D e s i g n

---

t_1 (1. day): Standard series of 22 intracutaneous test-
applications (Clinic for Dermatology, Dept.
of Allergy and Immunology)

t_2 (2. day): 1 day later inspection:
Blood-sample, (P)RIST and RAST
(Clinic of Dermatology)
Appointment for t_2 (following week)

t_3 (7. day): Psychological tests and anamnestic interview
(Clinic for Internal Medicine II, Dept.
Medical Psychology)

t_4 (14. day): Lungfunction-Test (if Whole Body-Pletismo-
graphy, so 1 provocation per day, and 4
sessions at least)
(Clinic for Internal Medicine I, Dept. of
Pulmonology)

t_5 (21. day): Last Lungfunction-Test, Polyphysiographic
registering of ECG, EMG, PGR, Respiration
under 5 different situations: a) relaxation,
imagination of b) anxiety, c) anger,
d) depression, and e) pleasant feelings
(Clinic of Internal Medicine II, Dept.
Medical Psychology)

Time differnces t_1 - t_5 varied between 3 - 9 weeks !

conditioned by a number of factors - infectious, immunological, endocrine, neural, social, psychological and hereditary factors all bear consideration individually and in combination. Therefore, a.br. as a psychoneuroimmunological illness offers a good paradigm for theory formation in medicine. The first scientists to attempt this were Dutch, Groen and Bastiaans (1955).

For the purposes of our study we used the following definition:

Bronchial asthma is a primarily functionally determined, reversible, respiratory disability, localized for the most part in the bronchioles and the smaller bronchi. Bronchial asthma is essentially a sudden increase in the intrabronchial airflow resistance due to obstruction of the respiratory passages,which is in turn caused by different pathological mechanisms.(Fig.1+2)

The newest hypothesis for the pathophysiology of bronchial constriction is a biochemical one. The latest results show the importance of "second messengers" such as cAMP and cGMP in intracellular activity. The existence of specific receptors on the surface of mastcells is noteworthy in considering the interrelatedness of the immune and endocrine systems. The existence of specific receptors with specific sites for histamine, adrenergic catecholamines and actylcholine has been established. They affect the functional activity of lymphocytes. The fact that cholinergic and ß-adrenergic receptor sites exist on certain lymphocytes is especially relevant to our consideration of links between CNS and the immune system. Szentivanyi and Filipp (1958) were among the first to study the role of the hypothalamus in anaphylaxis. Autonomic effect may be mediated by cyclic nucleotides at the cellular level. Increased intracellular concentration of cAMP after activation of ß-adrenergic receptors has been found to inhibit IgE-mediated release of histamine and other mediators of anaphylaxis from lungtissues (Orange and Austin, 1970), a point that might be of the outmost importance from the endocrinological point of view.

We consider a.br. to be a somatic illness whose onset and progress is more or less influenced by psychological factors, e.g., emotional conflicts. The behavior pattern of the patients

696

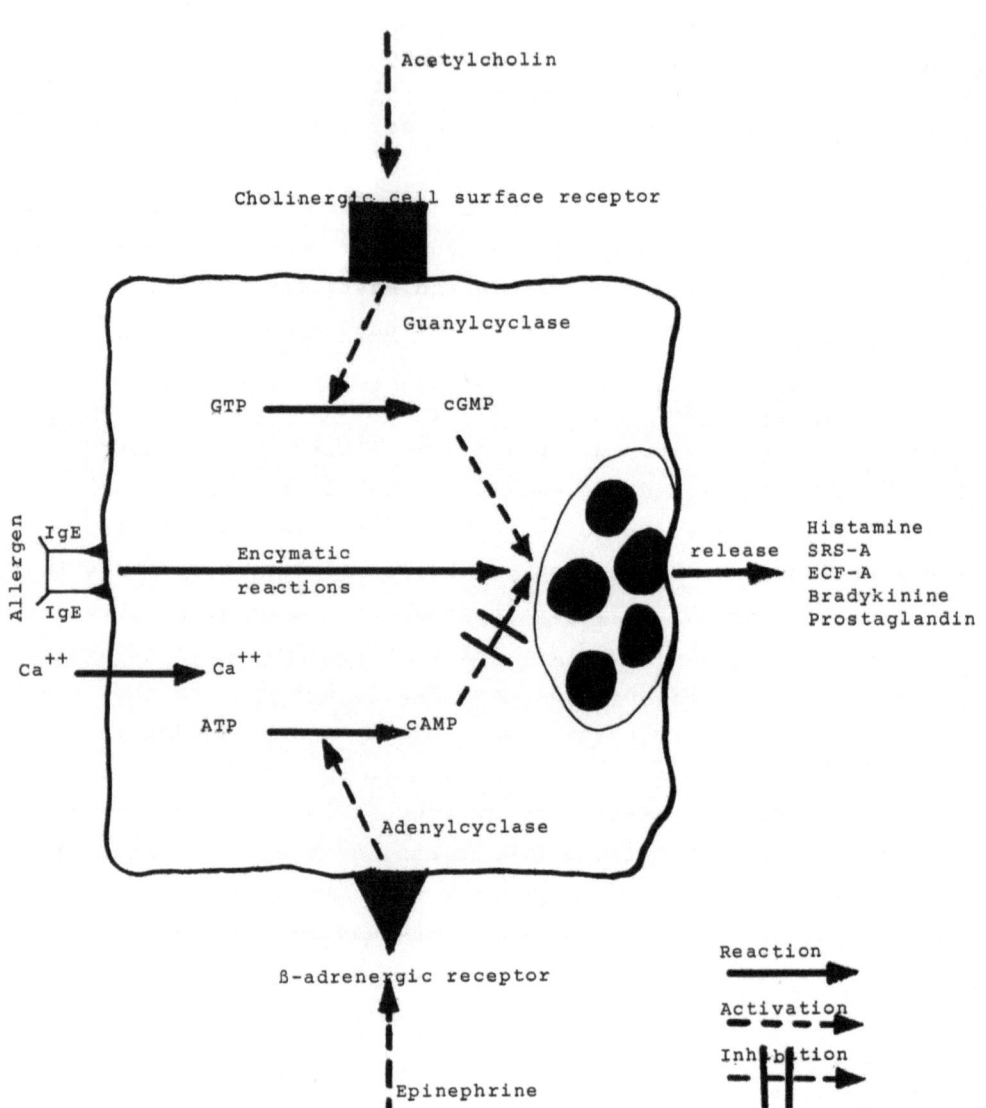

Fig. 1: Influences of acetylcholine and epinephrine
on the cAMP/cGMP system in the mastcelli.
In: Aas, K.: Allergy, 36, 3, 1981

Fig. 2: Basic mechanisms of cAMP and cGMP systems for the
tonus-regulation of smooth muscle (Histamine-mediated)
Adapted from: Aas, K.; Allergy, 36, 3, 1981

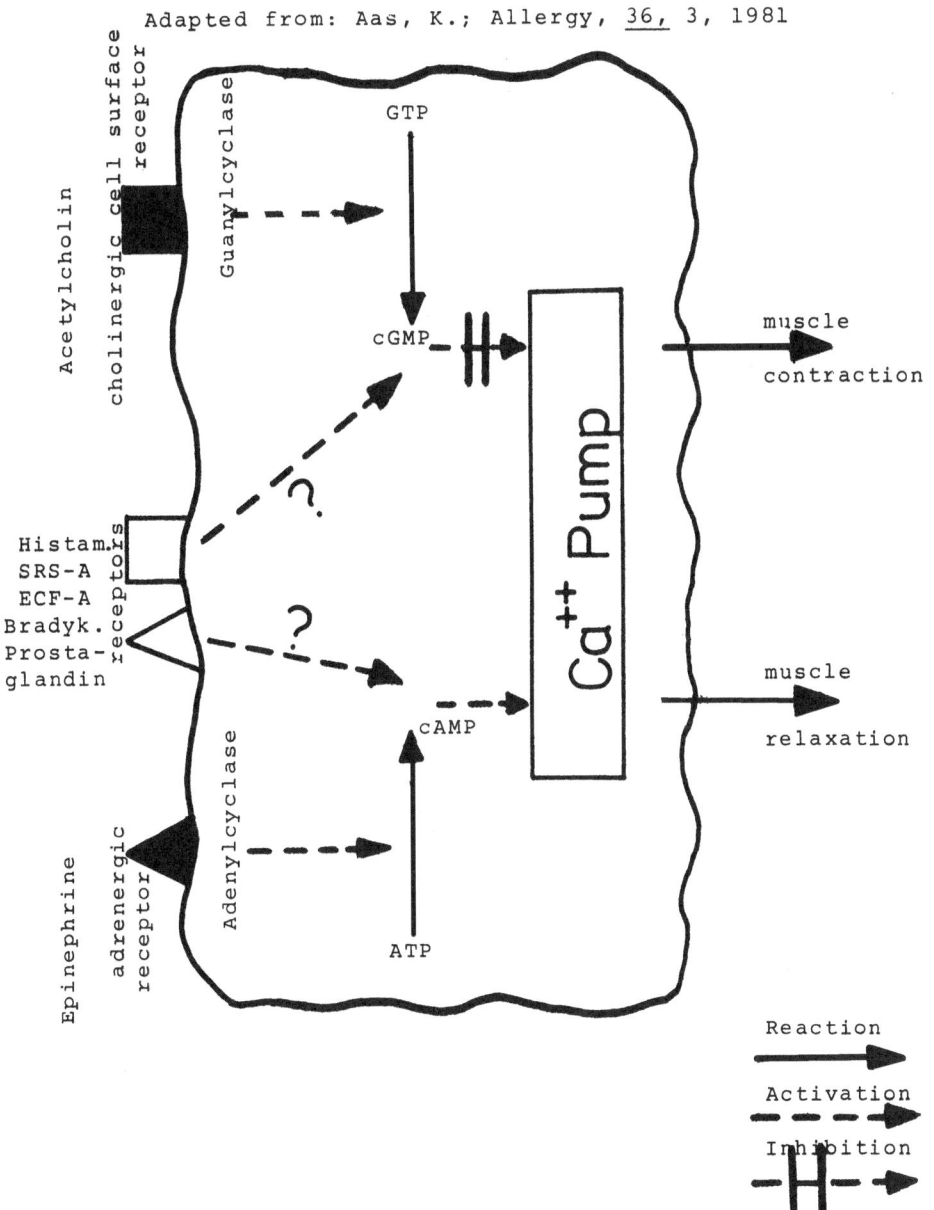

could be acquired through classical conditioning, and operant and cognitive learning processes. Changes in immune responses are associated with various affective disorders and states.

We agree with Solomon and Amkraut (1981) that "there is no asthmatic personality", but asthma patients have dependency wishes, and emotional conflicts, and the asthmatic attack occurs when these wishes are frustrated, or conflicts reactivated.

We could not find any psychological characteristics which differentiate asthma patients with diagnosed allergies from non-allergic asthmatics. The classification of a.br. in extrinsic and intrinsic does not differentiate between groupings of personality-traits or behavior-patterns of patients.

We have determined that the majority of patients with diagnosed allergic bronchial asthma, who, in their medical histories, attributed no significance to the allergen identified by tests in triggering symptoms, could identify more or less specific emotional conflicts as symptom provoking. In many cases an exact analysis of the situations in which attacks occurred revealed that the allergens provoked asthmatic reactions in the patients only in psychological conflict situations. I think this findings provides further evidence in support of Ader's (1980) theory that, "however pathogenic a given stimulus is presumed to be, the possibility that its effects are subject to the influence of psychosocial variables cannot be overlooked."

In the part of our study dealing with allergy and immunology we investigated the interrelations between psychological and allergic reactions in patients who were more or less strongly allergic. We also attempted to determine whether the allergy as a form of somatic hypersensitivity correspondy to a form of psychological hypersensitivity. We were also interested in a methodological problem, that is, which diagnostic procedure for allergies correlated best with psychological testing: intracutaneous, lung provocation, or immunological.

We found some significant intercorrelations between lung-function parameters, skin-tests (in vivo test) on the one hand

and psychometric measures on the other. (Tab. 3) Certain allergic diseases may be expressions of 2 or more types. Atopic disease is expressed as a.br. or atopic dermatitis. The patient may have 2 or more manifestations of the atopic state, but not necessarily at the same time. In contrast, all correlations with (P)RIST (in vitro-test) were insignificant. There was a slight to medium intercorrelation of the 3 allergy tests, but that is not surprising considering that the element common to the parameters used is the extent of atopic reactions (Tab.4 and Fig. 3)

Conclusions:

1. It appears unlikely to expect that it will be possible to coordinate laboratory and psychological testing of larger samples or reliably diagnosed patients for statistical evaluation. Hence it is necessary to plan single-case studies.

2. In single-case studies it would be possible to improve measurements of parameters such as breathing flow, which are subject to circadian rythms and mus be tested several times a day.

3. In addition to the patients's own subjective evaluation of their psychological state, evaluations by very well-trained psychodiagnosticians are necessary. Training in the recognition of relatively small psychopathological variations is an absolute prerequisite. This can only be achieved through uniform training of observers as, for example, through the use of video tapes and with international training workshops as for example in the use of the Burdock-Hardesty structured clinical interview.

TABLE 3: Significant correlations of allergic variables with FPI scales (similar to 16 PF) within two diagnostic groups (Spearman's Rho)

		Patients with Asthma br.	Patients with Dyshidrosis
FPI 1	Intracut.Test		
Nervousness	Inhalation	$-.20^{+}$	
FPI 2	Intracut.Test	$-.37^{++}$	
Spontaneous Aggressiveness	Inhalation	$-.20^{+}$	$-.50^{+}$
FPI 4	Intracut.Test	$.30^{+}$	
Excitability	Inhalation	$-.31^{++}$	
FPI 5	Intracut.Test		
Extroversion	Inhalation		$.35^{+}$
FPI 7	Intracut.Test		
Reactive Aggressiveness	Inhalation	$-.21^{+}$	
FPI 12	Intracut.Test	$.35^{+}$	
Masculinity	Inhalation		

No significant correlation between (P)RIST or RAST and psychodiagnostic variables with FPI scales was observed.

+   P   .50
++ p   .01

TABLE 4:   Number of negative test results compared
           with positive test results of varying
           number and strength in a standard series
           of 22 intracutaneous test

	Negative test results	Positive test results	N
Asthma [+]	14	30	44
Dyshidrosis[++]	1o	7	17

Legende: + Allergic Type I after Coombs and Gell
          ++ Allergic Type IV after Coombs and Gell

<u>Fig. 3:</u> The causes of slight correlations in
skin-tests, IgE and lung provocation
tests in asthmatics

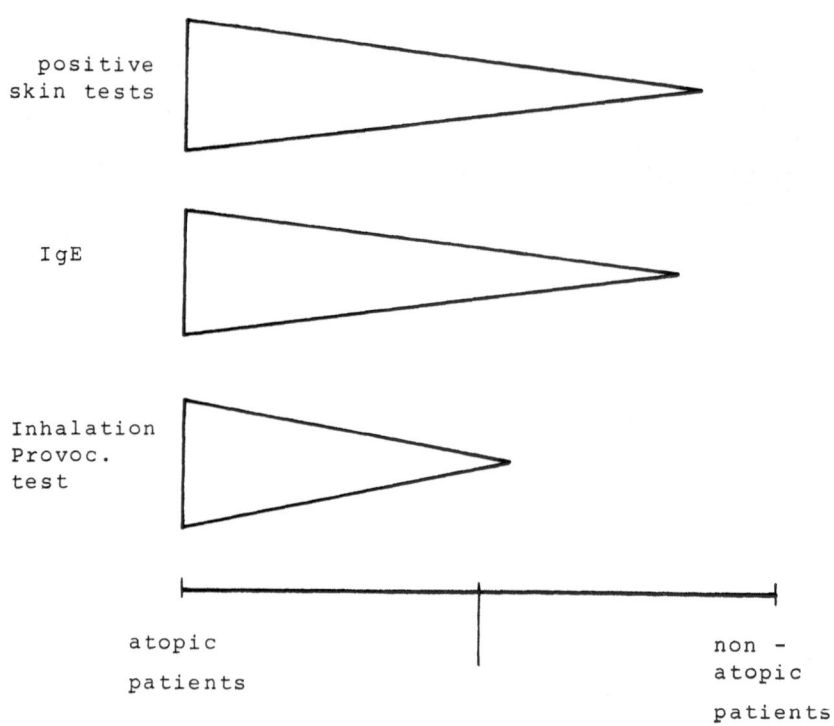

References:

Aas, K. - The radioallergosorbent test (RAST)
        Diagnostic and clinical significance.
        Ann.Allergy 33, 251, 1974

Ader, R. - Psychosomatic and psychoimmunologic research
        Psychosomatic Medicine, 42, 307-321, 1980

Bastiaans, J. and Groen, J. - Psychogenesis and psychotherapy
        of bronchial asthma
        Modern Trends in Psychosomatic Medicine
        Butterworth, London, 1954

Orange, P. and Austen, K.F. - Bronchial asthma: The possible role
        of the chemical mediators of immediate hypersensitivity
        in the pathogenesis of subacute chronic disease
        J.Exp.Med. 136, 556-567, 1972

Solomon, G. F. and Armkraut, A.A. - Psychoneuroendocrinological
        effects on the immune response
        Ann. Rev. Microbiol. 35, 155-184, 1981

Szentivanyi, A. - The beta-adrenergic theory of the atopic
        abnormality in bronchial allergy
        J.Allergy, 42, 2ol, 1968

FACTORS INVOLVED IN THE CLASSICAL CONDITIONING OF ANTIBODY RESPONSES IN MICE

R.M. Gorczynski, S. MacRae and M. Kennedy
Ontario Cancer Institute
500 Sherbourne Street
Toronto, Ontario
Canada M4X 1K9

Introduction

There is a great deal of evidence to support the contention that the immune response in mammals to foreign antigens is exquisitely regulated by the immune system itself, and by factors produced by cells associated directly or indirectly with the effects of immune stimulation (1). Amongst the former we could consider the intricacies of a network model for immuno-regulation (2), while amongst the latter for instance one could cite evidence for an effect of hormones. interleukins and products of arachidonic acid metabolism on immune responses (3-5). We (6) and others (7-11) have been interested, at the whole organism level, for any evidence which would support the notion that during immunomodulation, an organism can make some form of an association between non-antigenic environmental cues (a conditioned stimulus, CS) and a physiological stimulus known to perturb the immune system in a predictable way (an unconditioned stimulus, US). Thereafter presentation of the CS alone might then produce a physiological response (a conditioned response, CR) analogous to that initially evoked by the US itself (an unconditioned response, UR).

The studies described below examine evidence that, using a modification of a taste aversion protocol initially described by Ader and Cohen (7) psychological conditioning of antibody responses to SRBC can be observed. In this series of studies animals previously challenged with cyclophosph-amide (US) and saccharine (CS) subsequently showed immunosuppression when antigen (sheep erythrocytes - SRBC) was presented in the context of exposure to saccharine (CS). Despite attempts to control such variables as handling stress, age of mice, diurnal rhythmicity etc, the actual effect observed was not entirely predictable, conditioned suppression and enhancement being seen in individual experiments. However data obtained from animals condi-tioned after deliberate manipulation of their previous physical stress experience, or at different times of day, suggest a role for endogenous (hormonal?) factors associated with these states being of key importance in the development of the conditioned response measured. Prior adrenal-ectomy of the animals before initiation of conditioning trials abolished the conditioned immune response measured. In addition, we found that we were unable to develop a conditioned response in aged (24 months old) mice.

Most interesting of all perhaps was the data from a preliminary experiment in which we attempted to gain some independent behavioural measure which would be of predictive value in determining the subsequent conditioned immune response obtained. When animals were scored according to their activity in an open field trial and subsequently entered at random into a protocol to condition a suppressed antibody response to SRBC. those animals showing the lowest activity in the open field gave the greatest conditioned suppression and the greatest degree of correlation between the level of the conditioned response and the independently assessed activity level. The possible implications of these data in regard to our current knowledge of immune responses in general, and to future research in this field in particular, are discussed.

Materials and Methods

Mice: Eight week old male CBA mice were obtained from Jackson Laboratories. C57BL/6 mice (12 weeks and 24 months of age) were purchased from Charles River Laboratories. Mice were housed 7 per cage with food given ad libitum. All mice were adjusted to a restricted watering schedule (60 minutes per day - generally from 9:30 a.m. to 10:30 a.m.). The animal holding room itself was maintained on a regular 12 hour light - 12 hour dark cycle, with the light phase beginning at 6:30 a.m.

Sheep Erythrocytes: SRBC were obtained weekly from Woodland Farms, Guelph, Ontario. Cells were washed 3 times in PBS before use for injection or for plaque assay.

Cyclophosphamide: This was obtained from Bristol Laboratories of Canada (cytoxan) and was administered i.p. with 0.2 mls water at a dose of 100 mg/kg (2 mg for a 20 gram mouse).

Conditioning Protocol: After a series of preliminary investigations analysing the effect of varying the schedule of administration of cyclophosphamide and saccharine during the treatment and test phases of the experiment (below) we finally settled empirically on the following protocol. The basic design was adapted from that using the classical (Pavlovian) conditioning model described by Ader and Cohen (7), as modified by Cunningham et al#. Three pairings of cyclophosphamide, Cy, (US) and saccharine, Sacc, (CS) (in the drinking water), were given to mice at 14 day intervals, followed by challenge of all groups with 4 x 10^8 sheep erythrocytes (SRBC) i.p. with/without subsequent re-exposure to saccharine in the drinking water. The essential groups used were:

```
Group 1 (CS + US) x 3: SRBC + CS
Group 2 (CS + US) x 3: SRBC
Group 3 US x 3:        SRBC + CS
Group 4 CS x 3:        SRBC + CS
        treatment      test
```

Where mice were given the CS (saccharine) in the 'test' phase of the experiment the flavoured drinking water was given on the day of antigen challenge (day 0) and on days 2 and 4. This has proven an essential step in the elicitation of 'conditioned immunosuppression'. Mice in all groups were sacrificed for PFC assay 6 days after antigen challenge.

Plaque-Forming Cell (PFC) Assay: The assay for IgM-SRBC-PFC was performed as desribed by Cunningham and Szenberg (12). Guinea pig complement was obtained from Cedarlane Laboratories, Hornby, Ontario.

\# Cunningham, A.J., Kennedy, M., Poulos, C.X., Ciampi, A. and Gorczynski, R.M. (1983) Behavioural conditioning of antibody responses in mice. Submitted for publication.

Physical Stress: Mice receiving no previous deliberate physical or
psychological stress were subjected to a rotational physical stress
as described by Riley (13). Acute stress was defined arbitrarily as one
period of rotation at 45 r.p.m. for 45 minutes while chronic stress was
defined as 5 daily periods of rotation (each for 20 minutes).

Adrenalectomy: Mice were adrenalectomized under anaesthesia and subsequently
allowed a 14-day recovery period before entry into any experimental
protocols. After adrenalectomy (or sham surgery for the control animals)
all mice were retained on salt water (0.1% NaCl, 2 hours restricted
drinking per day).

Behavioural Assessment: Preliminary experiments (Cunningham, A.J.,
unpublished) have failed to detect any clear correlation between the degree
of taste aversion and conditioned immunosuppression/immunoenhancement in
mice entered into the experimental protocol discussed above. As an
alternative behavioural parameter we have assessed activity of mice in
an open field. Mice were allowed free movement for 5 minutes in an open
field on days -16, -8 and -1. An activity score was recorded automatically
on each occasion and the final activity assessed to each mouse was the
geometric mean of the three readings. No consistent pattern of variation
in activities over the three assessment periods has been observed amongst
a group of animals so tested   On day 0, 34 mice were randomly assigned
to receive either (Cy + Sacc) or (Cy + H$_2$O). After 3 such pairings all
mice were injected with SRBC and the (Cy + Sacc) group was further
subdivided into mice re-exposed/not re-exposed to saccharine. IgM-PFC
responses in the individual groups were ranked, as were the activity
measurements and a correlation analysis of the two rankings performed
(Kendall's coefficient of concordance). After this same ranking was
performed for the conditioned mice re exposed to saccharine, the mice
were further subdivided into two sub-groups based on their IgM-PFC responses
and a correlation analysis performed on the sub-groups (see Table 4).

Results

   Data in Table 1 are shown from eight independent experiments in which we
examined the antibody response (PFC per 1/100th spleen cell suspension
assayed on day 6 after immunization of mice) of conditioned and non-
conditioned and control animals. As reported elsewhere by Ader and Cohen
(7) it is possible using this approach to develop a conditioned suppression
when SRBC are given in association with the CS (saccharine) which had
previously been paired with the non-specific suppressive agent cyclophosphamide. However, in general our experience has been that despite rigorous
attempts to ensure uniformity in treatment schedules from experiment to
experiment on some occasions an actual conditioned enhancement of the
response occurred (a compensatory response?) e.q , experiments 4 and 7.
Clearly before any biological system can be explored in depth it must
demonstrate reproducible and predictable variation after any given
defined perturbation. We have thus investigated the possible reasons for
the fluctuations observed

   Anticipating that diurnal rhythmicity and/or previous physical
(emotional) stress experience may be a major contributor to the variability
measured we examined the conditioned antibody response in groups of
syngeneic age-matched mice whose initial challenge occurred (i) at
different times of day (7:00 a.m., 1:00 p.m.. 7 00 p.m.) or, (ii) at the

same time of day (1:00 p.m.) in groups of animals previously subjected to acute or chronic physical stress. In this latter case, following Riley

Table 1

Classical Conditioning of Antibody Response to Sheep Erythrocytes in Mice

Treatment regime[1]		IgM-PFC per 1/100th spleen cell suspension [2] (day 6)			
Experiment No.	Conditioning Test	Group 1 Sacc + Cy Sacc + PBS	Group 2 Sacc + Cy H_2O + PBS	Group 3 H_2O + Cy Sacc + PBS	Group 4 Sacc + PBS Sacc + PBS
1	CIS	39 (31-49)	65 (51-81)	68 (54-85)	410 (316-513)
2	CIS	27 (20-36)	51 (41·65)	57 (42-79)	536 (398-724)
3	NE	39 (30-50)	36 (29-46)	43 (32-58)	301 (240-380)
4	CIE	75 (58-100)	42 (33-52)	39 (31·49)	495 (380-631)
5	CIS	11 (8.7·14)	29 (23-36)	33 (25-44)	562 (447-708)
6	CIS	27 (20-36)	69 (48-100)	75 (58-100)	409 (316-515)
7	CIE	69 (55-87)	27 (21-34)	34 (25-46)	316 (251-398)
8	CIS	7.7 (5.0-11)	39 (30-50)	40 (32-50)	286 (229-363)

Footnotes to Table 1

1. Treatment schedules (see Materials and Methods) refer to the three sequential pairings of cyclophosphamide and saccharine given at 14 day intervals to 'condition' mice, or to the final trial when mice were given SRBC as antigen challenge along with saccharine flavoured drinking water to 'recall' the previous experience of cyclophosphamide. CIS indicates 'conditioned immunosuppression' and CIE 'conditioned immunoenhancement'.

2. Data are expressed as arithmetic mean (with range) of IgM-PFC per 1/100th spleen cell suspension averaged over the group (7 individual CBA mice were assayed per group shown).

(13), we have developed a rotational stress model in which animals are exposed daily for 5 days to 20 minutes rotation at 45 r.p.m (chronic stress) or once to 45 minutes at 45 r.p.m. (acute stress). Preliminary experiments (not shown) indicate that if challenged with SRBC immediately after conclusion of the final stress treatment mice suffering acute stress show a 2- to 3-fold decrease in IgM/IgG-splenic PFC 6 days later relative to unstressed control mice, while mice under the chronic stress treatment show a 1.5- to 2-fold enhanced IgM/IgG response. Note that in the particular experiment under consideration in Table 2 it is the <u>conditionability</u>

Table 2
Effect of Diurnal Rhythm and Stress on Classical Conditioning of Antibody Responses

Treatment[1]		PFC/100th spleen (day 6)[2]		
Experiment No.	Conditioning Test	Sacc + Cy Sacc + PBS	Sacc + Cy H_2O + PBS	H_2O + Cy Sacc + PBS
(CIS)	7:00 a.m.	7.6 (5.5-10)	29 (19-45)	33 (21-52)
1 (CIS)	1:00 p.m.	18 (13-25)	35 (25-48)	38 (25-58)
(NE)	7:00 p.m.	28 (18-45)	36 (25-52)	31 (20-49)
(CIS)	7:00 a.m.	21 (13 33)	69 (44-110)	58 (40-83)
2 (NE)	1:00 p.m.	45 (32-63)	60 (44-83)	63 (40-100)
(NE)	7:00 p.m.	69 (50-96)	58 (44-79)	64 (43-98)
(CIS)	7:00 a.m.	11 (6.9-17)	49 (40-60)	53 (40-69)
3 (CIS)	1:00 p.m.	29 (20-43)	52 (45-62)	60 (50-72)
(CIE)	7:00 p.m.	84 (63-100)	49 (40-60)	53 (45-62)
(CIS)	No deliberate stress	31 (25-38)	53 (38-74)	56 (50-63)
4 (CIS)	Acute physical stress	11 (8.7-14)	39 (25-62)	43 (32-58)
(NE)	Chronic physical stress	53 (42-66)	49 (40-62)	50 (35-71)
(CIE)	No deliberate stress	62 (49-78)	32 (26-41)	38 (32-48)
5 (CIS)	Acute physical stress	9.3 (7.9-12)	30 (25-38)	29 (23-36)
(CIE)	Chronic physical stress	71 (56-89)	42 (32-55)	46 (36-58)

Footnotes to Table 2

1. and 2. As for Table 1. Information in the first column refers either to the time of the initial pairing of cyclophosphamide/saccharine (experiments 1 to 3) or to the physical pretreatment of the CBA mice before they were entered into their first pairing trial (at 1:00 p.m.).

of the response which is under investigation, and mice are selected to differ in their 'behavioural' background state only at the time of initial pairing of saccharine/cyclophosphamide. Subsequent pairings of cyclophosphamide/saccharine occurred at 1:00 p.m. for all groups, and challenge with SRBC (and sacrifice) was also standard for all groups.

It is apparent from investigation of Table 2 that when stress or diurnal rhythm are deliberately manipulated quite predictable results are obtained. In particular mice entering a conditioning protocol at the onset of light (in a 12 hour light-dark environment cycle) invariably show a conditioned immunosupppression, as indeed do mice conditioned after deliberate acute physical stress. In contrast at appreciably later times of daylight in the periodic cycle, or after chronic stress treatment, no predictable conditioned suppression is seen (see also Table 1).

Table 3

Affect of Age or Adrenalectomy on Conditioned Immunosuppression of Antibody Responses

Treatment regime[1]		IgM-PFC per 1/100th spleen cell suspension (day 6)[2]			
		Group 1	Group 2	Group 3	Group 4
Experiment No.	Conditioning Test	Sacc + Cy Sacc + PBS	Sacc + Cy H_2O + PBS	H_2O + Cy Sacc + PBS	Sacc + PBS Sacc + PBS
1	CIS Sham adrenalect- omized	29 (23-36)	63 (50-79)	55 (44-69)	181 (167-197)
	NE Adrenalectomized	58 (46-72)	52 (42-66)	61 (50-76)	153 (138-170)
2	CIS Sham adrenalect- omized	31 (25-38)	74 (59-93)	83 (69-100)	258 (235-286)
	NE Adrenalectomized	69 (55-87)	81 (66-100)	75 (60-96)	261 (228-299)
3	CIS 3 month	23 (16-33)	47 (37-59)	51 (41-65)	291 (229-363)
	NE 24 month	81 (65-102)	69 (55-87)	74 (59-93)	260 (204-324)
4	CIS 3 month	14 (10-20)	39 (32-48)	40 (32-50)	306 (245-389)
	NE 24 month	72 (63-83)	81 (68-98)	75 (60-96)	219 (174-275)

Footnotes to Table 3

1. and 2. As for Table 1. Additional information in the first column
refers to the use of adrenalectomized or sham-adrenalectomized CBA mice
for conditioning (the surgery was performed 14 days before the first
pairing of cyclophosphamide and saccharine - experiments 1 and 2) or to
the use of different age C57BL/6 mice: 3 months/24 months respectively -
experiments 3 and 4   In all cases mice received their first conditioning
trial at 7:00 a.m. (see Table 2).

The data of Table 3 indicate preliminary studies aimed at examining
other parameters which may affect the conditionability of the immune
response to SRBC. Mice adrenalectomized 14 days before entry into a
conditioning trial showed no evidence for either a conditioned suppression
or a compensatory immunoenhancement. In contrast sham-adrenalectomized
mice showed a significant conditioned immunosuppression in the same
experiments (in this case, following the data shown in Table 2, the initial
trial for the animals was performed at 7:00 a.m.). In a similar vein when
aged animals were used we were again unsuccessful in establishing evidence
for a conditioned immune response pattern, despite the fact that mature
syngeneic sex-matched adults (12 weeks of age) were conditioned (immunosuppressed) in the same experiment.

In one final preliminary study we have asked whether there was any
independent behavioural measure which may be of predictive value in
determining the subsequent outcome of the conditioning trials performed.
As mentioned in the Materials and Methods to date we have no evidence that
the degree of taste-aversion can serve in this fashion.

Table 4
Correlation of Behaviour Measurement and Antibody Responses in Conditioned Mice

Treatment regime conditioning	test	No. of mice	Activity[1] score	IgM-PFC per[2] 1/100th spleen	Kendall's[3] coefficient of concordance
H_2O + Cy	Sacc + PBS	10	231 (182-288)	61 (50-76)	0.37
Sacc + Cy	H_2O + PBS	10	219 (178-269)	66 (52-83)	0.33
Sacc + Cy	Sacc + PBS	14	226 (182-275)	79;(252)[*][**] (66-96) 47 (38-60) **24;(195)[*] (12-48)	0.36 0.78

Footnotes to Table 4

1. The activity score shown is a group geometric mean (with range) obtained
from the average open-field trial activity scores (3 readings in a 5
minute test at 8 day intervals - see Materials and Methods) of the

individual mice in each group.

2. As for Table 1. For the final test group shown (conditioned with (Sacc + Cy); tested with Sacc) the PFC and open-field trial data for the individual mice were subsequently analysed, separating animals into conditioned immunosuppressed (6 mice) (G. mean PFC outside of range of control group - 1st row) or non-conditioned (8 mice). Group mean IgM-PFC per 1/100th spleen suspension(**) and group mean activity scores (*) for these subgroups are shown separately.

3. Kendall's coefficient of concordance between activity scores and IgM-PFC after ranking mice in each group according to these two parameters.

However, as shown in Table 4, it may prove possible to establish a reproducible pattern of correlation between the activity of mice in an open field and their subsequent conditioned immune response in the saccharine/cyclophosphamide protocol discussed. We found that if mice re-exposed to the CS (after previous pairings of CS + US) were categorized into conditioned immunosuppressed/non-suppressed animals according to their IgM-PFC responses, there was a highly significant correlation between the rank order of the PFC and open field trial scores for the conditioned immunosuppressed mice, and a lesser correlation for all other groups of mice. In addition we observed that the conditioned immunosupppressed mice had exhibited a lower activity in the open-field (relative to other mice tested) before any conditioning trials were begun.

Conclusion

Following the initial lead of Ader and Cohen (7) a number of groups have reproduced the observation establishing evidence for a classical condition- ing of an immune response to antigen challenge if that challenge takes place under environmental conditions which the animal has previously learned to associate with discrete (generally immunosuppressive) physiological states (8-9). Thus, pairing of cyclophosphamide with the novel taste of saccharine in the drinking water can lead to the subsequent immunosuppression of an antibody response to challenge with SRBC if this challenge itself occurs in association with saccharine in the water supply. While our initial experiments in mice using this protocol produced inconsistent results (in general either immunosuppression or a compensatory immunoenhancement being observed - Table 1), the problem of the lack of reproducibility was apparently resolved by paying due attention to the potential role of diurnal rhythm and exogenous stress factors in the experimental mice (Table 2). These findings, preliminary as they are, are consistent with the notion that the conditioned response finally elicited is probably a reflection of the natural physiological state of the animal at the time of initiation of the experiment.

There are a number of physiological means by which the central nervous system could modulate the immune response in a non-specific manner. Several potential neuroendocrine immune system interactions have been described (14), and there is a growing body of evidence to suggest that discrete subpopulations of immunocytes have receptors for neuro- transmitters, stimulation of which receptors alters immunological activity (15-16). While adrenocorticosteroid hormones have been reported not to play a central role in mediating the conditioned immunosuppression of antibody responses in the rat (17), the data of Table 3 suggests that adrenalectomy of mice in contrast abolishes the ability to condition immunosuppression using the cyclophosphamide/saccharine regime discussed here. The similar failure of aged or chronically stressed mice to show a conditioned immunosuppression, coupled with other observations that brain and peripheral neurotransmitter levels change under these conditions (18,19), indicates one of the many possible routes by which these phenomena may be explored at the mechanistic level.

Early studies from an alternative approach to explore the mechanisms involved are reported in Table 4. Here we have asked whether there are any independent behavioural correlates of the conditioned-immune response pattern measured in our mice. The data suggests that mice showing low activity in an open field are particularly likely to develop a conditioned immunosuppressed response to SRBC after the pairing of cyclophosphamide/ saccharine. It will, we think. be of interest to investigate what physiological features can be attributed to such mice which may help to explain this correlation.

References

1. Dutton, R.W. and Swain, S.L., Crit. Rev. Immunol. 3; 209, 1982.
2. Jerne, N.K., Annal. Immunol. (Inst. Pasteur) 125C: 373, 1975.
3. Gillis. S. and Mizell, S.B., Proc. Natl. Acad. Sci. U.S.A. 78: 1133, 1981.
4. Pierpaoli, W., Fabris, N. and Sorkin. E. In: "Cellular Interaction in the Immune Response". S. Cohen, G. Cudkowicz and R.T. McCluskey, Eds. S. Karger, Basel, 1971 pp. 25-30.
5. Henney, C.S., Bourne, H.R. and Lichtenstein, L.M., J. Immunol. 108: 1526, 1972.
6. Gorczynski, R.M., MacRae, S. and Kennedy, M., J.Immunol. 129: 704, 1982.
7. Ader, R. and Cohen, N., Psychosom. Med. 37: 333, 1975.
8. Rogers M.P., Reich, P., Strom, T.B. and Carpenter, C.B., Psychosom. Med. 38: 447, 1976.
9. Wayner, E.A., Flannery, G.R. and Singer, G., Physiol. Behav. 21: 995, 1978.
10. Bovbjerg, D., Ader, R. and Cohen, N., Proc. Natl. Acad. Sci. U.S.A. 79: 583, 1982.
11. Ader, R. and Cohen, N., Science 215: 1534, 1982.
12. Cunningham, A.J. and Szenberg, A., Immunology 14: 599, 1968.
13. Riley, V. Science 212: 1110, 1981.
14. Ahlquist, J., In: "Psychoneuroimmunology", R. Ader, Ed. Academic Press, New York 1981 p. 355.
15. Hall, N. and Golstein, A.L., In: "Psychoneuroimmunology", R. Ader Ed., Academic Press, New York, 1981, p. 521
16. Strom, T.B. and Carpenter, C.B., Transpl. Proc. 12: 304, 1980.
17. Ader, R., Cohen, N. and Grota, L.J., Int. J. Immunopharmacol. 1: 141, 1979.
18. Kubanis, P. and Zornetzer, S.F., Behav. Neurol. Biol. 31: 115, 1981.
19. Anisman, H., In: "Psychopharmacology of Aversively Motivated Behaviour" H. Anisman and Bignani, Eds., Plenum Press, New York, 1978, p. 119.

THE BONE MARROW, OUR AUTONOMOUS MORPHOSTATIC "BRAIN"

W. Pierpaoli
Institute for Integrative Biomedical Research
Ebmatingen, Switzerland

ABSTRACT

It is proposed here that the bone marrow (BM) and its microenvironment constitute an evolutionary ancient organic entity with self-sufficient and largely independent functions which are dominant with respect to other superimposed systems (nervous, endocrine) and that the BM contains our fundamental autonomous devices for controlling the homoeostatic mechanisms for maintenance of self-tolerance and defence (immunity), and repair (regeneration).

Transplantation of bone marrow (BMT) between genetically different partners is possible in mice and other species. Long-lived, histoincompatible, hemopoietic irradiation chimeras are obtained by a new method with unmanipulated BM. The allogeneic hemopoietic chimerism thus obtained cannot be passively transmitted, and cannot be eradicated even when immunocompetent cells with the same genetic character of the chimeric recipient are injected into the chimera. This demonstrates that the bone marrow is central to maintenance of immune identity and that, once established, the new donor marrow will control that no suicide mechanisms will initiate (graft versus host reaction) and that an efficient defence be exerted against viral or bacterial, infectious agents. Hemopoietic chimeras across the histocompatibility barrier are thus an ideal experimental model for investigating the function of the bone marrow in immunity, oncogenesis, regeneration, ageing. Its adoption for interdisciplinary studies in psychoneuroimmunology is suggested.

INTRODUCTION

The extreme flexibility of biological structures and functions is proven almost daily by the steady reshaping of ideas and experimental approaches and by the abrupt collapse of comfortable dogmas. This extends from molecular biology to clinical research and concerns indifferently nucleic acids, cells, organs and patients. This fundamental lack of expectations for "final", "clear-cut" results might be a proper prophylaxis especially in interdisciplinary research. This was and is presently my own attitude in illustrating the models adopted in my studies. Whenever I was looking for a connection between the "systems" in the body, I found it. Alternatively, in the course of the last 15 years, many organs, glands, hormones and cells types have attracted me and somehow all of them contributed to the present view of bone marrow (BM) function. I wish only to remind chronologically the bidirectional relationship between thymus and

adenohypophysis (Pierpaoli and Sorkin, 1967) and between different hormones,
the lymphatic tissue and the immune response (Pierpaoli et al., 1970); the
wasting syndromes and the multifunctional lymphocyte (Pierpaoli and Sorkin, 1972); the endocrine disorders preceding and accompanying oncogenesis
(Pierpaoli et al., 1974, Pierpaoli et al., 1977a);the programming function
of the thymus in early ontogeny on the developing neuroendocrine system
(Pierpaoli and Besedovsky, 1975; Pierpaoli et al., 1976); the irreversibly
modified central neuroendocrine regulation as a consequence of low caloric
diet before puberty (Pierpaoli, 1977); the link between immune tolerance
and reproduction shown by changes of sexual hormones levels in the blood
of mice made tolerant to other strains across the H-2 barrier by sequential
inoculation of allogeneic cells during perinatal time (Pierpaoli et al.,
1977b); the demonstration that the thymus is not relevant in the aging
process (Pierpaoli et al., 1977c); the early endocrine events following injection of allogeneic cells in mice and the consequent extensive study on
a possible pharmacological control of the immune response (Pierpaoli and
Maestroni, 1977, 1978). It was finally the adoption for bone marrow transplantation (BMT) of this last system for controlling immunity which led us
to the present model for BMT and to the studies on BM function (Pierpaoli
and Maestroni, 1980). We are just at the end and at a new beginning of a
tortuous road.It took in fact a few years to make us aware that dogmatic
and simplistic views dominate transplantation immunity and that lack of
comprehensive new approaches has prevented real progresses. The idea that
immunogenetic difference between donor and recipient causes the initiation
of a bidirectional reaction against the host and the recipient of bone
marrow is valid beyond any doubt. However, this unilateral view based on
accepted and documented studies has prevented to propose different inter-
pretations and to provide evidence for them. In fact, this process of reciprocal immune rejection might initiate only when we do not properly transplant the bone marrow and are unable to promote an harmonious and rapid engraftment of donor marrow in the new host (Pierpaoli and Maestroni, 1980;
Maestroni and Pierpaoli, 1980). Thus a method has been now adopted which
is based on the idea that BMT is mainly an hematological problem and that
a deeper knowledge of BM physiology is required for achieving chimerism
across the histocompatibility barriers (Maestroni and Pierpaoli, 1980;
Pierpaoli et al., 1981; Maestroni et al., 1982, Pierpaoli and Maestroni,
in press).

METHODS AND RESULTS

Our system for BMT has been amply reported and is still being further elaborated (Pierpaoli et al., 1981; Maestroni et al., 1982; Pierpaoli and Maestroni, in press). It is based on the concept that the BM must be trans- planted without affecting, with certain limitations inherent to the techni- que, its internal cellular and humoral regulation which is unique and typi- cal of the marrow to be transplanted. Thus, no manipulation or removal of immunocompetent, cytotoxic T cells is operated with the idea that donor T cells are also needed for a proper seeding and proliferation of donor mar- row in the immunodepressed, irradiated host. Our method is based on the pro- position that the initiation of a GVHD can be avoided with a proper tech- nique by which the "new" marrow is forced to take and to populate the bones before T cells initiate a response against the host (Pierpaoli and Maestroni, 1980 and in press). Thus, additional basic hematological know- ledge is needed rather than immunological subtlety. In fact, the results achieved with the accepted methods have been disappointing, and convincing evidence for long-lasting allogeneic chimerism across the H-2 barrier in mice has not been reported. We also doubt that all those cases in which long-lasting allogeneic chimerism has been claimed or achieved (Emeson and Weintraub, 1981; Aizawa et al., 1981; Krown et al., 1981; Onoé et al., 1981; Muto et al., 1981; Vallera et al., 1981; Norin et al., 1981; Tutschka et al., 1981; Strober et al., 1981; Mueller-Ruchholtz et al., 1981; Coico et al., 1982; Onoé et al., 1980) the chimeric mice were a suitable model for ana- lyzing adoption and maintenance of donor character and acquisition of immunocompetence. It is in fact impossible to accept the irrational belief that a wildly manipulated BM deprived of some of its basic cellular com- ponents (lymphocytes and T cells) be suitable to repopulate an irradiated host and to confer immune competence and resistance.

We consider the BM as an autonomous and self-regulating organ. However, its internal regulatory mechanisms are still unknown. Thus our praxis is still based on administration of intact BM to a lethally irradiated host. The BM is administered together with marrow-derived factors called marrow regulating factors (MRF), whose chemical composition and mechanism of action are still largely unexplored. In spite of large variability in sur- vival and the empirical, exploratory character of our system, many chime- ric mice are obtained across the H-2 barrier. Those chimeras live long and do not manifest any sign of GVHD. Their "immune" character has been and is being now investigated (Maestroni et al., 1982, Maestroni et al., in press). Thus work is now proceeding along two main lines: a) production of allogeneic chimeras across the H-2 barrier (H-2^d; H-2^b; H-2^k) with unmanipulated bone marrow and marrow-derived factors; b) analysis of their immune character, resistance to infections, tumors (spontaneous, induced, transplanted) and extensive investigations on the mechanisms underlying the chimeric tolerance (Maestroni and Pierpaoli, submitted,Pierpaoli and Maestroni,submitted).

One of the most striking features in our long-lived chimeras is the persistence of donor marrow chimerism when they are challenged with immunocompetent lymphocytes from the genetic type of the recipient, e.g. H-2^b lymphocytes into H-2^d —→ H-2^b chimeras. In addition, chimerism is not transferable (Maestroni et al., in press). It is remarkable that the allogeneic chimeras reject skin grafts from the same H-2 recipient type (e.g. skin from H-2^b donors grafted on H-2^d —→ H-2^b chimeras). This documented paradoxical phenomenon (Maestroni and Pierpaoli,submitted) demonstrates the extraordinary interest of chimerism for the study of transplantation biology.

CONCLUSIONS AND PROSPECTS

The striving for a "central" regulation of morphostasis (identity-defence-immunity) in adult life might lead us to the BM as an important source of information. It is clear that a "diffuse" neuroendocrine network is as valid as a "diffuse" lymphohemopoietic network and that no functional, causal or structural distinction or separation can be proposed. However, there might be an evolutionary priority by which a certain "concentration" or dominance of function is maintained or condensed in certain structures like the nervous brain and the "morphostatic brain" (Pierpaoli, 1981).

Studies on BM physiology are in their infancy. The enormous variety of the cellular components in the bone marrow has always been a fascinating source of quarrels and a playground for the classical hematologist. It has been a comfortable source of "B" and "T" cells for the immunologist and is an unexplored Continent for the neuroendocrinologist. Its lack of scientific "identity" as a separate field of research for the "marrowlogist" is in fact a reflex and expression of its enormous complexity, which has prevented a separation as a distinct discipline. Not marginally, work with BM is not practical because of its hiding in the bones. BM as an organ has eluded systematic exploration even at a basic physiological level, perhaps because its complexity is comparable to that of the CNS. The BM has often been regarded as a second-class immune organ. However, nowadays awareness is awakening that the BM might be a basic playground for a comprehensive interdisciplinary approach to medical problems. I believe that the BM is our major, phylogenetically most ancient organ which contains the intact contribution of millions of years of evolution under integrated form, and is our main device for regulatory automatisms. The BM is our independent, morphostatic brain and the center of identity-defence. The BM constitutes a landing point for the automatic mechanisms which regulate resistance to diseases, elimination of non-self, tolerance to self, surveillance of cell mitosis and regeneration. Aging, the most common disease, might also be a consequence of BM decay in time. How can we evaluate all this? An inter-disciplinary approach to the study of BM function is needed. We know very little of the bidirectional BM-CNS connections, of a possible causal re-lationship between psychic disturbances and BM dysfunction; of hematologi-cal, BM alterations as a consequence of psychic disorders. Somehow, the elusive multipotential "stem cell" in the BM could be compared, as far as flexibility and adaptability are concerned, to the cells of the central neuroendocrine center, the brain cortex. They can provide and adapt to many functions, not of psychic-adaptive, but of morphostatic-adaptive cha-racter. It is even feasible that the cells of the CNS lack the adaptability which is intrinsic in the differentiative capacity and in the mobility of the BM-derived cells.

The pioneer work of Calvo (1968) has demonstrated the extensive and complex innervation of the BM. Bundles of nonmyelinated axons can be vi-sualized around blood-forming cells in the BM. There exists a striking coincidence between final myelinization of the nerves in the BM of the rat and complete maturation of the cell-mediated immune capacity. Both events occur at two-three weeks of age. Also at this time the BM begins to res-pond to hormonal stimulation. We know that transplantation tolerance can be induced by neonatal challenge with allo-antigens, at a time when the neuroendocrine system is still organizing its final structure and function. All these facts have been taken into consideration in the choice of the model we selected for a study of the cellular mechanisms of tolerance.

Hemopoietic chimerism achieved by transplantation of BM certainly constitutes a basic tool for understanding how the genetically determined structures in the body recognize their "self" and react or adapt to noxious
agents. Inter-species and inter-strain chimerism is an ideal model for
studying how the "identity-defence" system of an individual can control
proliferation of a cancer cell or of a lethal virus.

Table 1

The bone marrow as a "target" for interdisciplinary studies. Bone marrow
transplantation and allogeneic chimerism as a basic model for such studies.

Disciplines	Some possible studies
Neurobiology	Innervation, repair and restoration of innerva- tion after damage (e.g. irradiation, BMT, drugs). Response to different stimuli (electrical, stress, hormones).
Hematology	Ultrastructure, reconstitution after damage, dif- ferentiation of stem cells, function of endothe- lial cells, connectival network, fat cells, etc.
Endocrinology	Activity of hormones and drugs, bidirectional link between the BM and endocrine glands, explor- ation of endogenous factors and their link to other systems and organs, etc.
Microbiology and Virology	State and role of BM in infectious diseases, re- sistance under conditions of allogeneic chimerism, cell transformation by viruses, etc.
Immunology, immuno- genetics, trans- plantation biology	The BM in immunodeficiency diseases. Origin of tolerance and its control. Ontogeny of self and control of non-self, link between the BM and sexual functions: reproduction.
Pharmacology	Action of drugs on structure, composition and function of the BM, possible "boomerang" effect of drug administration due to toxicity on the BM.
Oncology, aging	Aging of BM and oncogenesis, aging of BM and aging, factors and cells affecting the two patho- logical states.

Psychiatry, psychology	Bone marrow alterations under stress or altered, abnormal psychic conditions. The bidirectional link between brain and BM: attempts at affecting psychic disorders by "marrow therapy".

REFERENCES

Aizawa, S., Sado, T., Muto, M. and Kubo, E. 1981.
    Immunology of fully H-2 incompatible bone marrow chimeras induced in specific-pathogen-free mice: evidence for generation of donor- and host-H-2 restricted helper and cytotoxic T cells. J. Immunol., 127: 2426-2431.
Calvo, W. 1968.
    The innervation of the bone marrow in laboratory animals. Amer. J. Anat., 123: 315-328.
Coico, R.F., Krown S.E., Good; R.A. and Hoffmann M.K. 1982.
    Helper Cell factors restore antibody responses of allogeneic bone marrow chimeras: evidence for ineffective cellular interactions. J. Immunol., 128: 1590-1594.
Emeson, E.E. and Weintraub, F.M. 1981.
    Prevention of AKR leukemia by transplanting H-2-incompatible allogeneic bone marrow requires the maintenance of chimerism. Transplant. Proc.,13; 774-777.
Krown, S.E., Coico, R., Scheid, M.P., Fernandes, G. and Good R.A. 1981.
    Immune function in fully allogeneic mouse bone marrow chimeras. Clin. Immunol. Immunopathol.,19: 268-283.
Maestroni, G.J.M. and Pierpaoli, W. 1980.
    Factor(s) elaborated by bone marrow that promote persistent engraftment of xenogeneic and semiallogeneic marrow. J. Clin. Lab. Immunol., 4: 189-193.
Maestroni, G.J.M., Pierpaoli, W. and Zinkernagel, R.M. 1982.
    Immunoreactivity of long-lived H-2 incompatible irradiation chimeras $(H-2^d \longrightarrow H-2^b)$. Immunology,46: 253-260.
Maestroni, G.J.M., Pierpaoli, W. and Zinkernagel, R.M. in press.
    Allogeneic $H-2^d \longrightarrow H-2^b$ irradiation bone marrow chimeras.
Müller-Ruchholtz, W., Blank, M., Wottge, H.-U. and Müller-Hermelink, H.K. 1981.
    Specific immune suppression in adult mice across K-I-D fully allogeneic histoincompatible barriers induced by lymphocyte-free bone marrow cells. Transplant. Proc.,13: 603-607.
Muto, M., Sado, T., Aizawa, S. Kamisaku, H. and Kubo, E. 1981.
    Bone marrow transplantation across the major histocompatibility barrier in specific-pathogen-free mice: effects of intact versus T cell-depleted bone marrow on the expression of anti-host reaction in the recipient spleens. J. Immunol.,127: 2421-2425.
Norin, A.J., Emeson, E.E. and Veith, F.J. 1981.
    Long-term survival of murine allogeneic bone marrow chimeras: effect of anti-lymphocyte serum and bone marrow dose. J. Immunol.,126: 428-432.

Onoé, K., Fernandez, G. and Good, R.A. 1980.
Humoral and cell-mediated immune responses in fully allogeneic bone marrow chimeria in mice. J. Exp. Med.,151: 115-132.
Onoé, K., Yasumizu, R., Oh-Ishi, T., Kakinuma, M., Good, R.A. and Morikawa, K. 1981.
Restricted antibody formation to sheep erythrocytes of allogeneic bone marrow chimeras histoincompatible at the K end of the H-2 complex. J. Exp. Med.,153: 1009-1014.
Pierpaoli, W. and Sorkin, E. 1967.
Relationship between thymus and hypophysis. Nature,215: 834-837.
Pierpaoli, W., Fabris, N. and Sorkin, E. 1970.
Developmental hormones and immunological maturation. In: G.E.W. Wolstenholme and J. Knight (Editors), Hormones and the Immune Response, Ciba Foundation Study Group No. 36, Churchill, London, pp. 126-153.
Pierpaoli, W. and Sorkin, E. 1972.
Hormones, thymus and lymphocyte functions. Experientia,28: 1385-1389.
Pierpaoli, W., Haran-Ghera, N., Bianchi, E., Müller, J., Meshorer, A. and Bree, M. 1974.
Endocrine disorders as a contributory factor to neoplasia in SJL/J mice. J. Natl. Cancer Inst.,53: 731-744.
Pierpaoli, W. and Besedovsky, H.O. 1975.
Role of the thymus in programming of neuroendocrine functions. Clin. Exp. Immunol.,20: 323-338.
Pierpaoli, W., Kopp, H.G. and Bianchi E. 1976.
Interdependence of thymic and neuroendocrine functions in ontogeny. Clin. Exp. Immunol.,24: 501-506.
Pierpaoli, W., Haran-Ghera, N. and Kopp, H.G. 1977a.
Role of host endocrine status in murine leukaemogenesis. Br. J. Cancer., 35: 621-629.
Pierpaoli, W. 1977.
Changes of hormonal status in young mice by restricted caloric diet. Relation to lifespan extension. Preliminary status. Experientia,33: 1612-1613.
Pierpaoli, W., Kopp, H.G., Müller, J. and Keller, M. 1977b.
Interdependence between neuroendocrine programming and the generation of immune recognition in ontogeny. Cell. Immunol.,29: 16-27.
Pierpaoli, W., Hämmerli, M., Sorkin, E. and Hurni, H. 1977 c.
Role of thymus and hypothalamus in aging. In: U.J. Schmidt, G. Brüschke, E. Lang, A. Viidik, D. Platt, V.V. Frolkis, F.H. Schulz (Editors), Vth European Symposium on Basic Research in Gerontology. D. Straube, Erlangen, pp. 141-150.
Pierpaoli, W. and Maestroni, G.J.M. 1977.
Pharmacological control of the immune response by blockade of the early hormonal changes following antigen injection. Cell. Immunol.,31: 355-363.
Pierpaoli, W. and Maestroni, G.J.M. 1978.
Pharmacological control of the hormonally modulated immune response. Immunology,34: 419-430.
Pierpaoli, W. and Maestroni, G.J.M. 1980.
The Facilitation of enduring engraftment of allogeneic bone marrow and avoidance of secondary disease in mice. Cell. Immunol., 52: 62-72.
Pierpaoli, W., Maestroni, G.J.M. and Sache, E. 1981.
Enduring allogeneic marrow engraftment via nonspecific bone-marrow-derived regulating factors (MRF). Cell. Immunol., 57: 219-228.

721

Pierpaoli, W. 1981.
    Integrated phylogenetic and ontogenetic evolution of neuroendocrine and
    identity-defense, immune functions. In: R. Ader (Editor), Psychoneuro-
    immunology, Academic Press, New York, pp· 575-606.
Pierpaoli, W. and Maestroni, G.J.M. in press.
    Marrow regulating factors (MRF) and radiation chimeras: a model for
    bone marrow-directed immunity. In: Immuneregulation 81. Int. Workshop,
    Urbino, Italy, Plenum Press, New York.
Strober, S., King, D.P., Gottlieb, M., Hoppe, R.T. and Kaplan, H.S.1981.
    Induction of transplantation tolerance after total lymphoid irradia-
    tion: cellular mechanisms. Federation Proceedings,40: 1463-1465.
Tutschka, P.J., Hess, A.D., Beschorner, W.E. and Santos, G.W. 1981.
    Suppressor cells in transplantation tolerance. Transplantation,32:
    203-209.
Vallera, D.A., Soderling, C.C.B., Carlson, G.J. and Kersey, J.H. 1981.
    Bone marrow transplantation across major histocompatibility barriers
    in mice. Transplantation,31: 218-222.

IMMUNE REGULATION OF THE HYPOTHALAMIC-HYPOPHYSIAL-ADRENAL AXIS:
A ROLE FOR THYMOSINS AND LYMPHOKINES

Nicholas R. Hall, Joseph P. McGillis, Bryan L. Spangelo,
David L. Healy[1] and Allan L. Goldstein

Department of Biochemistry, The George Washington University College
of Health Sciences, Washington, D.C. 20037 and
[1]Pregnancy Research Branch, National Institute of Child Health and
Development, Bethesda, Maryland 20205

ABSTRACT
     Considerable evidence suggests that the central nervous system is
able to modulate the course of immunogenesis. It is also apparent that
certain products of the immune system are able to influence the hypothalamic-hypophysial-adrenal axis. Evidence is discussed to support the
concept that the thymosins and lymphokines constitute part of a bidirectional circuit through which the immune system regulates itself via a
neuroendocrine axis. The possibility that there exists a micro-environmental circuit consisting of a releasing factor and ACTH within the
immunologic compartment is also considered.

INTRODUCTION

     Various types of evidence support the concept that the central

nervous system is able to modulate the course of host defense. These in-

clude results from studies in which electrolytic or pharmacologic manipu-

lation of the central nervous system has been correlated with changes in-

volving immunologic function (Tyrey and Nalbandov, 1972; Stein et al.,

1976;; Korneva et al., 1978; Devoino et al., 1970; Cotzias et al., 1977).

Conversely, during the immune response, changes in the firing rate, mor-

phology, as well as the transmitter content of certain brain regions have

been reported (Besedovsky et al., 1977; Srebro et al., 1974; Vekshina and

Magaeva, 1974; Cotzias et al., 1977). The results of other studies

suggest that these changes are manifestations of neuroendocrine and/or

autonomic influences upon various immunologic compartments. During the

immune response, alterations in the levels of hormones that are controlled

by the hypothalamic-hypophyseal axis have been demonstrated (Besedovsky

et al., 1975). Furthermore, nerve terminals have been found within the

thymus gland (Bulloch and Moore, 1981).

     Recognition of immunogenic determinants and the subsequent differen-

tiation of lymphocytes into effector cells occurs primarily within the

peripheral immunologic tissues. Consequently, if afferent signals from

the central nervous system are to occur in the correct temporal sequence,

then the initiating signal would most likely originate from within this

immunologic compartment. This signal could be regarded as being functionally the same as that produced by numerous endocrine tissues that ultimately regulate their own activity depending upon the relative concentration of their product. If such a mechanism is involved in CNS modulation of immunity, then the immunologic event that is responsible for initiating or terminating the neural influence would have to be regarded as being of paramount importance.

NEUROENDOCRINE-IMMUNE INTERACTIONS

Several laboratories in Europe and in the United States are currently conducting research in order to identify and characterize the immunologic signal that is responsible for coordinating the neuroendocrine and autonomic events that modulate host defense. Most of this work has concentrated on the role that might be played by thymosins and lymphokines.

Certain of the products synthesized by lymphocytes undergoing mitogenesis have been found to stimulate an increase in serum corticosteroid levels in both rodents (Besedovsky et al., 1981) and in humans (Dumonde et al., 1982). These include products that may be responsible for the rise in corticosterone that occurs during the course of the primary immune response (Besedovsky et al., 1975). There is also evidence that the supernatants from dividing lymphocytes contain thymosin α_1-like immunocrossreactivity (McGillis et al, unpublished observation). Several studies implicate this peptide in modulating the hypothalamic-hypophyseal-adrenal axis.

Thymosin fraction 5, a partially purified preparation of peptides which includes α_1, has been found to stimulate serum corticosterone when injected either intraperitoneally or intravenously in several species including mice, rats, rabbits and monkeys (McGillis et al., 1982; Sivas et al., 1982; Healy et al., 1983). Several types of evidence suggest that this effect is mediated at the level of the hypothalamus and/or pituitary gland. First, there is no evidence that any of the thymosin peptides or lymphokines are able to directly stimulate adrenal fasciculata cells (Vahouny et al., 1983). Rat cells were incubated with purified thymosin or lymphokine preparations as well as heterogenous mixtures of each. At none of the concentrations evaluated was there either increased corticosteroid synthesis or C-AMP production. Also, Fraction-5 does not stimulate corticogenesis in cultured bovine adrenal cells (Healy, et al., submitted).

An alternate experimental approach has been to evaluate the effect of thymosin upon ACTH release when superfused over pituitary tissue. Using an RIA for ACTH, there was no stimulatory effect of several concentrations

of thymosin fraction 5 (McGillis et al., unpublished observation). Superfusion experiments with hypothalamic tissue in sequence with pituitary
cells are now underway. However, intracerebral injection of the peptides
using an in vivo murine model has proved to be effective (Hall et al.,
1982). Thymosin or control substances were injected into the lateral
cerebroventricle of mice through guide cannulas implanted in the cal-
varium. Within 3 hours after the injection there was a significant rise
in serum corticosterone. Thymosin Fraction-5 and thymosin α_1, were both
capable of elevating corticosterone levels. Control injections of either
the saline vehicle or thymosin β_4 were ineffective. Furthermore, the
effective dose of peptide administered into the brain was ineffective when
injected intraperitoneally. Intravenous injection of thymosin Fraction-5
has also been found to stimulate the adrenal axis in primates. Prepubertal
female macaque monkeys were fitted with a vest that allowed for injections
and blood collection to be carried out in an unrestrained, unanesthesized
state. There were no changes in plasma LH, FSH, prolactin, growth hormone
or thyroid stimulating hormone. However, there was a significant rise in
ACTH and β-endorphin as well as cortisol. These effects were both time
and dose dependent and did not occur following either vehicle or BSA control injections (Healy et al. 1983).

That the thymosin stimulated rise in glucocorticoids is sufficient to
alter the immune responsiveness of the individual was demonstrated in
another set of experiments. Chronically cannulated mice received an intracerebroventricular injection of thymosin peptides concommitant with a
sensitizing injection of sheep red blood cells. Five days later, each
animal's immune system was evaluated using a variety of functional assays.
It was found that antibody titers, as well as lymphocyte responsiveness to
T-cell mitogens, were significantly depressed in mice that were injected
with thymosin fraction-5 or thymosin α_1 but not when injected with the
vehicle alone or with thymosin α_7.

If thymosin α_1 is the component that is responsible for the corti-
cogenic effects of lymphokine preparations and heterogeneic thymic extracts, then it should be present in areas of the brain that regulate
adrenal-cortical activity. This has been demonstrated by the use of a
radioimmunoassay. Thymosin α_1-like immunocrossreactivity was found
throughout the CNS, however, the highest concentrations were detected in
the pituitary gland and the hypothalamus (Palaszynski, 1981; Hall et al.,
1982). Within the hypothalamus, the highest levels of the peptide were

found in the arcuate nucleus and the median eminence.

Finally, if thymosin α_1 is capable of modulating the synthesis and/or release of glucocorticoids, then it would be expected that a correlation would exist between this peptide and the steroid when the levels of either fluctuate. This has been demonstrated in a study comparing the circadian rhythm of corticosterone with fluctuations in thymosin α_1 (McGillis et al., 1983). Not only do thymosin α_1 levels exhibit cyclic changes over a 24 hour interval that are entrained to the onset of light, but the pattern is inversely related to the glucocorticoid rhythm. Whether one is in part responsible for the changes in the other remains to be demonstrated. However, the fact that a correlation does exist supports the hypothesis that there is a physiological relationship between the glucocorticoid axis and the immune system. It is also noteworthy that the time of low corticosterone and high thymosin α_1 corresponds to the time when a variety of immunologic parameters have been shown to be optimal (Fernandes et al., 1976; Pownall and Knapp, 1978).

Other products of the activated immune response have been suggested to play an immunomodulatory role via a neuroendocrine axis. According to the reflex hypothesis the immunogen stimulates receptors which results in an electrical signal being transmitted to the brain for the purpose of initiating immunomodulatory events (Gordienko, 1958). It has also been shown that C3a and C5a of the complement cascade can mimic the effects of dopamine in the hypothalamus (Seeman et al., 1978). There are other products such as interleukins, immunoglobulins and biologically active substances produced by subsets of T-lymphocytes that have the potential of serving as signals to initiate CNS events. So far, only the crude lymphokine containing supernatants and thymosin can be proposed to serve this role on the basis of experimental evidence. A common component of each is thymosin α_1 which may ultimately be the chemical signal that is responsible.

Activation of the hypothalamic-hypophyseal-adrenal axis does occur during the immune response which is further evidence in support of a functional relationship between these two systems. Changes in the morphology (Srebro et al., 1974), electrical firing rate (Besedovsky et al., 1977) and neurotransmitter content of subcortical nuclei implicated in regulating ACTH have been observed during the course of the immune response. This activity is thought to precede the dramatic rise in glucocorticoids during the logarithmic phase of antibody production (Besedovsky and Sorkin, 1977).

From a teleological perspective, it has been proposed that this activation results in negative feedback of immunologic functions (Besedovsky and Sorkin, 1977). The rise in steroids occurs at a time when lymphocytes have differentiated into effector cells and are performing their biological function. At this stage of clonal expansion in response to a specific immunogen, the sensitized T-lymphocytes have reduced susceptibility to the suppressive effects of glucocorticoids. However, T-cells with no or little affinity for the sensitizing immunogen would be suppressed by the steroids. Since these cells would not be required to participate in the reaction against the immunogen that indirectly resulted in the rise in steroid, the feedback inhibition could be regarded as a mechanism by which the immune system is finely tuned as proposed by Besedovsky and Sorkin (Besedovsky and Sorkin, 1977).

THYMOSIN AS AN ANTISTRESSOR

An alternative interpretation of the existing data, and one that should be pursued experimentally, is that the rise in corticosterone occurs for reasons other than modulation of lymphocytes. Associated with many infectious processes is a marked inflammatory response that results in part from the actions of complement proteins, a system that appears to be enhanced by ACTH administration (Pitner et al., 1951). Consequently, the steroid changes may modulate inflammatory processes with only in-direct influences upon lymphocyte functioning. Since elevated steroid levels are usually immunosuppressive, the role of lymphokines and thymosins might be to counteract steroid induced immunosuppression. There is some evidence in support of this hypothesis.

The results of several studies have suggested that thymosin treatment might enhance recovery from glucocorticoid induced immunosuppression. Enhancement was found in functional assays, thymocyte counts, concentra-tions of a T-cell associated enzyme (TdT), and gross tissue weights following immunosuppression induced by administration of exogenous corti-costeroid or by stress.

In one of the earlier studies (Thurman et al., 1977), it was found that thymosin treatment was able to induce a rapid recovery of Con-A responsivity of splenic lymphocytes. In the control mice, thymosin treat-ment consistently doubled the Con A response. However, biphasic effects were observed when thymosin was given to hydrocortisone treated mice. It increased the Con-A response if it was depressed, but decreased it if it was elevated. In a subsequent study, the daily injection of thymosin fraction 5 and two of its components, thymosin β_3 and β_4 both stimulated a marked increase in the enzyme while thymosin α_1 and β_1 were ineffective. In addition to TdT, thymosin fraction 5, β_3 and β_4 also partially normalized the loss of thymocytes that follows the injection of hydrocortisone. At a concentration of 100 µg, thymosin fraction 5 was most effective, while 1 µg each of thymosin β_3 and β_4 was effective. For all peptides, the ability to reverse the corticosteroid induced inhibition was associated with the lower concentrations.

In a preliminary study, we have found that thymosin fraction 5 can also reverse stress-induced changes of immunologic parameters. Male Swiss-Webster mice, 4 to 6 weeks of age, were subjected to rotation stress for 10 minute intervals once an hour for four consecutive hours. Animals were rotated at 45 RPM in their home cages. This paradigm was repeated daily for all experimental groups. Mice in control groups remained undisturbed. Preceeding each exposure to stress, all animals were injected i.p. with either 100 µg of thymosin Fraction 5 or phosphate buffered saline (vehicle) in a volume of 0.2 ml. Stress alone resulted in a slightly depressed stimulation-index for PHA but not for LPS. Thymus weight was decreased while spleen weight was elevated. In every case, treatment with thymosin reversed the stress-induced changes (Kim et al., unpublished observation). It is also noteworthy that these effects of thymosin were stress dependent. Unstressed mice that received thymosin exhibited the same changes induced by stress alone. This finding is consistent with the previously discussed observation that thymosin Fraction 5 and thymosin α_1 are able to cause elevations in serum glucocorticoid levels via a mechanism involving the hypothalamic-hypophyseal axis.

The mechanism(s) by which thymosin peptides counteract the immunosuppressive effects of either exogenous glucocorticoid administration or stress exposure is not known. One possibility is that thymosin modulates the binding affinity of glucocorticoids in lymphocytes. In a preliminary study, we have found that thymosin fraction 5 can inhibit the binding of ^{3}H-dexamethasone to cultured thymocytes in a dose dependent fashion (Goldstein et al., 1981). This observation is consistent with Bach's report that a related peptide, "Facteur Thymic Serique" (FTS), can also alter the binding of glucocorticoids to lymphocytes (Bach et al., 1975).

PROSPECTUS

Alternatively, recent evidence may necessitate a re-evaluation of the existing data and an interpretation that excludes the central nervous system. In the past, biological models have been formulated based upon state-of-the-art research. One such model system which forms the basis of our understanding of many endocrine functions is the classic hypothalamic-hypophyseal-target tissue axis which is regulated by long and short feedback circuits. Indeed, this is the model that currently best explains the data discussed in this review. However, with the advent of more precise analytical techniques, it is now apparent that peptides that were once thought to be produced within the boundaries of a single type of tissue can now have multiple sources. Two such peptides are ACTH and β-endorphin, the sources of which now appear to include lymphoid cells (Smith and Blalock, 1981). Consequently, changes in the endogenous levels of these peptides may be a direct consequence of immune activation rather than evidence for the existence of a neuroendocrine pathway. Although the ACTH that is produced has not been found to directly stimulate the adrenal fasciculata cells (Vahouny et al., 1983), it could have an as yet undefined role, during the course of immunogenesis. β-endorphin appears to exert a net stimulatory influence upon T-lymphocytes (Gilman et al., 1982) as well as natural killer cells (Mathews et al., 1983). This influence may be direct since receptors for β-endorphin are present on lymphocytes (Hazun et al., 1979).

Since there is a component of lymphokine and thymosin preparations that can stimulate ACTH and β-endorphin release, it is possible that a microenvironmental feedback circuit exists within the immunologic compartment. Similar circuits could also exist in other tissues. In some instances, the chemical sequence of the components might be identical, but the biological activities dependent upon the responsive cell type. With respect to the immune system, the source of the CRF-like peptide (i.e., a subset of lymphocytes and/or thymic epithelial cells) would share biological homology with the hypothalamus. The cells producing ACTH and β-endorphin would assume the role of the adenohypophysis while the target of the peptides' action would be expected to synthesize a product capable of regulating the levels of the CRF-like material. If a corticosteroid is eventually found to be produced by lymphoid-ACTH, it's suppressive effects include all of the cell types that are the putative sources(s) of lymphokines and thymosins (reviewed by Hall & Goldstein, 1983). If future data are consistent with the observation that immune ACTH fails to stimulate glucocorticoid release, a lymphokine or thymosin peptide could function in this capacity. There is also the possibility that certain thymosin and lymphokine peptides have biological functions beyond the immunologic compartment. Thus, immunoreactive thymosin α_1 in the brain might well be a neuronal or glial cell product with a non-immunologic function in the CNS.

These are questions that will undoubtedly have to be addressed as psychoneuroimmunology continues to evolve. Neuroscience and Immunology are perhaps two of the most dynamic and rapidly changing fields in medical science. Concepts that are formulated in this multidisciplinary subject will therefore have to remain flexible and be re-evaluated as new developments emerge within each of these component fields.

REFERENCES

Bach, J.F., Duval, D., Dardeene, M., Solomon, J.C., Tursz, T. and Fournier, C. 1975. The effects of steroids on T-cells. Transplant. Proc., 7(1):25-30.

Besedovsky, H.O., DelRey, A. and Sorkin, E. 1981. Lymphokine-containing supernatants from Con A-stimulated cells increase corticosterone blood levels. J. Immunol. 12:385-387.

Besedovsky, H., Sorkin, E., Felix, D. and Haas, H. 1977. Hypothalamic changes during the immune response. Eur. J. Immunol. 7:323-325.

Besedovsky, H.O., Sorkin, E., Keller, M. and Muller, J. 1975. Changes in blood hormone levels during the immune response. Proc. Soc. Exp. Biol. Med. 150:466-470.

Bulloch, K. and Moore, R.Y. 1981. Innervation of the thymus gland by brainstem and spinal cord in mouse and rat. Am. J. Anat. 162:157-166.

Cotzias, G.C., Miller, S.T., Tang, L.C., Papavasilious, P.S. and Wang, Y.Y. 1977. Levadopa, fertility and longevity. Science 196:549-551.

Devoino, L.V., Dremina, O.F. and Yu Ilyutchenok, R. 1970. The role of the hypothalmo-pituitary system in the mechanism of action of reserpine and 5-hydroxytryptophan on antibody production. Neuropharmacology 9:67-72.

Dumonde, D.C., Pulley, M.S., Hamblin, A.S., Singh, A.K. Southcott, B.M., O'Connell, D.O., Paradinas, F.J., Robinson, M.R., Rigby, C.C., Hallander, F., Schuurs, A., Verheul, H. and Van Vliet, E. 1982. Short-term and long-term administration of lymphoblastoid cell line lymphokine (LCL-LK) to patients with advanced cancer. IN: Lymphokines and Thymic Hormones: Their Potential Utilization in Cancer Therapeutics. (A.L. Goldstein and M.A. Chirigos, eds.) New York, Raven Press, pp. 301-318.

Fernandes, G., Halberg, F., Yunis, E.J. and Good, R.A. 1976. Circadian rhythmic plaque-forming cell response of spleens from mice immunized with SRBC. J. Immunol. 117:962-966.

Gilman, S.C., Schwartz, J.M., Milner, R.J., Bloom, F.E., Feldman, J.D. 1982. Beta-endorphin enhances lymphocyte proliferative responses. Proc. Natl. Acad. Sci. 79:4226-4230.

Gordienko, A.N. 1958. The mechanism of the formation of certain immuno-logical processes. IN: Control of Immunogenesis by the Nervous System (A.N. Gordienko, ed.) Rostov-on-Don, (Translated by The Israel Program for Scientific Translations). pp. 1-15.

Hall, N.R. and Goldstein, A.L. 1983. Endocrine regulation of host immunity: the role of steroids and thymosins. IN: Mechanisms of Immune Regulation. (M.A. Chirigos, ed.) Marcel Dekker. (in press).

Hall, N.R., McGillis, J.P., Spangelo, B., Palaszynski, E., Moody, T. and Goldstein, A.L. 1982. Evidence for a neuroendocrine-thymus axis mediated by thymosin peptides. IN: Current Concepts in Human Immuno-logy and Cancer Immunomodulation. (B. Serrou et al., eds.) Elsevier, pp. 653-660.

Hazum, E., Chang, K.J. and Cautrecasas, P. 1979. Specific nonopiate receptors for β-endorphins. Science 205:1033-1035.

Healy, D.L., Hall, N.R., Williams, R.F., Goldstein, A.L. and Hall, N.R. 1983. Thymosin elevates ACTH, β-endorphin and cortisol in prepubertal primates. The Endocrine Society (Abstract) (in press).

Healy, D.L., Hodgen, G.D., Schulte, H.M., Chrousos, G.P., Loriaux, D.L., Hall, N.R. and Goldstein, A.L. The thymus-adrenal connection: Evidence for a thymic corticotropin releasing factor in primates. (Submitted to Science).

Hu, S.K., Low, T.L.K. and Goldstein, A.L. 1981. Modulation of terminal deoxynucleotidyl transferase activity by thymosin. Molecular and Cellular Biochemistry 41:49-58.

Korneva, E.A., Klimenko, V.M. and Shhinek, A.K. 1978. Neurohumoral Regula-tion of Immune Homeostasis. Nauda, Leningrad, 1978.

Mathews, P.M., Froelich, C.J., Sibbett, W.L., Bankhurst, A.D. 1983. Enhancement of natural cytotoxicity by β-endorphin. J. Immunol. 130: 1658-1662.

McGillis, J.P., Hall, N.R. and Goldstein, A.L. 1983. Circadian rhythm of thymosin α_1 in normal and thymectomized mice. J. Immunol. (in press).

McGillis, J.P., Feith, T., Kyenne-Nyombi, E., Vahouny, G.V., Hall, N.R. and Goldstein, A.L. 1982. Evidence for an interaction between thymosin peptides and the pituitary-adrenal axis. Fed. Proc. 41:4918 (Abstract).

Palaszynski, E. 1982. Doctoral Dissertation, Department of Biochemistry, The George Washington University, Washington, D.C.

Pitner, G., Smith, L.C. and Colwell, C.A. 1951. Bact. Proc. 22:88 (Abstract).

Pownall, R. and Knapp, M.S. 1978. Circadian rhythmicity of delayed hyper-sensitivity to oxazolone in the rat. Clin. Sci. and Mol. Med. 54: 447-449.

Seeman, B., Schupf, N. and Williams, C.A. 1978. Mimicry of dopamine stimulation of hypothalamic pathways in C3a anaphylatoxin. Soc. Neurosci. Abstr. p. 414 1310.

Sivas, A., Vysal, M. and Oz, H. 1982. The hyperglycemic effect of thymosin f5, a thymic hormone. Horm. Metab. Res. 14:330-331.

Smith, E.M. and Blalock, J.E. 1981. Human lymphocyte production of corti-cotropin and endorphin-like substances: Association with leukocyte interferon. Proc. Natl. Acad. Sci. 78:7530-7540.

Srebro, A., Spisak-Plonka, I. and Szirmai, E. 1974. Neurosecretion in mice during skin allograft rejection. Agressologie 15:125-130.

Stein, M., Schiavi, R.C. and Camerino, M. 1976. Influence of brain and behavior on the immune system. Science 191:435-440.

Thurman, G.B., Rossio, J.L. and Goldstein, A.L. 1977. Thymosin-induced recovery of murine T-cells function following treatment with hydrocortisone acetate. 9:1201-1203.

Tyrey, L. and Nalbandov, V. 1972. Influence of anterior hypothalamic lesions on circulating antibody titers in the rat. Am. J. Physiol. 222:179-185.

Vahouny, G.V., Kyeyne-Nyombi, E., McGillis, J.P., Tare, N.S., Huang, K-Y., Tombes, R., Goldstein, A.L. and Hall, N.R., 1983. Thymosin peptides and lymphokines do not directly stimulate adrenal corticosteroid production in vitro. J. Immunol. 130:791-794.

Vekshina, N.L. and Magaeva, S.V. 1974. Changes in the serotonin concentration in the limbic structures of the brain during immunization. Bull. Exp. Biol. Med. 77:625-627.

STRESS AND IMMUNE RESPONSE:

PARAMETERS AND MARKERS

R. Ballieux

Department of Clinical Immunology

University Hospital

Catharijnesingel 101

3511 GV UTRECHT - The Netherlands

INTRODUCTION

The nervous system and the immune system show a certain degree of congruence (Jerne, 1967), which has stimulated a number of investigators to assume that the two systems are functionally connected (reviewed in Ader, 1982). As discussed in detail by Ader (1983) and Solomon (1983) elsewhere in this volume, the interaction between the mind, the endocrine system and the immune system may considerably influence the body's defence. Both animal studies as well as investigations in man have provided data concerning the role of psychosocial factors in resistance to disease. The biological consequences of e.g. emotional or anxiety stress have been described to result in adverse influences upon certain functions of the immune system. It is well established that this can be effected via the hypothalamic-hypophysical-adrenal axis (Besedovsky & Sorkin, 1977; Stein et al., 1982). Recent findings suggest also neuropeptides to be involved (Johnson et al., 1982a). However, also regulation of the activity of the central nervous system (CNS) by an ongoing immune response has been firmly documented (Besedovsky & Sorkin, 1977). In this respect recent observations that thymic hormones and lymphokines may influence neuroendocrine pathways are intriguing (Hall et al., 1983). Indeed, as Golub (1982) recently suggested 'we may just begin to see a regulatory and functional interdependance of the two systems'.

It is beyond doubt that the rapid developments in the disciplines of Immunology and of Brain and Behavior will lead to a better insight in the processes that govern the interactions between the CNS, the neuroendocrine system and the immune system. Although this ultimately will be of (great) importance for a better understanding of certain health consequences in Breakdown in Human Adaptation, a number of obstacles have to be passed before an integrative view will be developed. In the following paragraphs a biased selection of a few of these will be mentioned.

IMMUNOLOGICAL PARAMETERS AS STRESS MARKERS

As mentioned above the CNS, after appropriate stimuli, is able to modulate immune responses. This implicates that immunological parameters may potentially serve as stress markers. Indeed, the data presented by Ursin et al. (1983) in this volume indicate that low levels of immunoglobulins, although still within the normal range, relate to stressful working conditions. The results have been obtained in a population study. This brings up the question whether the lower levels of immunoglobulins result from changes in activity of the B lymphocyte population, in the rate of synthesis, metabolism and/or distribution of immunoglobulins over the intra- and extravascular spaces. Indeed, Dorian et al. (1982) showed in a recent study that during psychological stress, antibody synthesis as measured in vitro by a plaque forming cell assay, is impaired whereas the number of blood B cells was even increased. This finding reflects decreased B cell activity which ultimately might be caused by a change in the regulatory T cell circuits (Heijnen et al., 1980; Ballieux and Heijnen, 1983). Earlier investigations made in chickens have shown that 'stressful' social interactions result in an accelarated decline of established antibody titres to a bacterial antigen (Siegel and Latiner, 1975), probably reflecting decreased synthesis and/or increased metabolism of immunoglobulins.

Alternatively the possibility cannot be excluded that personality type is linked to genetically determined low responsiveness. In this respect it is interesting that an inverse correlation of intelligence and immunoglobulin levels has been documented (Roseman & Buckley, 1975). It is obvious that prospective studies on the relationship of stress and immunoglobulin levels which include pre-stress data, are of considerable interest. They potentially may indicate whether the wide variation in serum immunoglobulin values, as found in 'normal' reference populations (Zegers et al., 1975) is partly due to different stress levels in the individuals of the reference group.

In a number of investigations the proliferative response of blood lymphocytes after stimulation in vitro with mitogens such as Phytohaemagglutinin (PHA), Concanavalin A (ConA) and Pokeweed mitogen (PWM), has been applied as a functional immunological parameter. Although the clinical relevance of these types of assays often has been overestimated (Basten et al., 1982), an interesting fact emerges if one analyses the results of mitogen stimulation in stressful conditions. Ader (1983) rightly mentions that the results of stress on humoral and cell-mediated immune responses are suppressed, in particular when measured in mitogen stimulation tests. The nature of the stressful stimulation seems to dictate the duration of the suppression. Shortterm effects have been observed in acute or shortterm stress conditions in humans (e.g. Dorian et al., 1982) as well as in animals (Monjan and Collector, 1977). In contrast, studies in man following the loss of a loved object have established that bereavement is associated with a longterm suppression of lymphocyte stimulation responses (Bartrop et al., 1977; Schleifer et al., 1983). Whether the persistance of suppression in the latter situation reflects a fundamental different mechanism operating, compared to shortterm stress-induced impairment of lymphocyte responses, remains to be determined.

With the newer developments in the field of Immunology (see Cohen, 1983, in this volume) a more detailed analysis of lymphocyte distribution patterns and functions seems feasible. The data thus obtained may help to evaluate more precisely the potential value of immune function as stress marker. Related to this is the aspect of a quantitative correlation between the stress-stimulus and the impact on the immune response. In general it cannot be expected a simple relationship to exist, since a number of factors determine the outcome of the interaction between the 'stressor', the CNS, the neuroendocrine and the immune system. However, from careful designed experimental conditions it emerges that in animals (Keller et al., 1981), stress suppresses lymphocyte reactivity in proportion to the intensity of the 'stressor'. It might be worthwhile in relation to research in Breakdown in Human Adaptation, to analyse the quantitative relationships between stress and immune response. It is obvious that this has to be linked to quantification of psychosocial parameters of cognitive, emotional and behavioral nature as well as 'buffering factors' (e.g. coping skill and social support). The integrative nature of these studies might be a serious obstacle for rapid development in the field of Psychoneuroimmunology, but they are worthwhile to be undertaken.

QUANTIFICATION OF IMMUNOLOGICAL PARAMETERS: WHAT, WHEN AND HOW?

The subject to be dealt with in this paragraph is extremely complex. This is partly due to the lack of insight in stress induced processes resulting in alterations in immune reactivity. The relative ignorance in this area may lead to incorrect paradigms which then dictate inappropriate designs of experimental studies. Thus it has been shown that in human volunteers negative feelings of distress activate the pituitary-adrenal system (Lundberg and Frankenhaeuser, 1980). Stress-studies in humans as well as in animals have suggested that the release of ACTH in these conditions will increase corticosteroid levels which ultimately will result in suppression of a number of immune functions. In addition the symphatic nervous system may regulate immune reactivity via adrenal catecholamines (del Rey et al., 1981). Taken these data together it seems that adrenal activity is involved in stress induced alteration of immunity. However, in a recent paper Keller et al. (1983) report that adrenalectomy in the rat does not interfere with stress induced suppression of the PHA-response of lymphocytes in vitro. This finding implicates that the measurement of PHA-induced lymphocyte proliferation does not discriminate between immunosuppression caused by adrenal hormones and immunosuppression effected along non-adrenal pathways. It furthermore illustrates another aspect of complexicity. The outcome of the relatively single test of PHA-induced lymphocyte response is the result of rather complicated, cooperative type of interactions of monocytes and different types of thymus derived (T-) lymphocytes. An inadequate proliferative response, as measured by the uptake of radiolabeled thymidine, may among other things be the result of a diminished Interleukin production by monocytes or T lymphocytes, an increased release of prostaglandines by monocytes or a change in T subsets due to an altered recirculation pattern. Furthermore, the PHA responsiveness may differ when measured in (defibrinated) whole blood cultures or in cultures of isolated blood lymphocytes in control plasma. In the former case neuroendocrine hormones released in the bloodstream by stress (Rossier et al., 1977) may influence immune reactivity (Gilman et al., 1982; Johnson et al., 1982a). This is in particular relevant since neuroendocrine hormones may have lymphokine activity and therefore directly interfere with lymphocyte function (Johnson et al., 1982b).

To complicate the issue, several mitogens are currently used by investigators to analyse lymphocyte reactivity (see preceeding paragraph).

ConA and PWM are frequently used next to PHA. These mitogens also stimulate T cells but they seem to have some specificity for particular subsets. Therefore, differential responses can be expected in stressful
conditions. In addition, PWM stimulates a subset of B lymphocytes to
proliferate and differentiate into immunoglobulin producing plasmablasts.
Quantification of immunoglobulin-secreting cells after PWM stimulation of
lymphocytes in vitro may yield different results compared to proliferative assays when applied in stress-studies. Also antigen-specific responses may essentially differ in susceptibility to stress-induced immunoregulation compared to mitogen induced responses. This is illustrated by
the following example. Johnson et al. (1982a) examined the regulatory
effect of β-, γ- and α-endorphin on the in vitro antibody response to
sheep red blood cells of mouse spleen cells. It was found that α-endorphin was a potent inhibitor, whereas the suppression of the antibody
response by β- and γ-endorphin was minimal. In contrast, Gilman et al.
(1982) reported that α-endorphin as well as β-endorphin had no effect on
the response of lymphocytes to B cell mitogen (LPS), whereas only β-endorphin enhanced the proliferative response of T cells after stimulation
with ConA and PHA.

Another aspect, which should be of concern of workers in the field
of psychoneuroimmunology, is the relationship, if any, between measurements made in in vitro systems and in vivo tests. Is measurement of graft
rejection equivalent to the induction in vitro of specific cytotoxic T
cells? Does analysis of delayed type hypersensitivity by skin testing in
vivo overlap with lymphokine production (such as macrophage inhibiting
factor: MIF) in vitro. The answer to these question can be 'partly' or
'sometimes' if the tests are used to screen for changes in immune reactivity. However, as soon as a more detailed analysis has to be made of the
mechanisms underlying the change in immune response, the potentials of
the various in vivo tests and their 'analogs' in vitro are different.
Indeed, sometimes in vivo studies cannot be replaced by in vitro analysis. This is apparent e.g. from the research of Totman et al. (1980) on
life stress and susceptibility to experimental infection with rhinovi-
ruses. Measurement of interferon production in vitro cannot adequately
substitute for the factors affecting common cold in vivo. However, the
rapid development in the field of immunology with the accompanying sophistication of the methology (Webster, 1981), may allow for a proper
choice of analytic methods which are of relevance for the problem under

study. This implicates that in the field of Breakdown in Human Adaptation to stress, the answer to the question: what, when and how has to be formulated in close collaboration with and in a constant dialogue between the psychosocial, neuroendocrine and immunological specialists. This definitively will lead to impressive progress in the quantification of parameters.

REFERENCES

Ader, R. (editor), 1982.
    Psychoneuroimmunology. Academic Press, New York.
Ader, R., 1983.
    Psychoneuroimmunology. In: R. Ballieux, J. Fielding amd A. L'Abbate (eds), Breakdown in human adaptation to 'stress'. Towards a multidisciplinary approach. Vol. II. Martinus Nijhoff Publ. Comp., The Hague,
Ballieux, R.E. and Heijnen, C.J., 1983.
    Immunoregulatory T cell subpopulations in man: Dissection by monoclonal antibodies and Fc-receptors. Immunol. Rev., 74: 5-28.
Bartrop, R.W., Lazarus, L., Luckhorst, E., Kiloh, L.G. and Penny, R., 1977.
    Depressed lymphocyte function after bereavement. Lancet, i: 834-836.
Basten, A., McCaughan, G.W., Adams, E. and Callard, R.E., 1982.
    Human immunoregulation: a commentary. Immunol. Today, 3: 178-180.
Besedovsky, H. and Sorkin, E., 1977.
    Network of immune-neuroendocrine interactions. Clin. exp. Immunol., 27: 1-12.
Cohen, N., 1983.
    Immunology for nonimmunologists: some guidelines for incipient psychoneuroimmunologists. In: R. Ballieux, J. Fielding and A. L'Abbate (eds), Breakdown in human adaptation to 'stress'. Towards a multidisciplinary approach. Vol. II. Martinus Nijhoff Publ. Comp., The Hague,
Dorian, B., Garfinkel, P., Brown, G., Shore, A., Gladman, D. and Keystone, E., 1982.
    Aberrations in lymphocyte subpopulations and function during psychological stress. Clin. exp. Immunol., 50: 132-138.
Gilman, S.C., Schwartz, J.M., Milner, R.J., Bloom, F.E. and Feldman, J.E., 1982
    β-Endorphin enhances lymphocyte proliferative responses. Proc. Natl. Acad. Sci., 79: 4226-4230.
Golub, E.S., 1982
    Connections between the nervous, haematopoietic and germ-cell systems. Nature, 299: 483.
Hall, N.R., McGillis, J.P., Spangelo, B.L., Healy, D.L. and Goldstein, A.L., 1983.
    Immune regulation of the hypothalamic-hypophysial-adrenal axis: a role for thymosins and lymphokines. In: R. Ballieux, J. Fielding and A. L'Abbate (eds), Breakdown in human adaptation to 'stress'. Towards a multidisciplinary approach. Vol. II. Martinus Nijhoff Publ. Comp., The Hague,
Heijnen, C.J., UytdeHaag, F. and Ballieux, R.E., 1980.
    In vitro antibody response of human lymphocytes. Springer Semin.

738

Immunopathol., 3: 63-92.

Jerne, N.K., 1967
Antibodies and learning: selection versus instruction. In: C. Quarton, Th. Melnechuk, F.O. Schmitt (eds), The neurosciences "A study program". The Rockefeller University Press, New York, pp 200-205.

Johnson, H.M., Smith, E.M., Torres, B.A. and Blalock, J.E., 1982a.
Regulation of the in vitro antibody response by neuroendocrine hormones. Proc. Natl. Acad. Sci (USA), 79: 4171-4174.

Johnson, H.M., Farrar, W.L. and Torres, B.A., 1982b.
Vasopressin replacement of interleuking 2 requirement in gamma interferon production: Lymphokine activity of a neuroendocrine hormone. J. Immunol., 129: 983-986.

Keller, S.E., Weiss, J.M., Schleifer, S.J., Miller, N.E. and Stein, M., 1981.
Suppression of immunity by stress: effect of a graded series of stressors on lymphocyte stimulation in the rat. Science, 213: 1397-1400.

Keller, S.E., Ackerman, S.H., Schleifer, S.J., Schindledecker, R.D., Camerino, M.S., Hofer, M.A., Weiner, H. and Stein, M., 1983.
Effect of premature weaning on lymphocyte stimulation in the rat. Abstract for the annual meeting of the Am. Psych. Ass. In: Psychosomatic Medicine, 45: 75.

Lundberg, U. and Frankenhaeuser, M., 1980.
Pituitary-adrenal and sympathetic-adrenal correlates of distress and effort. J. Psychosom. Res., 24: 125-130.

Monjan, A.A. and Collector, M.I., 1977.
Stress-induced modulation of the immune response. Science, 196, 307-308.

del Rey, A., Besedovsky, H.O., Sorkin, E., da Prada, M. and Arrenbrecht, S., 1981.
Immunoregulation mediated by the sympathetic nervous system, II. Cell. Immunol., 63: 329-334.

Roseman, J.M. and Buckley, C.E.,III, 1975.
Inverse relationship between serum IgG concentrations and measures of intelligence in elderly persons. Nature, 254: 55-56.

Rossier, J., French, E.D., Rivier, C., Ling, N., Guillemin, R., Bloom, F.E., 1977.
Foot-shock induced stress increases ß-endorphin levels in blood but not in brain. Nature, 270: 618-620.

Schleifer, S.J., Keller, S.E., Camerino, M., Thornton, J.C. and Stein, M., 1983.
Suppression of lymphocyte stimulation following bereavement. J. Amer. Med. Assoc., in press.

Siegel, H.S. and Latimer, J.W., 1975
Social interactions and antibody titres in young male chickens (Gallus Domesticus). Anim. Behav., 23: 323-330.

Solomon, G.F., 1983
Emotions, immunity and disease: an historical and philosophical perspective. In: R. Ballieux, J. Fielding and A. L'Abbate (eds), Breakdown in human adaptation to 'stress'. Towards a multidisciplinary approach. Vol. II. Martinus Nijhoff Publ. Comp., The Hague.

Stein, M., Keller, S.E. and Schleifer, S.J., 1982.
The role of brain and the neuroendocrine system in immune regulation: potential links to neoplastic diseases. In: S.M. Levy (ed.), Biological mediators of behavior and disease: Neoplasia. Elsevier Science Publ. Co., Inc., New York.

Totman, R., Kiff, J., Reed, S.E., Craig, J.W., 1980.
    Predicting experimental colds in volunteers from different measures
    of recent life stress. J. Psychosom. Res., 24: 155-163.
Ursin, H., Mykletun, R., Isaksen, E., Murison, R, Vaernes, R., and Tøn-
der, O., 1983.
    Immunoglobulins as stress markers? In: R. Ballieux, J. Fielding and
    A. L'Abbate (eds), Breakdown in human adaptation to 'stress'.
    Towards a multidisciplinary approach. Vol. II. Martinus Nijhoff
    Publ. Comp., The Hague,
Webster, A.D.B. (guest editor), 1981
    The assessment of immunocompetence. In: Clinics in Immunology and
    Allergy, vol. I, no. 3, W.B. Saunders Company Ltd, London.
Zegers, B.J.M., Stoop, J.W., Reerink-Brongers, E.E., Sander, P.C., Aal-
berse, R.C. and Ballieux, R.E., 1975.
    Serum Immunoglobulins in healthy children and adults. Levels of the
    five classes, expressed in international units per millimetre. Clin.
    Chim. Acta, 65: 319-329.

PART 4

Breakdown in human adaptation and gastrointestinal
dysfunction: clinical, biochemical and
psychobiological aspects

edited by

J.F. Fielding

THE BRAIN AND THE GUT

Frank P. Brooks

GI Section/Department of Medicine, 574A Maloney Building, Hospital of the University of Pennsylvania, 3400 Spruce Street, Philadelphia, PA   19104

In order for events in the brain to lead to a breakdown in adaptation and gastrointestinal function, signals must reach the digestive tract by way of the vagus or sympathetic nerves, the release or failure to release of humoral substances from the brain into the systemic circulation or by a brain-induced effect on blood flow to the digestive tract. The effects on the gut may be entirely in terms of function - e.g. failure of the stomach to empty, or in terms of structural pathology - e.g. ulcerations. Delayed gastric emptying can produce the clinical syndrome of non-ulcer dyspepsia. Gastric mucosal ischemia in response to brain-initiated influences on the circulation can lead to acute gastric mucosal erosions and probably ulcerations. In experimental animal external factors acting on the brain can alter the migrating myoelectrical complexes in the upper digestive tract and modify intestinal transit times. Brain lesions can be followed by mucosal ulcerations where local resistance factors and/or excessive aggressive factors appear to be mediating the structural changes. The role of newer neurotransmitters and neuromodulators in contributing to the breakdown in adaptation or in preventing it is an active research area.

The anatomical relationships between the visceral brain and the digestive tract are well known. However it is only recently that Langley's view of the enteric nervous system as a kind of visceral brain has become popular again (Gershon, M.D. and Erde, S.M., 1981). The arrangement of afferent and efferent neurons with interneurons in the submucosal and myenteric plexuses is characteristic of a complex autonomous nervous system. Now some of the newer neurotransmitters such as neuropeptides and purines are turning up in both the brain and the enteric nervous system. Does this similarity imply functional relationships or does it merely reflect utilization of a common successful mechanism (Brooks, F.P., 1983)?

The principal connections between the brain and the gut are by way of the vagus nerves. Sympathetic (efferent) and splanchnic nerve connections connect nerve pathways in the spinal cord which reach the brain. Pelvic efferent nerves carry connections to the distal colon from the spinal cord and brain. The paraspinal and prevertebral ganglia serve as integrating points for sympathetic nervous activity to the gut (Szurszewski, J.A., 1977).

In the brain there is a series of levels of nervous control: the dorsal motor nuclei in the medulla, the mid brain reticular system, the hypothalamus, and the limbic system. Excitatory and inhibitory influences on functions of the digestive tract such as exocrine and endocrine secretion, absorption blood flow and motor activity can be exerted at each level.

The digestive tract is the site of many symptom complexes of unknown etiology (Lennard-Jones, J.E., 1983). Because of the relationship of external events or the benefits of cutting nervous connections it is a reasonable hypothesis that nervous control from the central nervous system is involved. Sometimes there is gross pathological changes in the digestive organs, e.g.: reflux esophagitis, peptic ulcer, gastric erosions, ischemic necrosis. In other digestive orders there are no demonstrable tissue changes but abnormalities in secretory, or motor function are readily demonstrated. Animal models may illustrate similar functional changes in response to manipulations in the central nervous system.

Table 1 lists seven digestive disorders where functional changes are prominent in the pathophysiology.

Neuromuscular and Neurosecretory Disorders of the Digestive Tract

1. Reflux esophagitis
2. Vagal drive in duodenal ulcer
3. Delayed gastric emptying
   Tachygastria
4. Biliary dyskinesia
5. "Nervous diarrhea"
6. Ischemic erosions
7. The aging colon and the irritable bowel syndrome

In reflux esophagitis the basic abnormality is gastroesophageal reflux of gastric content. This can be the result of transient inappropriate relaxation of the lower esophageal sphincter, low resting sphincter tone, or increased intragastric pressure (Dodds et al., 1982). Delayed emptying of solid gastric meals is a feature of some patients with reflux (McCallum et al., 1981). Failure of a reflex induced contraction of the lower esophageal sphincter has been proposed as a factor (Dodds et al., 1975).

Acid-pepsin hypersecretion in response to excessive efferent activity in efferent vagal neurons has been a subject of speculation for many years. Recently the acid secretory response to sham feeding (Feldman et al., 1980) and changes in immunoreactive pancreatic polypeptide in the plasma have been interpreted to indicate vagally-mediated hypersecretion in some patients with duodenal ulcer (Schwartz et al., 1979).

The availability of scintiscans of the abdomen and radionuclide markers of solids, liquids, and lipids make it possible to assess gastric emptying and intestinal transit times in intact human subjects (Carryer et al., 1982, Malagelada, J.R., 1979). Patients with symptoms of non-ulcer dyspepsia have been found to have marked delays in the emptying of solids and to respond to agents which accelerate gastric emptying (You et al., 1980). Some of these patients exhibit tachygastria - an accel-erated frequency of gastric pace-setter electrical potentials (Stoddard et al., 1981, Telander et al., 1980).

Some patients have abdominal pain similar to that seen with stones in the common duct but with normal size ducts and no stones. Measurements can now be made of pressures in the sphincters of the common bile duct, pancreatic duct and Sphincter of Oddi. Grossly elevated pressures with the clinical syndrome may be an indication for sphincterotomy (Hogan W., 1983).

There is a group of patients with watery diarrhea in whom no struc-tural pathology can be demonstrated. They have shortened small intestinal transit times (Kalser et al., 1956). In patients with the irritable bowel syndrome abnormalities in transit time were seen in both the large and small bowel (Read et al., 1980, Read, N.W., 1980). Drugs such as diphenoxylate and loperamide which prolong transit time may reduce the diarrhea.

Routine endoscopic studies of the stomach in patients with severe burns indicate that focal ischemia is the earliest manifestation of gastric erosions (Czaja et al., 1974). A new method of in vivo spectrophotometry showed similar changes in patients with burns or head injuries (Kamada et al., 1982).

The symptoms of abdominal pain related to defecation are common in both "normal" subjects and in patients seen in the office practice of gastroenterology. Recent studies indicate abnormalities in electrical activity in colonic smooth muscle as well as increased contractile activity in response to meals (especially fat) and cholinergic agonists (Snape et al., 1977, Sullivan et al., 1978). Elderly persons are particularly sus-ceptible to constipation and a clinical syndrome resembling idiopathic megacolon. Cerebrovascular ischemia and inactivity appear to be predis-posing factors (Andersson et al., 1979).

The common thread that runs through all of these disorders is a change in gastrointestinal function which either precedes structural lesions or exists in the absence of structural abnormalities.

Studies in man indicate that the brain can modify gastrointestinal motility and secretion. Labyrinthine stimulation delays the gastric emptying of a liquid meal and changes the duodenal contractile pattern from a fasted to a fed configuration (Thompson et al., 1982). REM sleep correlates with migrating myoelectrical complexes (phase III activity fronts) originating the stomach but not those originating in the jejunum (Evans et al., 1982). Insulin hypoglycemia has long been known to stimulate gastric acid secretion and large amplitude phasic gastric contractions. Cold applied to body surface elicits increases in mucosal blood flow in the colon (Forrester et al., 1981).

In our laboratory we have studied the effects of decortication and decerebration as well as those of electrolytic lesions of the lateral hypothalamus on gastric function in anesthetized cats. Decortication reduces the acid secretory phasic antral contractile response to electrical stimulation of vagal efferents but not gastrin release, while mid collicular decerebration markedly increases gastrin release from the antrum but has no effect on gastric acid secretion or gastric antral contractions (Feng et al., 1983). Bilateral electrolytic lesions of the lateral hypothalamus reduces the vagal response of acid secretion and antral contractions. Electrical stimulation in the lateral hypothalamus stimulates phasic antral contractions without an increase in gastric acid secretion (Feng et al., 1983). The response is abolished by bilateral cervical vagotomy. Electrical stimulation in the dorsal motor nucleus of the vagus almost always produces a doubling in acid secretion in about a third of the cats (Lombardi et al., 1982). Phasic antral contractions were seen in 5 of the 7 cats with a doubling of acid secretion but also in 17 of 30 experiments without doubling of acid secretion. Stimulation of the efferent cervical vagal trunks produced both an increase in acid secretion and phasic antral contractions. We conclude that there is an excitatory pathway for phasic antral contractions from the lateral hypothalamus to the dorsal motor nucleus of the vagus and hence to the antrum. There is some degree of organization within the dorsal motor nucleus of the vagus to selectively stimulate antral contractions and acid secretion. From experiments in cats giving intra-arterial eserine during electrical vagal stimulation we would conclude that the pattern of gastric motor response depends upon the amount of acetylcholine available at the myoneural junction and not on the parameters of electrical stimulation of the vagi (Lombardi et al., 1981).

In other laboratories the pathways from the amygdala, cingulate and orbital gyrus to the cat colon have been traced. Both augmentating and inhibitory influences on phasic contractions of the colon can be demonstrated (Rostad, H., 1973). Augmenting activity is mediated in the proximal colon by the vagi. Inhibitory influences travel in the splanchnics. In the distal colon the augmenting influences travel in the pelvic nerves and the inhibitory with the lumbar colonic nerves. In the dog the ending of phase III of the migrating motor complex is much more gradual than in denervated intestine. This may indicate a role for the central nervous system in the control of small intestinal motor activity (Nakaya et al., 1982).

Gastric blood flow is increased by vagal stimulation and reduced by sympathetic stimulation (Guth, P.H. and Smith, E., 1977). The effects of vagal stimulation on the lower esophageal sphincter may be mediated by vasoactive intestinal peptide while that on the pylorus may be the result of the release of enkephalins and substance P. Gastrin release in response to vagal stimulation may be the result of the release of bombesin as a neurotransmitter with a reciprocal inhibition of the release of somatostatin (Brooks, F.P., 1984).

These results establish that the brain can selectively influence motor activity, exocrine secretion, and the release of hormones affecting the function of the digestive tract. The specific mechanisms and their relation to behavior need to be elucidated. Understanding of normal mechanisms will permit the definition of abnormal responses originating in the brain.

Andersson, H., Bosaeus, I., Falkheden, T. and Melkersson, M., 1979. Transit time in constipated geriatric patients during treatment with a bulk laxative and bran: A comparison. Scand. J. Gastroent., 14: 821-826.
Brooks, F.P., 1983. Central nervous system and the digestive tract in functional disorders of the digestive tract. Edited by W.Y. Chey, Raven Press, New York, pp. 21-27.
Brooks, F.P., 1984. Section on digestive tract in Best and Taylor's Physiological Basis of Medical Practice. Edited by J.B. West, Williams and Wilkins, Baltimore (in press).
Carryer, P.W., Brown, M.L., Malagelada, J-R., Carlson, G.L. and McCall, J. T., 1982. Quantification of the rate of dietary fiber in humans by a

newly developed radiolabeled fiber marker. Gastroenterology, 82: 1389-1394.

Czaja, A.J., McAlhany, J.C. and Pruitt, B.A., Jr., 1974. Acute gastro-duodenal disease after thermal injury: an endoscopic evaluation of the incidence and natural history. New Eng. J. Med., 291: 925-929.

Dodds, W.J., Dent, J., Hogan, W.J., Helm, J.F., Hauser, R., Patel, G.K. and Egide, M.S., 1982. Mechanisms of gastroesophageal reflux in patients with reflux esophagitis. New Eng. J. Med., 307: 1547-1552.

Dodds, W.J., Hogan, W.J., Miller, W.N., Stef, J.J. and Arndorfer, R.C., 1975. Effect of increased intra-abdominal pressure on lower esophageal sphincter pressure. Am. J. Digest. Dis., 20: 298-308.

Evans, D.F., Foster, G.E. and Hardcastle, J.D., 1982. The motility of the human antrum and jejunum during the day and during sleep: an investigation using a radiotelemetry system. In Motility of the Digestive Tract. Edited by M. Wienbeck. Raven Press, New York, pp. 185-192.

Feldman, M., Richardson, C.T. and Fordtron, J.S., 1980. Effect of sham feeding on gastric acid secretion in healthy subjects and duodenal ulcer patients: evidence for increased basal vagal tone in some ulcer patients. Gastroenterology, 79: 796-800.

Feng, H-S., Aronchick, C.A. and Brooks, F.P., 1983. Decortication and de-cerebration effects compared on efferent electrical vagal stimulation of the stomach in cats. Clinical Research, 31 (in press).

Feng, H-S., Brobeck, J.R. and Brooks, F.P., 1983. Electrical stimulation and lesions of lateral hypothalamus and gastric function in cats. Submitted to 29th Congress of the International Union of Physiological Sciences, Sidney Australia.

Forrester, D.W., Davison, J.S., Spence, V.A. and Walker, W.F., 1981. Changes in human colonic mucosal-submucosal blood flow after body surface cooling. Gut, 22: 469-474.

Gershon, M.D. and Erde, S.M., 1981. The nervous system of the gut. Gastroenterology, 80: 1571-1594.

Guth, P.H. and Smith, E., 1977. Nervous regulation of the gastric microcirculation in nerves and the gut. Edited by F.P. Brooks and P.W. Evans, Charles B. Slack, Inc., Thorofare, NJ., pp. 365-376.

Hogan, W., 1983. Motility and biliary dyskinesia in functional disorders of the digestive tract. Edited by W.Y. Chey, Raven Press, New York (in press).

Kalser, M.H., Zion, D.E. and Bockus, H.L., 1956. Functional diarrhea: an analysis of the clinical and roentgen manifestations. Gastroenterology, 31: 629-646.

Kamada, T., Sato, N., Kawano, S., Fusamoto, H. and Abe, H., 1982. Gastric mucosal hemodynamics after thermal or head injury. Gastroenterology, 83: 535-540.

Lennard-Jones, J.E., 1983. Functional gastrointestinal disorders. New Eng. J. Med., 308: 431-435.

Lombardi, D.M., Feng, H-S. and Brooks, F.P., 1981. Phasic vs tonic contractions of cat stomach in response to electrical vagal stimulation. Fed. Proc., 40: 576.

Lombardi, D.M., Feng, H-S. and Brooks, F.P., 1982. Dissociation of secretory and motor responses to stimulation of the dorsal motor nucleus of the vagus in anesthetized cats. Gastroenterology, 82: 1120.

McCallum, R.W., Berkowitz, D.M. and Lerner, E., 1981. Gastric emptying in patients with gastroesophageal reflux. Gastroenterology, 80: 825-891.

Malagelada, J-R., 1979. Physiologic basis and clinical significance of gastric emptying disorders. Digestive Dis. and Sci., 24: 657-661.

Nakaya, M., Takeuchi, S., Aizawa, I., Suzuki, T. and Itoh, Z., 1982.

Involvement of the central nervous system in regulation of interdigestive contractions in the stomach in motility of the digestive tract. Edited by M. Wienbeck, Raven Press, New York, pp. 175-180.

Read, N.W., Miles, C.A., Fisher, D., Holgate, A.M., Kime, N.D., Mitchell, M.A., Reeve, A.M., Roche, T.B. and Walker, M., 1980. Transit of a meal through the stomach, small intestine and colon in normal subjects and its role in the pathogenesis of diarrhea. Gastroenterology, 79: 1276-1282.

Read, N.W., 1980. Disordered transit of a meal through the small and large bowel in irritable bowel syndrome. Gut, 21: A906.

Rostad, H., 1973. Colonic motility in the cat V peripheral pathways mediating the effects induced by hypothalamic and mesencephalic stimulation. Acta Physiol. Scand., 89: 154-168.

Schwartz, T.W., Stenquist, B., Olbe, L. and Stadil, F., 1979. Synchronous oscillations in the basal secretion of pancreatic polypeptide and gastric acid. Depression by cholinergic blockade of pancreatic polypeptide concentrations in plasma. Gastroenterology, 76: 14-19.

Snape, W.J., Carlson, G.M., Matarazzo, S.A. and Cohen, S., 1977. Evidence that abnormal myoelectrical activity produces colonic motor dysfunction in the irritable bowel syndrome. Gastroenterology, 72: 383-387.

Stoddard, C.J., Smallwood, R.H. and Duthie, H.L., 1981. Electrical arrhythmias in the human stomach. Gut, 22: 705-712.

Sullivan, M.A., Cohen, S. and Snape, W.J., 1978. Colonic myoelectrical activity in irritable bowel syndrome. New Eng. J. Med., 298: 878-883.

Szurszewski, J.H., 1977. Toward a new view of prevertebral ganglion in nerves and the gut. Edited by F.P. Brooks and P.W. Evers, Charles B. Slack, Inc., Thorofare, NJ., pp. 244-260.

Telander, R.L., Morgan, K.G., Kreulen, D.L., Schmatz, P.F., Kelly, K.A., Szurszewski, J.H., 1980. Human gastric atony with tachygastria and gastric retention. Gastroenterology, 79: 311-314.

Thompson, D.G., Richelson, E. and Malagelada, J-R., 1982. Pertubation of gastric emptying and duodenal motility through the central nervous system. Gastroenterology, 83: 1200-1206.

You, C.H., Lee, K.Y., Chey, W.Y., Menguy, R., 1980. Electrogastrographic study of patients with unexplained nausea, bloating and vomiting. Gastroenterology, 79: 311-314.

THE ROLE OF PSYCHIATRIC ASSESSMENT IN THE MANAGEMENT
OF FUNCTIONAL BOWEL DISEASE

David H. Alpers

Division of Gastroenterology
Washington University School of Medicine
660 S. Euclid Avenue
St. Louis, MO 63110 USA

The gastroenterologist is faced every day with patients who have a psychiatric illness as a primary significant diagnosis. This encounter occurs frequently because certain of the psychiatric illnesses can have gastrointestinal symptoms as a presenting complaints. Most important among these are depression and hysteria,which can present with anorexia, nausea, vomiting, a change in bowel habits, and abdominal pain. Table I lists the psychiatric disorders which can present with primary intestinal complaints. The psychiatric disorders which do not present with primary gastrointestinal complaints are also interesting, as they include some diagnoses considered to be associated with some intestinal disorders. Those psychiatric disorders not usually presenting with GI symptoms include schizophrenia, bipolar affective disorders (mania), phobic neurosis, obsessive compulsive neurosis, and personality disorders. Personality disorders are seen with functional GI disorders, but the association is often unrelated to the presenting complaints.

In addition to the frequency with which patients with psychiatric illness may present to a gastroenterologist, there has been other evidence suggesting a role for emotional factors in gastrointestinal disease.These data come from the psychoanalytic and epidemiologic literature. Some papers suggest that early problems with elimination and feeding lead to problems focussing on intestinal organs; e.g. psychogenic constipation (Buxbaum and Sodergren, 1977), rumination (Philippopoulos, 1973), and anorexia nervosa (Drossman et al, 1979). Other studies have documented an increased incidence of peptic ulcer in a high stress situation, such as in air traffic controllers (Cobb and Rose, 1973), or have suggested that a

Table I

Psychiatric disorders which can present with GI symptoms

Diagnosis	Common symptoms
Affective disorders	
Depression	Nausea, vomiting, anorexia, weight loss, abdominal pain, altered bowel habits
Somatoform disorders	
Somatization disorder (hysteria)	Nausea, vomiting, diarrhea abdominal pain
Psychogenic pain disorder	Abdominal pain
Anxiety disorders	
Generalized anxiety disorder	Diarrhea, dyspepsia
Eating disorders	
Anorexia nervosa	Aversion to food, weight loss
Bulimia	Vomiting, symptoms of depression
Alcoholism	Anorexia, weight loss

certain pattern of response to stress will predict the onset of peptic ulcer (Weiner et al, 1957). For these reasons it would be useful for the gastroenterologist to be able to assess patients objectively.

The medical history alone is not usually adequate for psychiatric assessment, as it does not focus on behavioral problems. The psychiatric history is very helpful in making psychiatric diagnoses, once some experience is obtained. The interview technique is practical and economical, and is subject only to concern for the validity of the data, since it is all self reported. We will return shortly to the use of the interview for making psychiatric diagnoses.

One frequent use of the interview made by gastroenterologists is to record recent life stresses, with the implication that these are related either to the onset or exacerbation of disease or symptoms. This tradition is reinforced by the few studies which suggest such a relationship (Cobb and Rose, 1973; Weiner et al, 1957). However, for most of the studies relating life stresses and gastrointestinal disorders, there are serious problems in study design (Cohen, 1979). Most studies rely upon retrospective self reporting of events and illnesses, and in many cases the hypotheses being tested were clear to the subjects. There are usually no guidelines regarding the type of events and illness reported. Undesirable events are stressed, yet no change might have an effect whereas change in life events might have been expected. Moreover, many of the groups studied vary in frequency of their reporting and in the readjustment required by each life event. Such factors have been shown to vary in different age and socioeconomic groups (Goldberg and Comstock, 1980). The life event may trigger a response which is equally or more dependent on the subject's personality than upon the life event itself. Thus, the event may alter illness behavior (i.e. response to the underlying illness) but not the incidence of disease. When effects of life events are seen, the magnitude of relationship to the occurrence of symptoms is small (Mendeloff et al, 1970). Finally, since the disease groups usually include only those who sought treatment and could be identified retrospectively, those who self select may have personality characteristics (e.g. anxiety) which could lead to selection by a study group. There is some relationship between stress and changing life events and illness, but it is a weak one, and depends as much or more on the individual's behavior and personality as on the event itself (Wershaw and Reinhart, 1974). For these reasons it seems prudent to examine patients with functional complaints by modern objective methods, whether they are psychometric or psychiatric.

Psychometric testing is infrequently used in the practice of gastroenterology, but has been used for research purposes. Some of the comprehensive personality profiles (e.g. MMPI) have been used to study ulcerative colitis (West, 1970). Objective rating scales have been applied to the study of peptic ulcer patients (Piper et al, 1981), and irritable bowel syndrome (Palmer et al, 1974). However, the literature of functional GI disorders is remarkable for the paucity of studies which have used objective psychological tests, tested for their validity and reliability. The other methods which have been used to assess behavior patterns in gastrointestinal disease include psychiatric interview (Feldman et al, 1967) and self administered questionnaire (Drossman et al, 1982). The advantages of these two methods are that they are practical, economical, and easy to use. However, they depend upon self reporting and are of unknown validity, unless the validity is tested for each study. The use of biofeedback has been restricted thus far to treatment of rectal sphincter function (Schuster, 1977), and has not been used for diagnosis or treatment of other intestinal disorders because no reliable intestinal signal has been developed to assess function of internal organs.

While the unstructured interview is of uncertain validity, it may still provide very useful information if proper attention is given to psychiatric symptoms and to the patient's response to them (coping behavior). Personality types can be identified (Kahana and Bebring, 1964) which, when exaggerated, provide the basis for the psychiatric diagnosis of personality disorder. Recognition of these personality types can be helpful in the management of any patient whether or not accompanied by organic disease. The major personality types seen with functional GI disorders (and their relative personality disorders) include the dependent (passive-aggressive), the orderly (obsessive-compulsive) and the dramatizing (hysterical).

Within the past few years increasing attention has been given to objectivity of diagnosis in psychiatry. Attempts have been made to define explicit diagnostic criteria (Wing et al, 1974; Spitzer and Endicott, 1975), and to develop supervised, structured psychiatric interviews which are designed to provide life time diagnosis (Feighner et al, 1972; Woodruff et al, 1974). These methods have been tested for

Table II

Differential diagnosis of abdominal pain

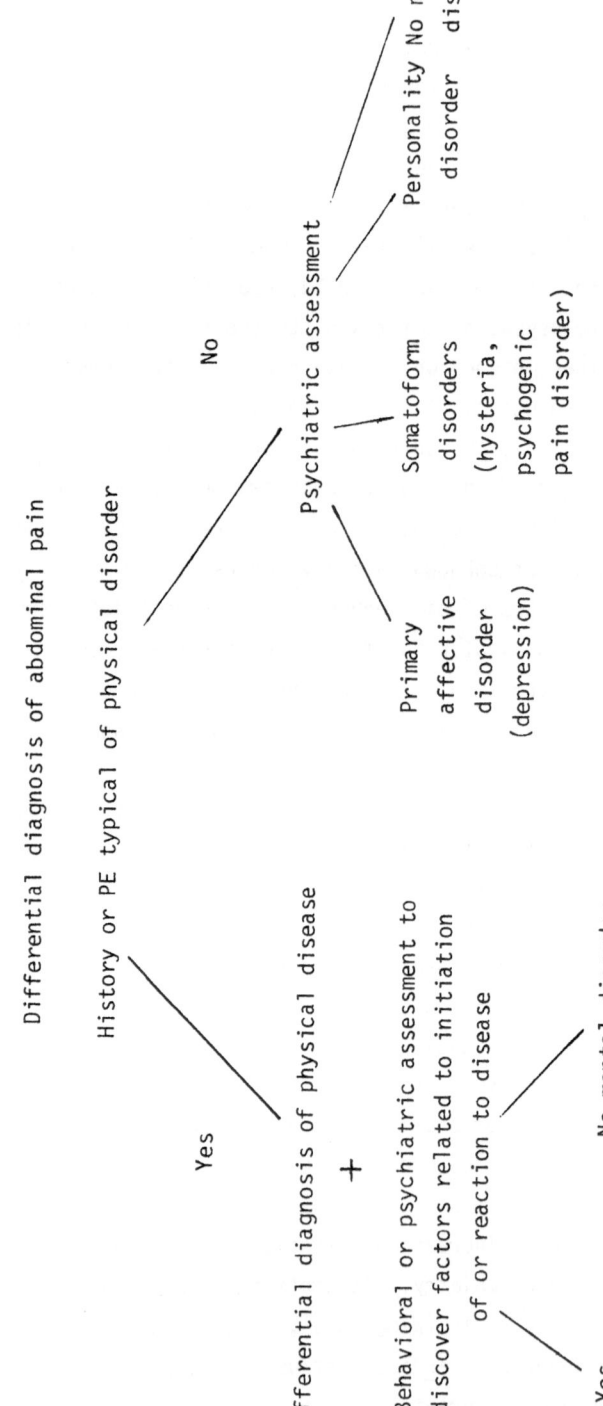

History or PE typical of physical disorder

Yes / No

Differential diagnosis of physical disease

Psychiatric assessment

+

Behavioral or psychiatric assessment to discover factors related to initiation of or reaction to disease

Primary affective disorder (depression)

Somatoform disorders (hysteria, psychogenic pain disorder)

Personality disorder

No mental disorder

Yes / No mental disorder

their reliability (Helzer et al, 1977) and validity (Helzer et al, 1978). They have been shown to provide psychiatric diagnoses which compare in reliability to other diagnostic studies such as EKG's and radiologic procedures (Keran, 1975). These methods have proved so successful that the criteria developed from them have formed the basis for the diagnostic criteria in the recent edition of the Diagnostic and Statistical Manual (DSM-III) of the American Psychiatric Association (DMS-III, 1980). There is increasing realization that specific diagnostic criteria are important not only for psychiatric research but also for sound clinical practice. This publication provides diagnostic criteria which can be used when taking a history and can allow the gastroenterologist to make diagnoses.

The usefulness of this approach can be seen if one takes a common symptom as an example, abdominal pain. Table II illustrates the process. Once it is determined whether or not the symptom suggests a physical disorder, the evaluation of psychiatric or behavioral status can assist in therapeutic decisions. It is important to keep in mind that the psychiatric disorders can occur in the presence of organic disease. In that instance attention may need to be given to both disorders.

We have studied patients with irritable bowel syndrome with structured research psychiatric interviews. Other functional disorders have been assessed by the use of nonstructured diagnostic interviews to make psychiatric diagnoses. As a result of these observations we have been able to propose a correspondence between functional GI disorders and psychiatric illness (Table III). However, not all functional GI disorders seem to be associated with psychiatric illness. These disorders would include aerophagia, stress induced diarrhea, proctalgia fugax, pruritus ani, and esophageal spasm. This is not to say that there is no psychiatric illness associated with these syndromes; merely that none has yet been demonstrated. Furthermore, some of these disorders (e.g. proctalgia fugax) have been associated with certain personality traits as determined from the MMPI, suggesting that these patients are anxious and perfectionistic (Pilling et al, 1965). One must also keep in mind that the observations made have documented an association, not implying causation. In addition, not all patients with the syndromes listed in Table III will have a psychiatric disorder. In our experience over half will, however.

The data involving patients with irritable bowel syndrome are the best defined either by clinical interview (Hislop 1971), psychological

Table III

Functional GI syndromes associated with psychiatric disorders

Syndrome	Primary complaints	Usual psychiatric disorders
Psychogenic vomiting	Nausea and vomiting	Depression
Dyspepsia	Epigastric discomfort	Anxiety neurosis Depression
Irritable bowel	A. Constipation and pain	Depression (A,B)
	B. Alternating diarrhea and constipation	Hysteria (A,B,C)
	C. Diarrhea	
Recurrent abdominal pain	Abdominal pain without other complaints	Psychogenic pain disorder

testing (Palmer et al, 1974), or by validated psychiatric interview
(Young et al, 1976). The patients studied by Young and coworkers all
had a chronic disorder involving mainly the colon and rectum manifested
by daily patterned irregularity of bowel function, with or without
abdominal pain. This study analyzed 29 subjects and 33 controls matched
for chronic illness and as closely as possible for age and sex. Psychiatric
disorders were found in 18% of controls, a figure which matches well with
other ambulatory populations. Patients with irritable bowel syndrome had
psychiatric disorders identified 72% of the time (21 of 29 subjects).
Half of these patients (34% of total) had either hysteria or depression.
Most of the rest (31% of total) had psychiatric illness which could not
be classified by research criteria, but by the more lenient classification
outlined by DSM-III could be diagnosed as hysteria or depression. Only
two of the 21 psychiatric diagnoses were for other illnesses (anxiety
neurosis and alcoholism). Only nine of the 21 psychiatric diagnoses were
correctly made by the internist caring for the patient. Seven times a
psychiatric diagnosis was made which was incorrect, usually depression,
when hysteria was the diagnosis. Of the eight patients with no
psychiatric disorder, four were thought to be psychiatrically ill by
their internist. The patients with unrecognized hysteria were taking
twice as many medications and were hospitalized twice as often as the
nonhysteric irritable bowel patients. These data convince us that
psychiatric illness is frequently encountered in irritable bowel patients,
and that making the correct diagnosis has important therapeutic
implications, either positive (use of antidepressants) or negative
(decreased use of medications).

The gastroenterologist who is unskilled in psychiatry will feel
uncertain in attempting to utilize any of the above information in his
practice. However, a transfer to a psychiatrist when no organic pathology
is found will be unsuccessful in most cases. The patient has selected
an internist or gastroenterologist for care, partly because of his own
self image, and partly because his (or her) behavior pattern is not
sufficiently disturbing to his family or coworkers to suggest the need
for psychiatric referral. Under such a circumstance the patient will not
respond well to the implication that his case needs psychiatric attention.
However, in our experience, such patients will invariably accept psychiatric
evaluation if the following procedures are followed: 1) the primary
physician must elicit the initial history and formulate some idea of the

problem himself. This will enable him to 2) discuss the problem with the patient and explain the need for skilled delineation of the problem so that proper therapy can be given. 3) The explanation must focus both on the physiological GI abnormality and on the psychiatric problem as separate but related issues. Therapy for both can be undertaken as needed. 4) Therefore, the physician must outline the treatment plan with himself as the prime manager. The psychiatrist can be used as a consultant and in some cases can follow the patient along with the gastroenterologist, but in most instances the gastroenterologist must remain the primary physician. 4) The psychiatrist to be used must be known to and trusted by the physician. Furthermore, he must be interested in seeing cases of functional GI disease. In this way an active dialogue will be established and the gastroenterologist will learn most about psychiatric diagnosis and eventually feel more comfortable with his ability to assess these disorders. 6) Finally, or perhaps first in importance, the gastroenterologist must develop an interest in psychiatry. His failure to do so is to close his eyes to a large percent of his patients who will enter his office. As he becomes more proficient in his recognition of psychiatric disorders, he will require less and less extensive medical investigations to make the diagnosis of functional GI disorders by ruling out unlikely organic diseases. The gastroenterologist can learn to deliver psychotherapy of a practical kind, centered largely around the delivery of emotional support for the patient by recognition of his psychiatric problem and by continuing to support his need to have the physiological basis of his complaints explained and respected.

References

Buxbaum, E. and Sodergren, S.S.,1977.A disturbance of elimination and motor development. The mother's role in the development of the infant. Psychoanal. Study Child., 32:195-214.

Cobb, S. and Rose, R.M., 1973. Hypertension, peptic ulcer and diabetes in air traffic controllers. J.A.M.A. 224:489-492.

Cohen, F., 1979. Personality, stress and the development of physical illness. In Health Psychology, B.C.Stone, F.Cohen and N.E.Adler, eds., Jossey-Bass, San Francisco.

Drossman, D.A., Ontjes, D.A. and Heizer, W.D., 1979. Anorexia nervosa. Gastroenterology 77:1155-1131.

Drossman, D.A., Sandler R.S., McKee, D.C., and Lovitz, A.J., 1982. Bowel patterns among subjects not seeking health care. Use of a questionnaire to identify a population with bowel dysfunction. Gastroenterology 83:529-534.

DSM-III: Diagnostic-statistical Manual of Mental Disorders, (3rd ed.),1980. J.B.W.Williams, ed., American Psychiatric Assoc., Washington, D.C.

Feighner, J.P., Robins, E., Guze S., 1972. Diagnostic criteria for use in psychiatric research. Arch. Gen. Psych. 26:57-63.

Feldman, F., Cantor, D., Soll, S., and Bachrach, W., 1967. Psychiatric study of a consecutive series of 34 patients with ulcerative colitis. Brit. Med. J. 3:14-17.

Goldberg, E.L. and Comstock, G.W., 1980. Epidemiology of life events: frequency in general populations. Am. J. Epidemiology 111:736-752.

Helzer, J.E., Robins, L.N., Taibleson, M.,et al,1977. Reliability of psychiatric diagnosis. I. A methodological review. Arch. Gen. Psych. 34:129-135.

Helzer, J.E., Clayton, P.J., Penbakian, R.,et al,1978.Concurrent diagnostic validity of a structured psychiatric interview. Arch. Gen. Psych. 35:849-853.

Hislop, I.G., 1971. Psychological significance of the irritable bowel syndrome. Gut 12:452-457.

Kahana, R.J. and Bebring, G.L., 1964. Personality types in medical management. In "Psychiatry and Medical Practice in a General Hospital", ed. N.E.Ginsberg, New York, International Universities Press, 108-123.

Keran, L.M., 1975. The realiability of clinical methods, data and judgements. N. Engl. J. Med. 283:642-646, 695-701.

Mendeloff, A.I., Mark, M., Siegel, C.I., and Lilienfeld, A., 1970. Illness experience and life stresses in patients with irritable colon and with ulcerative colitis. An epidemiologic study of ulcerative colitis and regional enteritis in Baltimore 1960-1964. N. Engl. J. Med. 282:14-17.

Palmer, R.L., Stonehill, E., Crisp, A.H.,et al, 1974. Psychological characteristics of patients with the irritable bowel syndrome. Postgrad. Med. J. 50:416-419.

Philippopoulos, G.S., 1973. The analysis of a case of merycsim: psychopathology and psychodynamics. Psychother. and Psychosom. 22:364-371.

Pilling, L.F., Swenson, W.M., and Hill, J.R., 1965. The psychologic aspects of proctalgia fugax. Diseases of the Colon and Rectum 8: 372-376.

Piper, D.H., McIntosh, J.H., Ariotti, D.E., Caloguiri, J.V., Brown, R.W., and Shy, C.M., 1981. Life events and chronic duodenal ulcer: a case control study. Gut 22:1011-1017.

Schuster, M.M., 1977. Constipation and anorectal disorders. Clin. Gastroenterology 6:643-658.

Spitzer, R.L. and Endicott, J., 1975. Schedule for affective disorders and schizophrenia (2nd ed.), New York.

Weiner, H., Thaler, M., Reiser, F., and Mirsky, I.A., 1957. Etiology of duodenal ulcer. I. Relation of specific psychological characteristics to rate of gastric secretion. Psychosom. Med. 19:1-10.

Wershaw, H.J. and Reinhart, G., 1974. Life change and hospitalization - a heretical review. J. Psychosom. Res. 18:393-401.

West, K.L., 1970. MMPI correlates of ulcerative colitis. J. Clin. Psychol. 26:214-219.

Wing, J.D., Cooper,J.E., and Sartorius, N., 1974. Measurement and classification of psychiatric symptoms. Cambridge Univ. Press, London, 10-20.

Woodruff, R.A., Goodwin, D.W., and Guze, S.B., 1974. Psychiatric diagnosis Oxford Univ. Press, Oxford.

Young, S.J., Alpers, D.H., Norland, C.C., and Woodruff, R.A., 1976.
    Psychiatric illness and irritable bowel syndrome: practical
    implications for the primary physician. Gastroenterology 70:162-166.

APPLICATION OF PSYCHOLOGICAL MEASURES IN EPIDEMIOLOGICAL STUDIES OF GASTROINTESTINAL DISEASE: A CRITICAL OPINION

J. Fielding
Department of Medicine and Gastroenterology
The Charitable Infirmary
Jervis St., Dublin 1, Ireland

To date gastroenterologists have employed psychological studies in defined diseased (syndrome rather than disease state) populations. It is not acceptable that these post-morbid studies can be applied to population screening programmes. An ideal opportunity for an interchange of current thinking should enable us to move forward in defined studies employing defined methods on defined populations.

The psychological studies undertaken recently in the irritable bowel syndrome point out the pitfalls that clinicians have fallen into, and hopefully one can learn the psychoanalytical weaknesses in such studies. The irritable bowel syndrome is the commonest disorder seen by gastroenterologists and is the stress associated gastrointestinal disorder par excellance.

The first clinical weakness is that only those whose symptoms have lead their possessors to seek medical assistance are defined as suffering from the irritable bowel syndrome although we know that 30% of healthy people have similar symptoms (Thompson and Heaton, 1980). Second there is a widespread pratice to separate patients accordine to varying symptoms. Thus with regard to bowel habit patients are often subdivided into those whose major complaint is constipation and those whose major complaint is diarrhoea. Yet, this could be a chicken and egg situation. For example, of three patients with alternating constipation and diarrhoea one may complain of constipation, one of diarrhoea and only one of both. Moreover, only detailed enquiry into both the frequency and nature of stool may bring the true situation to light. It seems that there are there are two major factors operative here: a) patients complain of what they consider to be abnormal and this consideration may be based on a combination of personal, family, social and religious attributes; b) patients complain of their current problems and may forget their different problems of the recent or remote past. What I wish to emphasize is that we do not know if the different medical definitional approaches are more doctor or patient related. With the humility of any biased observer I am convinced that most of the separation is doctor related and therefore artificial. Third, the approach to diagnosis is positive in some centres whilst others have the negative "all investigations normal" approach. This in itself must influence patient attitude and responses. Thus, how can we apply even the most sophisticated of tests and hope to necessarily arrive at clinical meaningful answers.

I now wish to turn to the psychological tests that have recently been employed in the irritable bowel syndrome. Two European studies have employed the Crown Crisp experiental index (Palmer et al., 1984 and Ryan et al., in press). One has shown patients to come half way between normals and psychoneurotics and to be significantly different from both with regard to free floating anxiety, phobic anxiety obsessional traits, somatic anxiety and depression. Hysterical traits were not different from controls (Palmer et al., 1974).

The second study showed the same trends but patients were only significantly different from normals with regard to free floating anxiety and somatic anxiety. Moreover the patients were not significantly different from controls (Ryan et al., in press). This second group of workers found the same patients emminently normal on testing with the Cattel 16 PF test. They felt that these results suggested that patients premorbid anxiety levels might be normal and that it was insufficient to look only at the postmorbid state.

A locus of control study showed that the site of major symptomatology was not influenced by the locus of control score but that patients with lower scores had a more favourable prognosis (Ryan et al., in press).

In one American study four fifths of irritable bowel syndrome patients had identifible psychiatric illness and in contradistinction to the European studies there were five patients diagnosed as hysterics. Moreover, the psychiatric illness anteceded or coincided with the onset of symptoms in over half the patients and the authors felt that the irritable bowel syndrome symptoms may have been part of the psychiatric illness in these patients (Young et al., 1976).

A study employing a modified Hopkins Symptoms Check list noticed significantly higher scores in most IBS patients than the published normal levels for the test on all the global psychopathology scales and in the following sub scales; somatization of affect, interpersonal sensitivity, depression, anxiety and hostility. Predominant bowel habit deviation did not influence the results obtained.

The current state of relative ignorance is apparent in both the clinical and psychological assessment fields with regard to the irritable bowel syndrome. The current state of the art is no better with regard to other stress associated gastrointestinal disorders.

References

Palmer RL, Stonehill E, Crisp AM, Waller Sheila L, Misiewicz JJ, 1974. Psychological characteristics of patients with the irritable bowel syndrome. Postgrad Med J, 50: 416-419.

Ryan WJ, Arthurs Yvonne, Kelly MG, Fielding JF. Personality and the irritable bowel syndrome. Irish Jour Med Sci, in press.

Ryan WJ, Dolphin Thelma, Fielding JF. The locus of control: its significance in the irritable bowel syndrome. Dig Dis Sci, in press.

Thompson WG, Heaton KN, 1980. Functional bowel disorderd in apparently healthy people. Gastroenterology, 79: 283-288.

Whitehead WE, Engel BT, Schuster MM, 1980. Irritable bowel syndrome. Physiological and psychological differences between diarrhoea predominant and constipation predominant patients. Dig Dis Sci, 25: 404-413.

Young SJ, Alpers DH, Norland CC, Woodruff RA Jr., 1976. Psychiatric illness and the irritable bowel syndrome. Practical implications for the primary physician. Gastroenterology 1976, 70: 162-166.

STRESS-RELATED NICOTINE ABUSE AND DISORDERS

OF THE GASTROINTESTINAL TRACT

John R. Bennett,
Hull Royal Infirmary,
Anlaby Road,
Kingston upon Hull HU3 2JZ,
United Kingdom.

Abstract

Smoking is not categorised into causation, and the mechanisms
by which it causes changes in the gut are rarely identifiable.
Smoking and alcohol drinking are usually found together and the
effects of one may be confused with the other, or may act jointly.
There are positive correlations between smoking and carcinomas of
lip, mouth, oesophagus and possibly stomach, but alcohol is also
implicated.   It is uncertain whether smoking is a causative
factor for peptic ulcers but it does interfere with healing.
Effects of smoking on gastric secretion are confusing, but it may
inhibit pancreatic bicarbonate secretion.   Smoking may lead to
gastro-oesophageal reflux, to duodeno-gastric reflux, and thus to
gastritis.   Oral infections may be encouraged by smoking, but aphthous
ulcers in the mouth, and ulcerative colitis have been reported as less
common or more likely to heal in cigarette smokers.

INTRODUCTION

From the outset I am going to make the assumption that all smoking
equates with nicotine abuse, and that all smoking is stress-related.
I realise that these assumptions can be challenged, and cannot be
absolutely true, (for example, some smoking is indubitably for reasons
of fashion and social trend) - but I am not aware of epidemiological
studies dividing smoking-related disease into sub-categories of types
of smoker, and when one sees how crude most data related to alimentary
disorders and smoking are, then any sub-division would be statistically
ludicrous.

There is a general **assumption** among physicians that smoking is bad
for the gut, and cessation of smoking is usually advised for any smoker
with abdominal complaints.   Facts to support this prejudice are
relatively few.

Inhaled smoke, which can more readily be shown to ravage the
respiratory tract, has no direct contact with the gut, so any
relationship is more indirect.   The two mechanisms are by smoke

dissolved (or suspended) in swallowed saliva, and by drugs absorbed
from inhaled smoke (mainly nicotine) having neural effects.
Demonstrating with certainty that smoking causes pathological changes,
or alters some physiological process is not easy. Smoking is a
repetitive but intermittent process; pharmacological effects on neural
transmitters may be rapid, but other chemical action may be delayed;
dosage required to produce measurable changes is difficult to
determine, partly because nicotine has a double effect, first
stimulating and later depressing autonomic ganglia; dose-variation is
hard to determine in individuals, and varies considerably between
individuals. A final flaw is the possibility that smokers may have
characteristics which predispose them both to acquire the smoking
habit and to the risk of certain diseases, (Yerusilalmy and Palmer
1959) possibly from physiological causes (Lilienfeld 1959).
Related to this is the well known tendency of habitual smokers also
to be regular drinkers of alcohol, the effects of one possibly being
confused by the effects of the other (Heath 1958). I shall try and
deal critically with such evidence as is available on a variety of
associations with alimentary disorders.

MALIGNANT NEOPLASMS

LIP AND MOUTH

Numerous anecdotes support the belief that smoking is a cause of
cancer of lip and mouth, especially when short-stemmed clay pipes are
used, but here thermal trauma may be as important as tobacco itself.
In the main, large epidemiological studies, buccal and pharyngeal
neoplasms are grouped together, occasionally lumped with oesophageal
ones. Excess deaths from cancer of the lip, mouth and pharynx in
cigarette smokers have been separately identified (Levin 1962,
Hammond 1964) though alcohol is also strongly incriminated (Wynder et
al 1957, Vincent and Marchetta 1963, Schwartz et al 1966, Schottenfeld
et al 1974). Probably alcohol and tobacco are independently
associated with cancer of the mouth and pharynx (Zeller 1965).
The recurrence rate of cancer of the mouth, pharynx, and larynx after
'clinical cure' was higher in smokers than in non-smokers, (Moore 1971,
Silverman and Griffith 1972) though another study found no difference
in recurrence rates (Castigliano 1968). Oral leukoplakia is known
to be pre-malignant and occurs more commonly in smokers (Waldron and

Shafer 1960). Pipe smoking may be particularly harmful, and a specific lesion known as stomatitis nicotina is often seen in heavy pipe smokers. Factors other than nicotine are probably involved: rabbits' gingiva develop leukoplakia when tobacco smoke is blown on to them, but not if tobacco extract or nicotine is applied, (Roffo 1930) while vitamin B deficiency makes the mouse ear susceptible to tobacco smoke (Kreshover and Salley 1957).

OESOPHAGEAL CARCINOMA

Several studies show a higher death rate from oesophageal cancer in smokers (Hammond 1964, Hammond and Horn 1958, Dorn 1962, Doll and Hill 1964), whether they smoked cigarettes, pipes, or cigars, and an excess of smokers in people dying of carcinoma of oesophagus and mouth (Levin 1962). Doll and Hill (1964) pointed out that the relationship might not be a direct causal one, as the rise in oesophageal cancer deaths from 1942 to 1962 was only 8% compared with 325% for lung cancer.

Other workers found more smokers in patients with oesophageal cancer than in control populations (Ahlbom 1937, Sadowsky 1953, Schwartz et al 1957). The relationship is stronger in males (Wynder et al 1957, Staszewski 1969, Wynder and Bross 1961). Alcohol consumption was also higher in the cancer patients (Wynder and Bross 1961). A careful study in Brittany looked at both alcohol and tobacco risks adjusting one for the other (Tuyns et al 1977). An independent dose-response was observed for each. Among light drinkers a 5-fold increase was found in heavy smokers compared with light or non-smokers. Heavy smokers and heavy drinkers had a 44 times increased risk.

A necropsy study of oesophageal epithelial histology showed that smokers had a high incidence of abnormalities, some of which were considered pre-malignant (Auerback et al 1965).

It seems that smoking may be a risk factor for oesophageal carcinoma, but many other aetiological agents must be incriminated.

GASTRIC CARCINOMA

Most major studies found no increase of deaths due to gastric neoplasms in smokers, though Dorn's study of 1962 indicated a risk in cigarette smokers of 1.86 times that for non-smokers (compared with 2.18 for cancer of mouth and oesophagus). In Poland an excess of male smokers was found in those dying from gastric carcinoma, predominantly at the cardiac end (Staszewski 1969).

A recent French study (Hoey et al 1981) shows a relative risk for gastric cancer of 4.8 in smokers, compared with 6.9 for drinkers, but in those who both drank and smoked the risk was 9.3.

COLON CARCINOMA

A report of an increased risk of rectal (but not colonic) cancer in doctors who smoked heavily (Doll and Peto 1976) was dismissed by one of its authors as a statistical fluke (Doll 1977), being unconfirmed by previous studies (Hammond and Horn 1958, Staszewski 1969, Kahn 1966, Hammond 1966).

PANCREATIC CARCINOMA

Although earlier studies (Hammond and Horn 1958, Dorn 1962, Doll and Hill 1964) showed no significant excess of pancreatic cancer in smokers, Hammond (1964) found that cigarette smokers (not pipe or cigar smokers) had a higher death rate from pancreatic cancer. This was supported by his smaller 'matched pair' analysis. There is support for this view in subsequent studies (U.S. Public Health Services 1971; U.S. Public Health Services 1973; Wynder et al 1973). However, the fact that alcohol is a known cause of chronic pancreatitis and the epidemological flaw of the difficulty of making positive diagnoses of pancreatic cancer leads to hesitation about accepting smoking as an important cause.

LIVER CANCER

While several series report an increased incidence of liver cancer in smokers no distinction is drawn between primary and metastatic growths. No worthwhile conclusions can be drawn.

PEPTIC ULCER

It cannot be said unequivocally that there is a causal association between smoking and peptic ulceration, although a variety of studies have suggested that there is an association, and that there are possible physiological explanations. The evidence is not simple, however. One immediate difficulty is that many epidemiological studies refer to "peptic ulcer", even though duodenal and gastric ulcers may have quite different causative mechanisms. Another was quizzically stated by Kramer (1979) "How can a disease whose cause is unknown be attributed to a cause whose cause is unknown"?

DEATH RATES

There is an increased death rate from peptic ulcers in smokers (Hammond and Horn 1958, Dorn 1962, Doll and Hill 1964, Doll and Peto 1976). Curiously the mortality was greatest in moderate or light smokers rather than in heavy smokers. Hammond (1964) found that the risk of dying from a gastric ulcer was significantly higher in cigar and pipe smokers as well as in cigarette smokers.

INCIDENCE OF PEPTIC ULCERS IN SMOKERS

Several studies have reported peptic ulcers to be commoner in various groups of smokers, the incidence varying considerably, however. (Jedrychowski and Popiela 1974, Friedman et al 1974 : Edwards et al 1959 : Gillies and Skyring 1969 : Paffenbarger et al 1974 : Doll and Peto 1976). Smokers tend to develop gastric ulcers on the lesser curvature (Thomas et al 1980).

INCIDENCE OF SMOKING IN PEPTIC ULCER PATIENTS

Although the first study (Barnett 1927) showed that 82% of males with peptic ulcer were smokers compared with 72% of a control population- an insignificant difference, Trowell (1934) found that more patients with duodenal ulcers smoked than did controls, and more of them inhaled the smoke. In patients with acute bleeding from, or perforation of, a peptic ulcer, an increased number of cigarette smokers was found (Allibone and Flint 1958), while in patients admitted to the Central Middlesex Hospital for treatment of peptic ulcers there was a significant excess of smokers among the gastric ulcer patients of both sexes, and in male duodenal ulcer patients (Doll et al 1958). Massachusetts physicians with peptic ulcers more often smoke than do their colleagues without ulcers, and those with duodenal ulcers started smoking at an earlier age (Monson 1970). Men with gastric ulcers under 55 are more commonly smokers (and drinkers) (Thomas et al 1980), though the most recent Australian study (Piper et al 1982), while confirming smoking as a risk factor for gastric ulcers in men and women (especially if analgesic takers), discounts alcohol.

SMOKING AND ULCER HEALING

The plethora of reports of clinical trials of ulcer-healing have yielded considerable, but sometimes confusing, information about the effects of smoking.

In an early study of the effect of an antacid regime on the symptoms of peptic ulcer it was found that non-smokers responded better and had fewer exacerbations than smokers (Batterman and Ehrenfeld 1949). However, a follow-up study of patients after acute perforation of an ulcer showed no improvement of symptoms in those who stopped smoking, (Jamieson et al 1946) although severe symptoms tended to be associated with smoking - this suggested an indirect relationship.

The classic study of gastric ulcer healing, by Doll and his colleagues in 1958, in which ulcer size was measured radiologically, also showed no alteration in symptoms among those advised to stop smoking. However, the reduction in ulcer size was significantly greater in those advised not to smoke, though the number of ulcers which healed completely was almost identical in the two groups. Moreover, patients who never smoked anyway showed no greater healing than smokers not advised to stop. Eloshoff and Grossman (1980) summarised eight different studies, all showing a <u>trend</u> towards better healing in non-smokers, a magisterial response to those who claimed that as individual studies showed no conclusive statistical differences, they should be considered as showing no relationship (Porro et al 1980).

An interesting study in 1981 (Korman et al 1981) showed that the healing effects on duodenal ulcers of cimetidine and of an antacid were much reduced in smokers, while a recent study from Germany illustrates the problem starkly (Gugler et al 1982). A double blind comparison of two H_2 receptor antagonists in duodenal ulcer healing, cimetidine and ranitidine, was carried out in two centres. The results in one centre showed 88% healing with cimetidine, 63% with ranitidine; in the other centre the reverse was found, 60% and 92%. In both centres, however, success correlated inversely with smoking (p <0.01), the smokers being unevenly distributed in the two centres. This may be related to the effects of smoking on anti-secretory drugs referred to later. (Boyd, Wilson, Wormsley 1983).

EXPERIMENTAL EVIDENCE

Animal Experiments

Dogs who inhale cigarette smoke through traceostomies do not develop peptic ulcers, but if they are also given histamine injections they develop ulcers more often than dogs given histamine alone. (Toon et al 1951). Rats whose oesophagus is perfused with acid alone sometimes develop duodenal ulcers, but the addition of subcutaneous nicotine considerably increases the risk, though nicotine alone does not cause ulcers. (Robert et al 1971). The addition of nicotine also increases the likelihood of duodenal ulcers developing in rats subjected to carbachol-pentagastrin infusions (Robert 1972).

Interesting and paradoxical effects have been observed on gastric secretion in animals. Tobacco smoke condensates and nicotine depress gastric juice volume, acid concentration and output, and peptic activity in rats (Thompson et al 1970, Thompson 1970, Thompson 1971) both basally and when treated with maximal or submaximal doses of histamine or pentagastrin. However, administration of nicotine for two weeks stimulates volume, acid, and

pepsin output.   In cats, an infusion of nicotine at 400 μg/hour
decreased a 'near maximum' gastric response to pentagastrin, whereas
half that dose of nicotine has no effect.   However, if peptic ulcers
are produced artificially by pentagastrin infusion the addition of
nicotine at the lower dose produces more ulcers, while at the higher
rate it causes fewer ulcers (Konturek et al 1971).

In a study on dogs (Konturek et al 1971) it was found that nicotine
injections affected neither basal nor half-maximal gastric secretions.
However, they inhibited a near-maximal secretion of pancreatic juice
(both volume and bicarbonate concentrations) and diminished volume and
bicarbonate content of bile.   This confirmed the original
observations made by Edmunds in 1909.

IN MAN

In a search for the possible mechanism whereby smoking could cause
dyspepsia, attention has predictably focused on gastric secretion.
The conclusions are more memorable for their disparity than for any
light they cast on the association.   Every reported study has used a
different technique, itself a reflection on the difficulty of
reproducing a physiological situation which can be accurately measured.

An early test meal method showed that in patients with 'functional
gastric disturbances' and with duodenal ulcers smoking tended to produce
'hyperacidity' (Gray 1929).

Basal acid output was not significantly changed after smoking four
to seven cigarettes in Schnedorf and Ivey's (1939) study, but later work
showed an increased output in 40% of control subjects and in 85% of peptic
ulcer subjects after one cigarette (Steigmann et al 1954).   Piper and
Raine found in 1959 that four to six cigarettes in an hour increased the
volume and acid output of basal secretions but when the study was
repeated in 1974 (Whitecross) no significant effect of smoking was
observed;   the authors suggest a change in cigarette composition as the
explanation.   Using an 'alcohol test meal' two cigarettes were found to
increase acid concentrations in both normal subjects and those with
peptic ulcers (Ehrenfeld and Sturtevant 1939).   Cooper and Knight
(1956) used modern techniques to measure acid output basally during
smoking and during stimulation with meat broth and insulin in patients
with duodenal ulcers.   Basal output (volume, pH, acid concentration,
and pepsin concentration) was the same in smokers and non-smokers, and

subsequent changes were no greater during and after smoking than in controls. Debas and his colleagues (1971) used the method of continuous stimulation by pentagastrin to produce 50% of maximal acid output. Three cigarettes smoked in an hour did not affect it. Wilkinson and Johnston (1971) also used a submaximal infusion of pentagastrin (though they did not relate the level to the patient's maximal output) and showed that smoking one or two cigarettes caused a fall in acid and pepsin secretion. Intravenous nicotine produced a similar reduction. Murthy et al (1977) found a diphasic effect of smoking on basal acid secretion, a rise followed by a fall.

Taylor's work in Liverpool on pepsin fractions suggests that, even if there is no consistent change in total pepsin output, pepsin I, present in increased quantities in the gastric juice of patients with gastric and duodenal ulcers (Taylor 1970) may be secreted in excess by smokers with ulcers when stimulated by pentagastrin or histamine (Walker and Taylor 1979).

No change in volume or constituents of gastric mucus was induced by smoking (Cooper et al 1957, Whitecross 1974).

While the effects of smoking on gastric secretion, natural or artificially stimulated, leaves us with a confused picture and no indication of this as a likely contribution to the causation of peptic ulcers, a recent study in Dundee may explain why smokers' ulcers heal less well than non-smokers. Smoking markedly reduced the diminution of nocturnal acid secretion caused by both histamine H_2 antagonists and an anticholinergic, though it did not block their inhibition of pentagastrin-stimulated secretion. (Boyd, Wilson, Wormsley (1983). This may explain the curious results of the German study in which two populations with different smoking habits produced entirely different results in a randomised trial of a histamine H_2 antagonist on ulcer healing (Gugler et al 1982).

PANCREATIC SECRETION

When it seemed that no consistent effect of smoking on gastric acid secretion could be shown, the suggestion was made that smoking might inhibit pancreatic secretion, thus diminishing bicarbonate neutralisation of acid in the duodenum. Subsequent studies have confirmed this (Bynum et al 1972, Bochenek 1973, Brown 1976, Murthy et al 1977, Murthy et al 1978), and in most animal experiments nicotine reduces pancreatic secretion (Konturek et al 1971, Konturek et al 1972, Solomon et al 1974).

It is tempting to see this as a ready explanation for duodenal ulcer problems in smokers, but the evidence is incomplete as Wormsley has pointed out in detail (Wormsley 1978). Moreover, it cannot resolve the gastric ulcer/smoking question.

GASTRITIS AND GASTRIC ULCER

In their careful study of histologically demonstrated atrophic gastritis, Edwards and Coghill (1966) found that, in people over 50, gastritis was much commoner in heavy cigarette smokers. Roberts (1972) found a similar, but less marked, trend. In shorter-term experiments, photographs of the gastric mucosa were taken before and during cigarette smoking (Hoon 1969). Four of six subjects showed blanching of the mucosa, though 'Tom's' stomach had not shown any change on smoking in Wolf and Wolff's (1947) famous experiments.

It is thought likely that chronic gastritis is a prelude to chronic gastric ulceration, and that bile reflux from the duodenum is one potential causative factor in chronic gastritis. Bile reflux seems to be increased by smoking (Whitecross et al 1974, Read and Grech 1973), perhaps by altering antro-duodenal motility, and diminishing pyloric sphincter pressure (Valenzuela et al 1976). Evidence that this is a significant causative relationship is lacking. Yeomans et al (1981 found no increase in gastric bile acids after volunteers smoked one cigarette.

GASTROINTESTINAL MOTILITY

OESOPHAGUS

Smoking reduces the resting tone of the gastro-oesophageal sphincter (Dennish and Castell 1971, Stanciu and Bennett 1972), and leads to increased acid gastro-oesophageal reflux, probably by blocking the cholinergic control mechanisms as in vitro (Ellis et al 1960, Misiewicz et al 1969). Clearing of an acid load from the oesophagus is also impaired by smoking (Kjellen and Tibbling 1978).

STOMACH

No consistent change in gastric motility has been observed with smoking. Radiological examination has suggested paralysis of peristalsis (Danielopolu et al 1925), increased peristalsis (Adler 1925), or no change (Gray 1929 a & b). Balloons in the stomach detected increased, decreased, and unchanged motility in equal proportions of patients studied by Batterman (1955), though in another study a fundal balloon recorded no change in motility on smoking (Cooper et al 1958). One isotopic study of gastric emptying using a gamma camera suggested that smoking accelerates liquid emptying (Grimes and Goddard 1978), while another indicated delayed emptying of solids (Harrison and Ippoliti 1979).

COLON

Smoking has frequently been observed to produce a call to stool, though the motility change has not been positively identified. In patients with the irritable bowel syndrome smoking may precipitate colonic hypermotility (Connell et al 1965).

COMMENT

The lack of firm data on the effects of smoking on most aspects of gastro-intestinal motility is disappointing. It presumably reflects both the difficulties in measuring motility accurately, and the problems of assessing the effect of a mixture of drugs in variable dosage - which smoking effectively is.

NON-MALIGNANT ORAL DISEASE

Aphthous ulcers have been observed to be less common in smokers (Shapiro et al 1970), and even to heal better when smoking was resumed (Bookman 1960, Dorsey 1963). The deposition of tar on the teeth of smokers is commonly observed. Dental calculus is commoner in smokers (Pindborg 1949, Alexander 1970), particularly in the supragingival area (Kowalski 1971).

The evidence regarding smoking as a cause of periodontal disease is not uniform. Some workers found a positive relationship (Arno et al 1958, Brandtzaeg et al 1964, Solomon et al 1968), while others have not (Alexander 1970, Ludwick and Massler 1952, Lilienthal et al 1965).

Three cases are reported in which cessation of smoking alone led to rapid remission of severe oral moniliasis (Beasley 1969), and acute oral infections such as Vincent's gingivitis are more likely to occur in smokers (Ludwick and Massler 1952).

CIRRHOSIS

Deaths from hepatic cirrhosis are more frequent in cigarette smokers than in non-smokers (Vincent and Marchetta 1963, Doll and Hill 1964), but this is probably due to the strong association between heavy drinking and heavy smoking.

ULCERATIVE COLITIS

Perhaps the most curious aspect of the relationship between smoking and the alimentary tract is the negative correlation with ulcerative colitis as with oral aphthous ulcers. Like the dog that failed to bark, this may be an important clue to the relationship between smoking and gut disease.

Harries, Baird and Rhodes (1982) reported from Cardiff that a higher proportion of colitics were non-smokers compared with Crohn's disease patients and controls from a fracture clinic. Anecdotal, but striking support came in correspondence (de Castella 1982; Roberts and Diggle 1982), and in a brief report of a survey in   (Bures et al 1982).
Finally a report from the Boston Collaborative Drug Surveillance Program confirms the findings (Jick and Walker 1983).

A variety of hypotheses has been canvassed to explain the findings, though the only guide to mechanism was one case report (Roberts and Diggle 1982) in which nicotine chewing gum suppressed

REFERENCES

Adler, E. 1925. Der Zigarettenmagen. Med.Klin. 27: 1005-1006.

Ahlbom, H.E. 1937. Pradisponierende Faktoren fur Platten-epithelkarzinom in Mund, Hals and Speiserohre. Acta radiol. 18: 163-185.

Alexander, A.G. 1970. The relationship between tobacco smoking calculus and plaque accumulation and gingivitis. Dent.Health. 9: 6-9.

Allibone, A. and Flint, F.J. 1958. Bronchitis, aspirin, smoking and other factors in the aetiology of peptic ulcer. Lancet. 2: 179-182.

Arno, A., Waerhaug, J., Lovdal, A. and Schei, O. 1958. Incidence of gingivitis as related to sex, occupation, tobacco consumption, tooth brushing and age. Oral Surg. 11: 587-595.

Auerbach, O., Stout, A.P., Hammond, E.C. and Garfinkel, L. 1965. Histologic changes in oesophagus in relation to smoking habits. Arch.environ.Health. 11: 4-15.

Barnett, C.W. 1927. Tobacco smoking as a factor in the production of peptic ulcer and gastric neurosis. Boston Med.Surg.J. 197: 457-459.

Batterman, R.C. and Ehrenfeld, I. 1949. The influence of smoking upon the management of the peptic ulcer patient. Gastroenterology. 12: 575-585.

Batterman, R.C. 1955. The biological effects of tobacco. Edited by Wynder, E.L. Little, Brown, Boston. 133-50.

Beasley, J.D. 1969. Smoking and oral moniliasis. J.Oral.Med. 24: 83-86.

Bochenek, W.J. and Kuronczewski, R. 1973. Effects of cigarette smoking and volume of duodenal contents. American J.Dig.Dis. 18: 729-733.

Bookman, R. 1960. Relief of ulcer sores on resumption of cigarette smoking. Calif.Med. 93: 235-236.

Boyd, E.J.S., Wilson, J.A., Wormsley, K.G. 1983. Smoking impairs therapeutic gastric inhibition. Lancet. 1: 95-97.

Brandtzaeg, P. and Jamison, H.C. 1964. Study of peridontal health and oral hygiene in Norwegian Army recruits. J.Periodont. 35: 302-307.

Brown, P. 1976. The influence of smoking on pancreatic function in man. Med.J.Australia. 2: 290-293.

Bures, J., Fixa, B., Komarkova, O., Fingerland, A. 1982. Non-smoking: a feature of ulcerative colitis. Brit.Med.Journal. 285: 440.

Bynum, T.E., Solomon, T.E., Johnson, L.R. and Jacobson, E.D. 1972. Inhibition of pancreatic secretion in man by cigarette smoking. Gut. 13: 361-365.

Castigliano, S.G. 1968. Influence of continued smoking on the incidence of second primary cancers involving mouth, pharynx and larynx. J.Amer.Dent.Assoc. 77: 580-585.

Connell, A.M., Jones, F.A. and Rowlands, E.N. 1965. Motility of the pelvic colon. Gut. 6: 105-112.

Cooper, P. and Knight, J.B. 1956. Effect of cigarette smoking on gastric secretions of patients with duodenal ulcer. New Engl.J. Med. 255: 17-21.

Cooper, P., Saltz, M., Harrower, H.W. and Burke, D.H. 1957. Effect of cigarette smoking on dissolved gastric mucins and viscosity of gastric juice. Gastroenterology. 33: 959-967.

Cooper, P., Harrower, H.W., Stein, H.L. and Moore, G.F. 1958. The effect of cigarette smoking on intra gastric balloon pressure and temperature of patients with duodenal ulcer. Gastroenterology. 35: 176-182.

Danielopolu, D., Simici, D. and Dimitriu, C. 19   Action du tabac sur la motilite de l'estomac etudiee chez l'homme a l'aide de la methode graphique. C.R.Soc.Biol. 92: 535-538.

Debas, H.T., Cohen, M.M., Holubitsky, I.B. and Harrison, R.C. 1971. Effect of cigarette smoking on human gastric secretory responses. Gut. 12: 93-96.

de Castella, H. 1982. Non-smoking: a feature of ulcerative colitis. Brit.Med.J. 284: 1706

Dennish, G.W. and Castell, D.O. 1971. Inhibitory effect of smoking on the lower esophageal sphincter. New Engl.J.Med.

Doll, R., Jones, F.A. and Pygott, F. 1958. Effect of smoking on the production and maintenance of gastric and duodenal ulcers. Lancet. 1: 657-662.

Doll, R. and Hill, A.B. 1964. Mortality in relation to smoking: ten years' observations of British doctors. Brit.Med.J. 1: 1399-1410, 1460-1467.

Doll, R. and Peto, R. 1976. Mortality in relation to smoking: twenty years' observations on male British doctors. Brit.Med.J. 4: 1525-1536.

Doll, R. 1977. Cancer of the large bowel: general epidemiology. Topics in Gastroenterology. 5: 3. Edited by Truelove, S.E.

Dorn, H.F. 1962. Death rates and causes of death of smokers and non-smokers. "Tobacco and Health". 172-190. Edited by G. James and T. Rosenthal. Thomas, Springfield, Illinois.

Dorsey, C. 1963. More observations on relief of aphthous stomatitis on resumption of cigarette smoking. Calif.Med. 101: 377-378.

Elashoff, J.D. and Grossman, M.I. 1980. Smoking and duodenal ulcer. Gastroenterology. 79: 181.

Edmunds, C.W. 1909. The antagonism of the adrenal glands against the pancreas. J. Pharmacol.exp.Ther. 1: 135-150.

Edwards, F., McKeown, T. and Whitfield, A.G.W. 1959. Association between smoking and disease in men over sixty. Lancet. 1: 196-200.

Edwards, F.C. and Coghill, N.F. 1966. Aetiological factors in chronic atrophic gastritis. Brit.Med.J. 2: 1409-1415.

Ehrenfeld, I. and Sturtevant, M. 1939. The effect of smoking tobacco on gastric acidity. Amer.J.Med.Sci. 201: 81-86.

Ellis, F.G., Kauntze, R. and Trouce, J.R. 1960. The innervation of the cardia and lower oesophagus in Man. Brit.J.Surg. 47: 466-472.

Friedman, G.D., Siegelaub, A.B. and Seltzer, L.C. 1974. Cigarettes, alcohol, coffee and peptic ulcer. N.Engl.J.Med. 290: 469-473.

Gillies, M.A. and Skyring, A. 1969. Gastric and duodenal ulcer: the association between aspirin ingestion, smoking and family history of ulcer. Med.J.Aust. 2: 280-285.

Gray, I. 1929. Gastric response to tobacco smoking. Amer.J.Surg. 7: 484-493.

Grimes, D.S. and Goddard, J. 1978. Effect of cigarette smoking on gastric emptying. Brit.Med.J. 2: 460-461.

Gugler, R., Rohner, H.G., Kratochvil, P., Brandstatter, G. and Schmitz, H. 1982. Effect of smoking on duodenal ulcer healing with cimetidine and oxmetidine. Gut. 23: 866-871.

Hammond, E.C. and Horn, D. 1958. Smoking and death rates - report on forty-four months of follow-up of 187,783 men. I.J.Amer.Med. Assoc. 166: 1154-1172. II. J.Amer.Med.Ass. 166: 1294-1308.

Hammond, E.C. 1964. Smoking in relation to mortality and morbidity. Findings in first thirty-four months of follow-up in a prospective study started in 1959. J.Nat.Cancer.Instit. 32: 1161-1188.

Hammond, E.C. 1966. Epidemiological study of cancer and other chronic diseases. National Cancer Institute Monograph. 19: 727. U.S. Govt.Printing Office, Washington D.C.

Harries, A. D., Baird, A., Rhodes, J. 1982. Non-smoking: a feature of ulcerative colitis. Brit.Med.J. 284: 706.

Harrison, A. and Ippoliti, A. 1979. Effect of smoking on gastric emptying. Gastroenterology. 76: 1152-(A).

Heath, C.W. 1958. Differences between smokers and non-smokers. Arch.Intern.Med. 101: 377-388.

Hoey, J., Montvernay, C. and Lambert, R. 1981. Wine and Tobacco: risk factors for gastric cancer in France. J.Epidemiol. 113: 668-674.

Hoon, J. R. 1969. Intragastric photographic observations of the effects of smoking on gastric mucosa. Gastrointestinal Endoscopy. 15: 172-174.

Jamieson, R.A., Illingworth, C.F.W. and Scott, L.D.W. 1946. Tobacco and ulcer dyspepsia. Brit.Med.J. 2: 287-288.

Jick, H., Walker, A.M. 1983. Cigarette smoking and ulcerative colitis. N.E.J.M. 308: 261-262.

Jedrychowski, W. and Popiela, T. 1974. Association between the occurrence of peptic ulcers and tobacco smoking. Public Health. 88: 195-200.

Kahn, H.A. 1966. An epidemiological study of cancer and other chronic diseases. National Cancer Institute Monograph 19. U.S. Govt. Printing Office, Washington DC.

Kjellen, G. and Tibbling, L. 1978. Influence of body position, dry and water swallows, smoking and alcohol on esophageal acid clearing. Scand.J.Gastroent. 13: 283-288.

Konturek, S.J., Radecki, T., Thor, P., Dembinski, A. and Jacobson, E.D. 1971. Effect of nicotine on gastric secretion and ulcer formation in cats. Proc.Soc.exp.Biol.Med. 138: 674-677.

Konturek, S.J., Solomon, T.F., McCreight, W.G., Johnson, L.R. and Jacobson, E.D. 1971. Effects of nicotine on gastro-intestinal secretion. Gastroenterology. 60: 1098-1105.

Konturek, S.J., Dale, J., Jacobson, E.D. 1972. Mechanism of nicotine-induced inhibition of pancreatic secretion of bicarbonate in the dog. Gastroenterology. 62: 425-529.

Korman, M.G., Shaw, R.G., Hansky, J., Schmidt, G.T. and Stern, A.I. 1981. Influence of smoking on healing rate of duodenal ulcer in response to cimetidine or high dose antacids. Gastroenterology. 80: 1451-1453.

Kowalski, C.J. 1971. Relationship between smoking and calculus deposition. J.dent.Res. 50: 101-104.

Kramer, P. 1979. Smoking and ulcers: true, true and related. Gastroenterology. 76: 1083-1084.

Kreshover, S.J. and Salley, J.J. 1957. Predisposing factors in oral cancer. J.Amer.dent.Assoc. 54: 538.

Levin, M.L. 1962. Smoking and cancer: retrospective studies and epidemiological evaluation. In Tobacco and Health. Edited by G. James and T. Rosenthal 163-171. Thomas, Springfield, Illinois.

Lilienfeld, A.M. 1959. Emotional and other selected characteristics of cigarette smokers and non-smokers as related to epidemiological studies of lung cancer and other diseases. J.Nat.Cancer Inst. 22: 259-282.

Lilienthall, B. Amerena, V. and Gregory, G. 1965. An epidemiological study of chronic periodontal disease. Arch.oral Biol. 10: 553-566.

Ludwick, W., Massler, M. 1952. Relation of dental caries experience and gingivitis to cigarette smoking in males 17-21 years old. J.dent.Res. 26: 261.

Misiewicz, J.J., Waller, S.L., Anthony, P.P. and Gummer, J.W.P. 1969. Achalasia of the cardia: pharmacology and histopathology of isolated cardiac sphincteric muscle from patients with and without achalasia. Quart.J.Med. 38: 17-30.

Monson, R.R. 1970. Cigarette smoking and body form in peptic ulcer. Gastroenterology. 58: 337-344.

Moore, C. 1971. Cigarette smoking and cancer of the mouth, pharynx and larynx. J.Am.Med.Assoc. 218: 553-558.

Murthy, S.N.S., Dinoso, V.P., Clearfield, H.R. and Chey, W.Y. 1977. Simultaneous measurement of basal pancreatic, gastric acid secretion, plasma gastrin and secretin during smoking. Gastroenterology. 73: 758-761.

Murthy, S.N.S., Dinoso, V.P., Clearfield, H.R. 1978. Serial pH changes in the duodenal bulb during smoking. Gastroenterology. 75: 1-4.

Paffenbarger, R.S., Wing, A.L., Hyde, R.T. 1974. Chronic disease in former college students with early precursors of peptic ulcer. Amer.J.Epidemiol. 100: 307-315.

Pindborg, J.J. 1949. Tobacco and Gingivitis: Correlation between consumption of tobacco, ulceromembranous gingivitis, and calculus. J.Dent.Res. 28: 461-463.

Piper, D.W. and Raine. J.M. 1959. Effect of smoking on gastric secretion. Lancet. 1: 696-698.

Piper, D.W., McIntosh, J.H., Greig, M., Shy, C.M. 1982. Environmental factors and chronic gastric ulcer. Scand.J. Gastroenterology. 17: 721-729.

Porro, G.B., Petrillo, M., Grossi, E., Lazzaroni, M. 1980. Smoking and duodenal ulcer. Gastroenterology. 79: 180-181.

Read, N.W., Grech, P. 1973. Effect of cigarette smoking on competence of the pylorus: preliminary study. Brit.Med.J. 3: 313-316.

Robert, A., Stowe, D.F., Nezamis, J.C. 1971. Possible relationship between smoking and peptic ulcer. Nature. 233: 497-498.

Robert, A. 1972. Potentiation. by nicotine, of duodenal ulcers in the rat. Proc.Soc.exp.Biol.Med. 139: 319-322.

Roberts. D.M. 1972. Chronic gastritis, alcohol and non-ulcer dyspepsia. Gut. 13: 768-774.

Roberts, C.J. and Diggle, R. 1982. Non-smoking: a feature of ulcerative colitis. Brit.Med.J. 285: 440.

Roffo, A.H. 1930. Leucoplasie experimentale produite par le tabac. Rev. de Stomatol. 32: 699.

Sadowsky, D.A.,Gilliam, A.G. and Cornfield, J. 1953. Statistical association between smoking and carcinoma of lung. J.Nat.Cancer Inst. 13: 1237-1258.

Schnedorf, J.G. and Ivy. A.C. 1939. The effect of tobacco smoking on alimentary tract. J.Amer.Med.Assoc. 112: 898-903.

Schottenfeld, D., Gantt, R.C., Wynder, E.L. 1974. The role of alcohol and tobacco in multiple primary cancers of the upper digestive system, larynx and lung: a prospective study. Preventive Medicine. 3: 277.

Schwartz, D., Denoix, P.F. and Anguera, G. 1957. Recherche des localisations du cancer associees aux facteurs tabac et alcool chez l'Homme. Bull.Ass.franc.Cancer 44: 336-361.

Schwartz, D., Lasserre, O., Flamant, R., Lellough, J. 1966. Alcohol et cancer. European J. of Cancer, 2: 367.

Shapiro, S., Olson, D.L., Chellemi, S.J. 1970. The association between smoking and aphthous ulcers. Oral.Surg. 30: 624-630.

Silverman, S., Griffith. M. 1972. Smoking characteristics of patients with oral carcinoma and the risk for second oral primary carcinoma. J.Amer.Dental Assoc. 85: 637.

Solomon, H.A., Priore, R.C., Bross, I.D.J. 1968. Cigarette smoking and periodontal disease. J.Amer.Dent-Assoc. 77: 1081-1084.

Solomon, T.E., Solomon, N., Shanbour, L.L. 1974. Direct and indirect effects of nicotine on rabbit pancreatic secretion. Gastroenterology. 67: 276-283.

Stanciu. C., Bennett, J.R. 1972. Smoking and gastro-oesophageal reflux. Brit.Med.J. 2: 793-795.

Staszewski, J. 1969. Smoking and cancer of the alimentary tract in Poland. Brit.J.Cancer. 23: 247-253.

Steigman, F., Dolehide, R.H. and Kaminski, L. 1954. Effects of smoking tobacco on gastric acidity and motility of hospital controls and patients with peptic ulcer. Amer.J.Gastroenterology. 22: 399-409.

Taylor, W.H. 1970. Pepsins of patients with peptic ulcers. Nature. (London) 227: 76-77.

Thomas, J., Grieg, M., McIntosh, J., Hunt, J., McNeil, D., Piper, D.W. 1980. The location of chronic gastric ulcer. Digestion. 20: 79-84.

Thompson, J.H., Spezia, C.A. and Angulo, M. 1970. Chronic effects of nicotine on rat gastric secretion. Experientia. 26: 615-617.

Thompson, J.H. 1970. Effect of nicotine and tobacco smoke on gastric
    secretion in rats with gastric fistulas. Digestive Diseases. 15:
    209-217.
Thompson, J.H. 1971. Tobacco smoke and gastric secretion. Lancet. 2:
    1040-1041.
Toon, R.W., Cross, F.S., Wangensteen, O.H. 1951. Effect of inhaled
    cigarette smoke on production of peptic ulcer in the dog.
    Proc.Soc.exp.Biol. (N.Y.) 77: 866-869.
Trowell, O.A. 1934. The relation of tobacco smoking to the incidence
    of chronic duodenal ulcer. Lancet. 1: 808-809.
Tuyns, A.J., Pequingot, G., Jensen, O.M. 1977. Le cancer de l'oesophage
    en Ille et Vilaine en function des niveaux de consommation
    d'alcool et de tabac. Bull.Cancer (Paris) 65: 1.
U.S. Public Health Service. 1971. The Health Consequences of Smoking.
    DHEW Publication No. 71: 7513 p298.
U.S. Public Health Services. 1973. The Health Consequences of Smoking.
    DHEW Publication No. 74: 8704.
Valenzuela, J.E., Defilippi, C., Csendes, A. 1976. Manometric studies
    on the human pyloric sphincter. Gastroenterology. 70: 481-483.
Vincent, R.G., Marchetta, F. 1963. The relationship of the use of
    tobacco and alcohol to cancer of the oral cavity, pharynx or
    larynx. Am.J.Surg. 106: 501-505.
Waldron, C.A. and Shafer, W.G. 1960. Current concepts of leukoplakia.
    Int.Dent.J. 10: 350-357.
Walker, V. and Taylor, W.H. 1979. Cigarette smoking, chronic peptic
    ulceration and pepsin I secretion. Gut. 20: 971-976.
Whitecross, D.P., Clarke, A.D., Piper, D.W. 1974. The effect of
    cigarette smoking on human gastric secretion. Scand.J.Gastroent.
    9: 399-403.
Wilkinson, A.R. and Johnston, D. 1971. Inhibitory effect of cigarette
    smoking on gastric secretion stimulated by pentagastrin in man.
    Lancet. 2: 628-632.
Wolf, S. and Wolff, H. 1947. Human gastric function. Oxford Univ.
    Press.
Wormsley, K.G. 1978. Smoking and Duodenal Ulcer. Gastroenterology.
    75: 139-152.
Wynder, E.L., Hultberg, S., Jacobsson, F., Bross, I.J. 1957.
    Environmental factors in cancer of upper alimentary tract.
    Cancer. 10: 470-487.
Wynder, E.L., Bross, I.J., Feldman, R.A. 1957. A study of etiological
    factors in cancer of the mouth. Cancer. 10: 1308-1323.
Wynder, E.L. and Bross, I.J. 1961. A study of etiological factors in
    cancer of the esophagus. Cancer. 14: 389-413.
Wynder, E.L., Mabuchi, K., Fortner, J.G. 1973. Epidemiology of cancer
    of the pancreas. J.Nat.Cancer.Instit. 50: 645.
Yerusilalmy, J. and Palmer, L.F. 1959. On the methodology of
    investigations of etiologic factors in chronic diseases.
    J.Chron.Dis. 10: 27-40.
Zeller, A.Z. and Terris, M. 1965. The association of alcohol and
    tobacco with cancer of the mouth and pharynx. Amer.J.publ.Health.
    55: 1578-1585.
Yeomans, N.D., Williams, D.R., McKinnon, M.A., McLeish, A.R.,
    Smallwood, R.A. (1981). Effect of cigarette smoking on
    duodenogastric reflux of bile acids. Austr. & N.Z. J.Med. 11,
    347-350.

USE OF QUANTITATIVE METHODS FOR THE STUDY OF PSYCHOLOGICAL FACTORS IN ULCER PATIENTS

G.C. Lyketsos

Dromokaition Mental Hospital, Athens, Greece

SUMMARY

The PDS is a useful instrument for the measurement of personality deviance as enduring psychological factors of vulnerable human adaptation. Quantitative methods for the study of anxiety, depression and other symptoms, like Sad, DSSI or PSE, need to be combined in order to detect whether another recent breakdown of human adaptation is involved.

Correlated biochemical or other equivalent laboratory methods need to be devised in order to fill the gap between measurements of the psychic and the biological components of human adaptation.

RESULTS

It has been claimed that personality types and specific psychological conflicts determine psychosomatic or psychophysiological disorders. Elaborate psychoanalytic research, however, has failed to measure these varaibles quantitatively (From-Reichmann, 1973; Alexander, 1968; Dunbar, 1968).

Recent development in the classification of mental disorders (DSM-III 1980) no longer includes the terms psychophysiological or psychosomatic. Instead it refers to psychological factors affecting physical condition. A temporal relationship between the psychological stimuli and the initiation or exacerbation of a physical condition is stated to be a pathognomic diagnostic criterion. The psychological factors are assumed to act as a stress, a more or less recent interference in the adaptational process of the individual.

During the past few years several questionnaires and rating scales have been devised to measure life events and life changes as psychological stress (Holmes and Rahe, 1967; Paykel et al., 1971; Horrowitz et al.,

1977; Hurst et al., 1978). Investigators have correlated various measures of life change with bodily symptoms and illnesses of all types (Theorel et al., 1975; Hurst et al., 1978). Particular attention has been paid to the construction and inclusion of a time dimension into the life event on the assumption that the more recent the event, the greater its effect (Horrowitz et al., 1977).

A few sophisticated methods have also been devised to measure the association of personality types with the physical illness, as for example the association between type A personality and coronary disease (Jenkins, 1976). In 1967 Caine, Foulds and Hope published a manual of a Hostility and Direction of Hostility Questionnaire "designed to sample a wide range of possible manifestations of (a drive) aggression, hostility or punitiveness". The HDHQ measures intro- and extra-punitiveness, a form of behaviour associated with introjective and projective mechanisms. They were able to classify psychotics according to the dimensions derived from the scale (1967, Fig. 1) : melancholics in the plus (+) or inward-directed hostility, with high general hostility, and clear-cut paranoids in the minus (-) or outward-directed hostility, also with high general hostility. Later (1976) Foulds developed a modification of the HDHQ, a self-rating scale measuring Personality Deviance (PDS), the emphasis here was on measuring personality characteristics enduring throughout life (traits) rather than state-related symptomatic behaviour. This is why all questions in this new scale begin with the phrase : "Most of my life ...".

The personality Deviance Scale contains 36 items, of which 12 measure extra-punitiveness (E), 12 intro-punitiveness (I), and 12 dominance (Do). Previous studies (Foulds, 1976) have shown that these three attributes relate to enduring personality traits. Thus Foulds added to intro- and extra-punitiveness an important third dimension, a subscale of dominance (Appendix A, Bedford and Foulds, 1978a).

There are six questions in this subscale to assess straightforward uninhibited hostile acts, and six to assess domineering social behaviour. A high score in this subscale indicates high dominance, and a low score

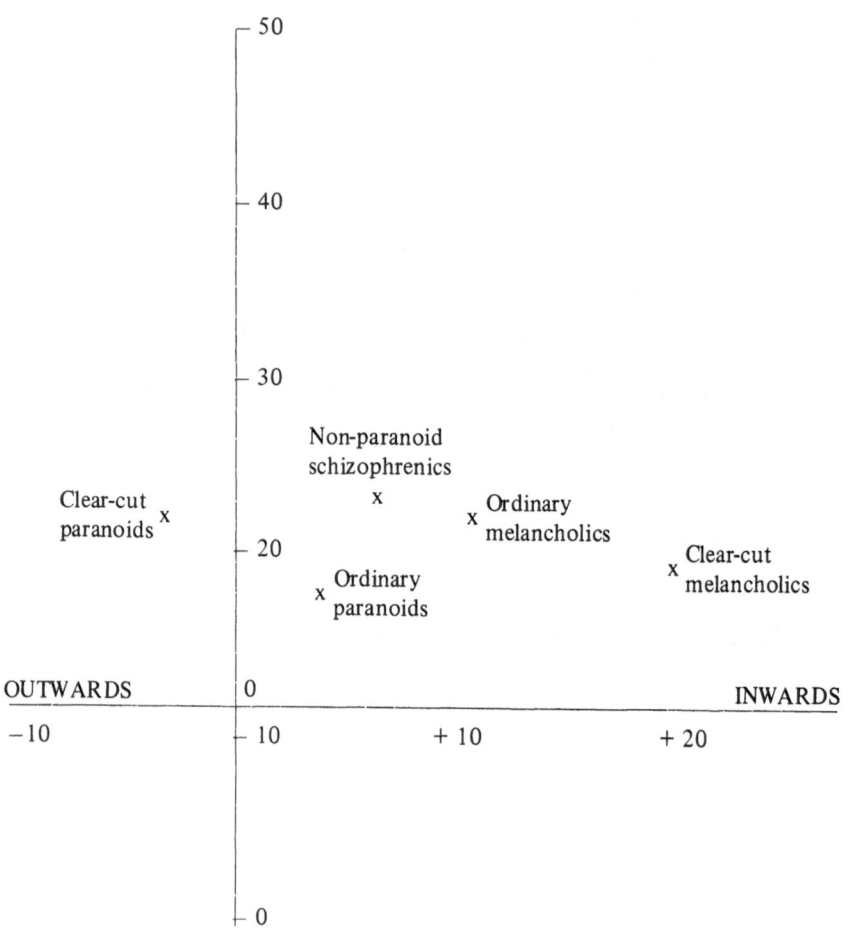

FIG. 1 - Clinical diagram for psychotics (HDHQ)

high submissiveness. In psychodynamic terms it indicates a life-style or a balanced adaptation which the individual maintains throughout his life in defence of his dominance-dependence conflict.

All 12 questions in the subscale of dominance refer to hostility being "consummated" (Alexander, 1950) by active behaviour during life. The 24 questions in the subscales measuring extra-punitiveness and intro-punitiveness refer to hostility maintained at the level of thoughts, feeling, preferences, beliefs or attitudes, not "consummated" in active behaviour during life.

Foulds has also developed (1976) a State of Anxiety and Depression (Sad) self-rating questionnaire, suitable both for evaluating treatments and for screening out psychopathology in the general populations (Bedford and Foulds, 1978b). Sad was derived empirically from the Delusions-Symptoms-Signs Inventory (DSSI, Foulds and Bedford, 1978c), and logically from the hierarchical model of psychopathology (Bedford and Foulds, 1975; Foulds and Bedford, 1975; Foulds, 1976). According to this model dysthymic states (in 93% of the cases, anxiety and depression) underlie all symptomatology of recent psychopathology. Scores above cut-off levels in the dysthymic scales indicate current psychopathology or "at risk" conditions. This is in agreement with epidemiological findings by the PSE(Wing, 1976), and with psychodynamic views according to which anxiety and/or depression is the prelude of defensive symptomatology (Laughlin, 1967). The Sad contains 14 items, 7 of which relate to axiety (A), and 7 to depression (D). Previous studies have shown both these questionnaires to be valid and reliable (Foulds, 1976; Foulds and Bedford, 1977a; Foulds and Bedford, 1977b).

Both the PDS and the Sad have been translated into Greek by three psychiatrists (Lyketsos et al., 1978). Normative data have been established in a general population sample (Lyketsos et al., 1978), and comparisons made between a Greek and a British sample (Lyketsos et al., 1979). Both the PDS and the Sad, measuring life-long psychological adaptation and recent symptomatic adaptation respectively, are suitable for the assessment of psychological factors associated with physical disease. In fact, psychological factors are dichotomized into personality characteristics and dysthymic symptoms, although they presumably act concurrently.

In a recent study (Lyketsos et al., 1982) we have used both these scales to assess psychological factors in duodenal ulcer and hypertensive patients. The aim of the study was to compare two specific psychosomatic groups with a general physical illness control group. All patients suffering from other psychosomatic or mixed illnesses were excluded. Several clinical parameters were taken into consideration - severity of illness, the presence or absence of other psychosomatic illness, neurotic symptoms - and careful questioning was carried out to determine the presence or absence of psychological and/or physical stress preceding the onset of the illness.

Table I shows the demographic data of the three groups. There were no statistical differences between the patients of the three groups concerning age and residence. As expected there were significant differences in the sex distribution of ulcer patients with a preponderance of males. There was also a bias towards social class 3 in both ulcer and hypertensive groups, whereas the controls were equally distributed in the three socio-economic groups (see Table I).

TABLE 1 - Description of subjects

	Age		Sex		Socio-economic level			Residence	
	≤30	≥31	M	F	1	2	3	Urban	Rural
Ulcers (N-27)	4	23	23	4	2	8	17	25	2
Hypertensives (N-24)	2	22	13	11	1	6	17	19	5
Controls (N-29)	9	20	15	14	11	11	7	28	1
χ^2	4.36 (df-2)		7.77[+] (df-2)		18.33[++] (df-4)			4.71 (df-2)	

+ $p < 0.05$
++ $p < 0.01$

The personality and state scores at admission and discharge for the experimental and the control groups are shown in Table II. The normative data are listed in the last row. Comparison of hypertensives and ulcer patients at admission (row 3 and row 1) by independent t-tests (df 49) showed that there were now differences in any of the personality variables (i.e. extra-punitiveness, intro-punitiveness and dominance), but the hypertensives were significantly more anxious (t - 2.20, p < 0.05) and significantly more

TABLE 2 - The personality and state (anxiety and depression) scores for the four groups on admission and discharge

		E	Do	I	A	D
Ulcers	Adm. (1)	27.3 ± 3.8	28.8 ± 4.5	23.6 ± 3.5	4.3 ± 3.6	2.4 ± 2.4
(N-27)	Dis. (2)	26.7 ± 2.6	28.3 ± 5.4	20.1 ± 4.7	3.4 ± 3.4	1.5 ± 1.7
Hypertensives	Adm. (3)	26.3 ± 4.3	27.9 ± 5.5	22.3 ± 4.4	7.3 ± 5.6	5.4 ± 5.5
(N-24)	Dis. (4)	26.4 ± 3.9	28.5 ± 6.4	22.2 ± 6.2	6.5 ± 5.1	4.0 ± 4.3
Controls	Adm. (5)	28.1 ± 3.5	32.3 ± 4.4	21.0 ± 5.1	2.1 ± 2.9	1.8 ± 2.9
(N-29)	Dis. (6)	28.2 ± 3.9	31.9 ± 3.8	20.5 ± 4.1	2.1 ± 3.0	1.8 ± 3.1
Norms (22)	(7)	28.5 ± 4.9	32.7 ± 5.0	22.1 ± 4.9	2.4 ± 2.9	1.5 ± 2.3

E: extrapunitiveness; Do: dominance; I: intrapunitiveness; A: anxiety; D: depression

depressed ($t - 2.65$, $p < 0.02$) than the ulcer group.

At discharge (row 4 and row 2) the personality variables remained undifferentiated and the state differences persisted, i.e. the hypertensives were significantly more anxious ($t - 2.49$, $p < 0.02$) and more depressed ($t - 2.62$, $p < 0.02$) than the ulcer group.

When compared with controls at admission (row 3 and row 5), the hypertensives did not differ in extra-punitiveness and intro-punitiveness but they were significantly less dominant ($t - 3.23$, $p < 0.005$, df - 51) more anxious ($t - 4.02$, $p < 0.001$) and more depressed ($t - 3.05$, $p < 0.005$).

At discharge (rows 4 and 6), the same differences occurred at a significant level, that is the hypertensives were still less dominant ($t - 2.44$, $p < 0.02$), more anxious ($t - 3.63$, $p < 0.001$) and more depressed ($t - 2.03$, $p < 0.05$) than controls.

When the ulcer group was compared with controls at admission (rows 1 and 5), the same pattern emerged, except for the difference in depression. The ulcer patients were less dominant than controls ($t - 3.08$, $p < 0.005$, df - 54) and more anxious ($t - 2.45$, $p < 0.02$).

At discharge (rows 2 and 6), the difference in anxiety disappeared,

with only one significant difference left; the ulcer patients, like the hypertensive patients, were still less dominant than controls (t - 2.86,
p < 0.01) at discharge.

As can be seen in Table II, the normative scores obtained from a previous study in Greece (Lyketsos et al., 1978) for both personality and
dysthymic state variables are very similar to those obtained in the control
group here.

DISCUSSION

Although psychodynamic theory implies a specific and elaborate chain of
psychological and physical events (Appendix B) which differentiate ulcer
and hypertensive patients, it refers to a common conflict: namely aggressive
and hostile versus dependent tendencies (Alexander, 1950). Increased dependence or lower dominance appears to be the preferred long-standing adaptation of these patients' personalities in this conflict. The PDS proved to
be a satisfactory measure for the assessment of these enduring dependent
characteristics. Both ulcer and hypertensive patients scored significantly
lower in Foulds' subscale of dominance in comparison with control groups
and norms, that is, in Alexander's terms, they were less prone to consummate
their hostility in actions. Thus their hostility found expression in disfunction of the organs (Alexander, 1952). In a previous study (1977c) Foulds
and Bedford reported that intro-punitiveness and extra-punitiveness differentiated normals from neurotics and psychotics, while dominance did not.
We seem to have identified in these two psychosomatic groups a population
which is significantly more submissive than normals, in the sense that they
fall short in the consummation of their hostility (action). It remains for
further research to evaluate whether dominance differentiates normals from
other psychosomatics.

Psychodynamic theory also implies a rise in anxiety as part of the
chain of events which ultimately leads to these psychosomatic illness. In
our recent study (Lyketsos et al., 1982) anxiety accompanied the onset or
the exacerbation of the ulcer, as was indicated by significantly high

anxiety scores on admission. This means that Sad detected anxiety in ulcer patients although the psychosomatic patient "may not be consciously aware of his emotional state" (DSM-II 1968). "Alexithymic" is the term best illustrating the inability of psychosomatic patients to express their feelings in words (Sifneos, 1973).

The disappearance of anxiety at discharge is a confirmation that Sad assessed anxiety as a temporary state rather than as an enduring trait. However, the temporal relationship between anxiety and the onset or exacerbation of the ulcer was not determined. It could be a conflict-anxiety - which led to an equivalent physical expression, namely gastric hyperfunction and ultimately the ulcer - or a dysthymic reaction to the threat of onset or exacerbation of the ulcer.

Another disadvantage of Sad is that it measures only the presence or the absence of dysthymic states which according to the hierarchical model of Foulds (1976) underlie any personal illness, in other words, any current psychopathology. It does not relate them to other symptoms or syndromes, nor does it classify them in categories. "The scale can be used to ascertain the degree of anxiety and depression once the ulcer, the hypertension, the asthma or whatever has been identified" (Foulds, 1976, p. 113). In our group of hypertensives the significantly high scores of both anxiety and depression registered at admission persisted at the discharge of the patients, indicating an enduring association of these dysthymic states with hypertension. But what is their relationship ? Foulds (1966) suggested that somatization of symptoms might be a substitute form of intro-punitiveness, and it is well known that some hypertensives are clinically diagnosed as masked depressions and treated as depressives. For this reason, in our current studies of psychosomatic disorders (bronchial asthma, alopecia, urticaria, etc.), we have added Wing's Present State Examination (1976) in order to delineate further the contribution of psychopathological symptoms in psychosomatic illness.

Another disadvantage of our recent study was that stress as a psychosocial factor was only associated with ulcer as present or absent during an unstructured interview. In our current studies we have added a modified social adjustment scale (Holmes and Rahe, 1967), in order to quantify psychosocial stress associated with somatic breakdown. Such a strategy, however, leaves another question open : how far stress associated with a somatic illness can be quantified when there are examples (Alexander, 1950) in which the death of a spouse - the highest score in Holmes' adjustment scale (Holmes and Rahe, 1967) - had a beneficial therapeutic effect in a case of psychosomatic rheumatoid arthritis ! In this case it is clearly stated, however, that the deceased husband was not an object of dependency. It remains to be seen whether normative scoring of recent life events will probe reliable.

In the present study no significant correlations were found between clinical or laboratory severity of the ulcer and any of the personality or state variables. However, biochemical or endochrine correlates may yet be detected. Recently several studies (Persky et al., 1971; Kreuz and Rose, 1972; Ehrenkraz et al., 1974; Meyer-Bahlburg et al., 1974; Rose et al., 1974; Edwards and Rowe, 1975; Rose, 1975; Monti et al., 1977; Persky et al., 1977; MacCulluch and Waddington, 1981) have attempted unsuccessfully to correlate aggression with steroid hormone (testosterone, oestrogen and progesterone levels). This research did not measure enduring personality characteristics. To overcome this difficulty in a pilot study of a male sample of volunteers we correlate the PDS scores with testosterone stability, oestradiol, stress influenced hormones, FSH, LH, and somatometric evaluations. If strong correlations with aggression are found, we plan to apply these measurable endocrine procedures to other breakdowns of human adaptation.

REFERENCES

Alexander F. (1950) Psychosomatic medicine. Norton, New York

Alexander F. and Ross H. (1952) Dynamic psychiatry. J.P. Lippincott Co., University of Chicago Press

Bedford A. and Foulds G.A. (1975) Humpty Dumpty and psychiatric diagnosis. Bulletin of the British Psychological Society, 28, 208-211

Bedford A. and Foulds G.A. (1978a) Personality Deviance Scale (Manual) NFER Publishing Co.

Bedford A. and Foulds G.A. (1978b) State of Anxiety and Depression (Manual) NFER Publishing Co.

Bedford A. and Foulder G.A. (1978c) Delusions-Symptoms-Signs Inventory. NFER Publishing Co.

Caine T.M., Foulds G.A. and Hope K. (1967) Manual of the Hostility and Direction of Hostility Questionnaire. University of London Press Ltd.

DSM-II (1968) Diagnostic and Statistical Manual of Mental Disorders. Ed. by American Psychiatric Association

DSM-II (1980) Diagnostic and Statistical Manual of Mental Disorders. Ed. by American Psychiatric Association

Dunbar F. (1968) Psychosomatic diagnosis. New York, Hoeber

Edwards D.A. and Rowe F.A. (1975) Neural and endocrine control of aggressive behaviour. In: Hormonal Correlates of Behaviour. Vol. 1, ed. Eleftheriou and Sprott, New York, Plenum Press, 275-304

Ehrenkraz J., Bliss E. and Sheared M.H. (1974) Plasma testosterone : correlation with aggressive behaviour and social dominance in man. Psychosom. Med., 36, 496-475

Foulds G.A. (1966) Psychic somatic symptoms and hostility. Brit. J. Soc. Clin. Psychol., 5, 185-189

Foulds G.A. and Bedford A. (1975) Hierarchy of classes of personal illness. Psychol. Med. 5, 181-192

Foulds G.A. and Bedford A. (1977a) Hierarchies of personality deviance and personal illness. Brit. J. Med. Psychol., 50, 73-78

Foulds G.A. and Bedford A. (1977b) Self-esteem and psychiatric syndromes. Brit. J. Med. Psychol. 50, 237-242

Foulds G.A. and Bedford A. (1977c) Personality and coping with psychiatric symptoms. Brit. J. Psychiat., 130, 29-31

From-Reichmann F. (1937) Contribution to the psychogenesis of migraine. Psychoanal. Rev., 24, 26-33

Holmes T.H. and Rahe R.H. (1967) The social readjustment rating scale. J. Psychos. Res., 71, 213-218

Horrowitz M., Schaefer C., Hiroto D., Wilner N. and Levin B. (1977) Life event questionnaire for measuring presumptive stress. Psychosom. Med., 39, 413-417

Hurst M.W., Jenkins C.D. and Rose M. (1978) The assessment of life change stress. A comparative and methodological inquiry. Psychosom. Med. 40, 121-125

Jenkins C.D. (1976) Medical Progress. Recent evidence supporting psychological and social risk factors for coronary disease. N. Eng. J. Med., 294, 987-1033

Kreuz L.E. and Rose R.M. (1972) Assessment of aggressive behaviour and plasma testosterone in a young criminal population. Psychosom. Med., 34, 321-332

Laughlin H. (1967) Clinical features in personality types and character reactions. In: The Neuroses by H. Laughlin, 245-294, Butterworths, Washington

Lyketsos G.C., Mouzakis D. and Beryanaki N. (1978) Dominance, hostility and dysthymic symptoms in a sample of the Greek population. Hyppocrates, Athens University Press, 6, 415-426

Lyketsos G.C., Blackburn I.M. and Mouzakis D. (1979) Personality variables and dysthymic symptoms : a comparison between a Greek and a British sample. Psychol. Med., 9, 753-758

Lyketsos G.C., Arapakis G., Psarras M., Photiou I. and Blackburn I.M. (1982) Psychological characteristics of hypertensive and ulcer patients. J. Psychosom. Res., 26, 255-262

MacCulluch M.J. and Waddington J.L. (1981) Neuroendochrine mechanisms and the aetiology of male and female homosexuality. Brit.J.Psych. 139, 341-345

Meyer-Bahlburg H.F., Boon D.A. and Sharmer M. (1974) Aggressiveness and testosterone measures in man. Psychosom.Med., 36, 267-274

Monti P.M., Brown W.A., and Corriveau D.P. (1977) Testosterone and components of aggressive and sexual behaviour in man. Am.J.Psych., 134, 692-694

Paykel E.S., Prusoff B.A. and Uhluhuth G.H. (1971) Scaling of life events. Arch. Gen. Psycho., 25, 340-347

Persky H., Smith K.D. and Basu G.K. (1971) Relation of psychologic measures of aggression and hostility to testosterone production in man. Psychosom. Med., 33, 265-277

Persky H., O'Brien C.P., Fine E., Howard W.J., Kahn M.A. and Beck R.W. (1977) The effect of alcohol and smoking on testosterone function and aggression in chronic alcoholics. Am.J.Psych., 134, 621-625

Rose R.M., Bernstein I.S. and Gordon T.P. (1974) Androgen and aggression : a review and findings in primates. In: Holloway (ed) Primate aggression, territoriality and xenophobia, New York, Academic Press, 275-304

Rose R.M. (1975) Testosterone, aggression and homosexuality : a review of the literature and implications for future research. In: E.J. Sachar (ed) Topics in psychoendochrinology, Grüne and Stratton Inc., New York

Sifneos P. (1973) The prevalence of "alexithymic" characteristics in psychosomatic patients. Psychoth. Psychosom., 22, 255

Theorel T., Lind E., and Floredus B. (1975) The relationship of disturbing life changes and emotions to the early development of myocardial infarctions and other serious illness. Int. J. Epidemiology, 4

Wing J. (1976) A technique for studying psychiatric morbidity in in-patient and out-patient series and in general population samples. Psychol. Med. 6, 665-671

APPENDIX A

Personality Deviance Scale (PDS) Items by Scale
(All items are preceded by "Most of my life")

Extrapunitive Scale
Booklet Number

1.	I should have liked to get my own back on someone	HT(L)
3.	I have thought that people will tell the truth, even if it gets them into trouble	DO(R)
7.	I have felt like telling people to go to blazes	HT(L)
9.	When someone has been particularly helpful, I've wondered what real reason lays behind it	DO(L)
13.	I have felt the urge to smash things	HT(L)
15.	I have believed that people are pretty reliable	DO(R)
19.	I have wanted to give someone a piece of my mind	HT(L)
21.	I have felt that people would tell lies to get ahead	DO(L)
25.	I should have liked to pick a quarrel with someone	HT(L)
27.	I have felt that people are out for what they can get	DO(L)
31.	I have felt like blaming others when things have gone wrong	HT(L)
33.	I have thought one can safely trust people	DO(R)

Intropunitive Scale

4.	I have felt as capable as other people	LSC(L)
6.	I have preferred to take a lot of advice before doing anything	DEP(R)
10.	I have had confidence in myself	LSC(L)
12.	I have wanted plenty of support from people	DEP(R)
16.	I have been very unsure of myself	LSC(R)
18.	I have liked to be told what needs doing	DEP(R)
22.	I have given up doing something because I thought too little of my own ability	LSC(R)
24.	I have been content to lean on other people for emotional support	DEP(R)
28.	I have felt that, even when difficulties were piling, I would overcome them	LSC(L)
30.	I have preferred to find out for myself what is to be done	DEP(L)
34.	I have felt pretty useless	LSC(R)
36.	I have needed a lot of help from other people	DEP(R)

Dominance Scale

2.	I have been content to act in a very humble way	MIN(L)
5.	When I've wanted to have a row with someone, I have done so	HA(L)
8.	When in a group, I have been quite content to be led	MIN(L)
11.	When I've disliked someone, I have shown it	HA(L)
14.	I have been content to be dominated by someone else	MIN(L)
17.	When I've been angry with someone, I've bottled it up	HA(R)
20.	I have preferred to let people have their own way	MIN(L)
23.	Even when crossed, I've let people get away with it	HA(R)
26.	I have been happy to play second fiddle	MIN(L)
29.	When I've thought I was justified in losing my temper, I have done so in no uncertain terms	HA(L)
32.	I have preferred to stay in the background	MIN(L)
35.	When I've felt like blaming someone to their face for something that has gone wrong, I have done so	HA(L)

APPENDIX B

Specific Dynamic Patterns in Gastric Hyperfunction

I

Frustration of oral-receptive longings → oral-aggressive response → guilt → anxiety → overcompensation for oral aggression and dependence by actual successful accomplishments in responsible activities → increased unconscious oral-dependent cravings as reaction to excessive effort and concentration → gastric hypersecretion.

II

Prolonged frustration of oral-receptive longings → repression of these wishes → gastric hypersecretion.

STRESS, THE IMMUNE SYSTEM, AND GI FUNCTION

A.S. Peña

Department of Gastroenterology

University Hospital,

Leiden, The Netherlands

ABSTRACT

The gastrointestinal part of the immune system differs from the rest of this system in several characteristics, for example, a different distribution of immunoglobulins - with predominance of secretory IgA, the presence of Peyer's patches with specific cells for antigen uptake, a capacity for the induction of tolerance, the presence of cytotoxic-suppressor lymphocytes, and an active traffic of cells to all the mucosas of the organism.

The mucosal immune system makes a major contribution to the survival of the organism and, in common with the rest of the immune system, is subject to a complex form of regulation. Four regulatory levels can be distinguished: a) the genetic: immune response genes located within the major histocompatibility complex and genes linked to blood groups and secretor status; b) the cellular: receptors on the lymphocyte surface; c) the hormonal : most of the pituitary hormones are represented in the GI tract; and d) the neuronal level.

In recent years several studies have shown that stress and behavioral factors can modulate both the humoral and the cell-mediated immunity at the cellular, hormonal, and neuronal levels. There is, however, a scarcity of data on the effects of stress on the modulation of gastrointestinal immunity. Since several diseases of the GI tract carry a higher risk of malignancy and major abnormalities of immune regulation, more effort should be made to study the effect of behavioral therapy in relation to the prevention of a further deterioration of the gastrointestinal immunity.

INTRODUCTION

When coupled with adaptive failure, certain types of experimental and naturally occurring stress are associated with immunosuppression (Locke 1982). Immunologic incompetence may play a role in the pathogenesis of infectious diseases, autoimmune diseases, and cancer (Soloman et al. 1974).

The immune apparatus of the gastrointestinal tract plays a major role in the survival of the organism. Evidence of immunological dysfunction has been documented in pernicious anaemia, coeliac disease, and inflammatory bowel disease, i.e., both ulcerative colitis and Crohn's disease. A higher risk of malignancy appears to accompany all of these conditions. Furthermore, behabioral factors can modulate the immune response at the cellular, hormonal, and neuronal levels (Borysenko and Borysenko 1982).

The present communication reviews briefly the available data supportting the notion that stress and inadequate coping lead to alterations in the humoral and cellular immune functions. As will be clear from this

review, the available data are scarce but encouraging. Prospective studies
on relationships between immune parameters and stress in different populations are needed.

Clinical observations linking stress and illness

The role of psychological factors in the onset and course of infectious and autoimmune diseases as well as allergy and cancer has been reviewed (Rogers et al. 1979). When the magnitude of a recent life change
is taken into account, it appears that stress contributes to the development of disease, as shown, for example in the Boston study on the effect
of bereavement (Parkes and Brown 1972).

Experimental observations in animals

Ader and Cohen (1975) have shown that behavioral conditions can be
used to suppress humoral and cell-mediated immune responses. They injected
cyclophosphamide into rats which had just consumed a saccharin solution,
and found that subsequent exposure to the saccharin concurrently with an injection of sheep erythrocytes led to a reduced serum antibody response.
Later, Bovbjerg et al. (1982) used the same protocol and found that condi-
tional immunosuppression also affected the popliteal graft-versus-host response. More recently, Ader and Cohen (1982) showed that the rate of development of proteinuria was significantly retarded and mortality reduced
in female New Zealand hybrid mice by classical conditioning of immunosuppression. So far, no parameters of gastrointestinal immunity have been
measured, although some experimental work linking stress with parasitism in
the gastrointestinal tract has been reported (Hamilton 1974), and crowding
is known to increase susceptibility to Salmonella typhimurium (Edwards and
Dean 1977).

Recent studies on relationships between stress and human function

Bereavement, sleep deprivation, and pre-examination stress all markedly
depressed lymphoblast transformation (for a review, see Locke 1982).

Stress and gastrointestinal immune function

Salivary immunoglobulin A (S-IgA) is considered part of the body's
time of defense against viral infections. McClelland et al. (1980) has reported that subjects with low concentrations of S-IgA had more severe
illnesses during the last months of the investigation than did those with
high concentrations of S-IgA. S-IgA was significantly lower during three

high stress periods than in either of two low-stress periods in dental students followed prospectively over a ten-month period(Jemmott et al.1981,Locke 82) Patients with acute necrotizing ulcerative gingivitis reported more negative life events in the preceding 12 months, as well as more anxiety and more emotional stress, compared with matched controls (Cohen-Cole et al. 1981, cited by Locke 1982).

More recently, McClelland et al. (1982) have shown that among prisoners high concentration of S-IgA were associated with reports of fewer respiratory infections, and those with high reported stress and the highest levels of reported illness had the lowest concentrations of S-IgA.

These studies support the concept that stress can impair gastrointestinal function but by no means provide proof, since several methodological questions may be raised as to both the timing of sampling and the measurement of secretory IgA.

Stress, immunity, and duodenal ulcer

The term ulcer disease covers an extremely heterogeneous group of disorders with a variety of genetic and environmental etiologies (Rotter and Rimoin 1977). Rotter and Heiner (1982) recently reviewed the findings in animals and man that suggest the existence of an immunological form of duodenal ulcer. The patients with this postulated form of duodenal ulcer are possibly characterized by the presence of antibodies to secretory IgA, increased frequency of allergic disorders, childhood duodenal ulcer, and possibly an increased frequency of certain HLA antigens. Antibody to parietal and gastrin-producing cells may mediate hyperfunction analogous to the situation in autoantibody-stimulated thyrotoxocosis. Stress may play a major role in the pathogenesis of these particular forms of duodenal ulcer.

Mediating mechanisms in immunomodulation

Several mechanisms have been postulated (Lock 1982):

. classical neuroendocrine mechanisms (hypothalamic-pituitary-adrenal axis),
. receptors on lymphocytes and granulocytes for neurotransmitters,
. direct autonomic innervation of lymphoid organs, and
. β-endorphin-induced degranulation of mast cells (histamine activating suppressor cells).

CONCLUSIONS

Stress and behavioral factors influences the immune response, but data concerning these effects on the immunity of the gut are scarse. This area of research should be stimulated, because a better understanding of immuno-modulation will contribute to the prevention of autoimmune disease and malig-nancy in patients with diseases of the gastrointestinal tract.

REFERENCES

Ader, R., and Cohen, N., 1975. Behaviorally conditioned immunosuppression. Psychosom. Med. 37:333-340.

Ader, R., and Cohen, H., 1982. Behaviorally conditioned immunosuppression and murine systemic lupus erythematosus. Science 215: 1534-1536.

Bovbjerg, D., Ader, R., and Cohen, N., 1982. Behaviorally conditioned suppression of a graft-versus-host response. Proc. Natl. Acad. Sci. 79: 538-585.

Borysenko, M., and Borysenko, J., 1982. Stress, behavior, and immunity: animal models and mediating mechanisms. Gen. Hosp. Psychiatr. 4:59-67.

Edwards, E.A., and Dean, L.M., 1977. Effects of crowding of mice on humoral antibody formation and protection to lethal antigenic challenge. Psychosom. Med. 39:19-24.

Locke, S.E., 1982. Stress, adaptation, and immunity: studies in human. Gen. Hosp. Psychiatr. 4:49-58.

McClelland, D.G., Davidson, R.J., Floor, E., and Saron, C., 1980. Stressed power motication, sympathetic activation, immune function and illness. J. Hum. Stress 6:11-19.

McClelland, D.G., Alexander, C., and Marks, E., 1982. The need for power, stress, immune function, and illness among male prisoners, J. Abn. Psychol. 91:61-70.

Parkes, C.M., and Brown, R.J., 1972. Health after bereavement - a controlled study of young Boston widows and widowers. Psychosom. Med. 34:449-461.

Rogers, M.P., Dubey, D., and Reich, P., 1979. The influence of the psyche and the brain in immunity and disease susceptibility: a critical re-view. Psychosom. Med. 41:147-164.

Rotter, J.I., and Rimoin, D.L., 1977. Peptic ulcer disease - a heterogeneous group of disorder? Gastroenterology 73:604-607.

Rotter, J.I., and Heiner, D.C., 1982. Are there immunological forms of duodenal ulcer? J.Clin. Lab. Immunol. 7:1-6.

Solomon, G.F., Amkraut, A.A., and Kasper, P., 1974. Immunity, emotions and stress, with special reference to the mechanisms of stress effects on the immune system. Psychosom. 23:209-217.

CLINICAL RECOGNITION OF STRESS RELATED GASTROINTESTINAL DISORDERS IN ADULTS

J.F. Fielding,
Department of Medicine and Gastroenterology,
The Charitable Infirmary,
Jervis St., Dublin 1., IRELAND

ABSTRACT

There is no agreement as to what constitutes the irritable bowel syndrome. Little has been written on its clinical recognition. This chapter is a predominantly personal view awaiting the agreement or disagreement of others.

INTRODUCTION

It is only within the past decade that attention has focused on the positive clinical recognition of the irritable bowel syndrome. Those who have written on it have done sofrom their own "point of departure". I make no pretence that mine is necessarily any better than that of others; with the humility of the true clinical investigator I am convinced that it is no worse! It is within such limitations that this chapter is written.

First one needs to define what one means by stress related gastrointestinal disorders and here it is probably easiest to start by stating what one does not include in such a definition. I exclude those clinically defined gastrointestinal states, such as duodenal ulceration and ulcerative colitis, whose overt presentation is often precipitated by stress. I also exclude the gastrointestinal consequences of stress induced by the abuse of caloric and dietary intake, ethanol consumption, nicotine inhalation and intravenous injection of opiates. One is left therefore with that group of disorders which although they account for some three quarters of all referrals to gastroenterologists are bedeviled by a lack of both agreed terminology and of pathophysiological understanding except at a superficial level. Whilst in the past these disorders were referred to as "psychosomatic" as distinct from "organic" we at least today recognise the organic nature of the abberations in motor sensory and or secretory function that may occur. I believe these varying disorders have much in common and prefer to recognise this homogeneity in diversity by referring to them as variants of the irritable bowel syndrome rather than by the more common attitude of referring to them primarily by the more historic terms which recognise the primacy and nature of the main symptoms such as nervous dyspepsia or spastic colon. I recognise three major clinical subdivisions of the irritable bowel syndrome (Fielding, 1979).

1. ORO–PHARYNGO–UPPER OESOPHAGEAL

Anatomically this group would more correctly be referred to as oropharyngeal but as the crico-pharyngeus muscle is really the upper oesophageal sphincter it seems reasonable to use the fuller title for this group of disorders. Abnormal motor function is the commonest patient percieved abnormality in this group leading to globus sensation. Its clinical recognition requires no more than an accurate history. This is often overlooked as the patient may complain of dysphagia and, as is often the case with all variants of the irritable bowel syndrome, only detailed enquiry will reveal the true nature of the complaint. Sensory deviations include ex - cessive perception of the glossal papillae, painful swallowing (which may be referred to as a sore throat) and a belief in a malodourus breath. Examination of the tongue, observation of swallowing and smelling the exhaled breath will enable the true nature of the complaint to be readily recognised in the large majority of cases.

2. OESOPHAGO–GASTRO–DUODENAL

Abnormal motor function may lead to angina pectoris and or nausea which may on occasion be accompanied by vomiting. For the former the age, female sex and association of chest tightness with swallowing should enable the diagnosis to be suspected in the large majority of cases. The hallmark of the nausea that these people get is that in the vast majority of instances it is unaccompanied by vomiting and there is no evidence of dehydration or weight loss. In those patients who do have vomiting it is an infrequent occurence in comparison with nausea. Cyclical vomiting may represent an extreme variant of this disorder. Another consequence of disordered motor function is excess belching of wind, which may also be contributed to by the excess swallowing of air with food. A history of rapid eating usually points to the latter. Sensory disturbance is probably at least in part either directly or indirectly secondary to motor dysfunction. This may give rise to a full or bloated feeling in the epigastrium during or after eating, or epigastric pain during or after eating. If pain occurs it is usually colicky or spasmodic and if it radiates it does so from right to left. Moreover, the latter pain is often associated with, and relieved by, defaecation.

In addition patients with this subgroup of the irritable bowel syndrome not infrequently have an excessively palpable though nontender colon; most often in the left iliac fossa, next most often in both iliac fossae. Thus although these patients are said to be difficult to differentiate from patients suffering from peptic ulceration and or gastritis I suggest that proper attention to history and examination allows a clinical differentiation in a large majority of patients.

3. COLONIC (IRRITABLE COLON SYNDROME)

This is by far the commonest and most important subdivision. Indeed, some are really referring to this entity when they speak of the irritable bowel syndrome, whilst at the same time pointing out the widespread nature of the disorder. I prefer my approach as I believe it gives a more unifying concept of the varying ways in which the disorder may affect different parts of the gastrointestinal tract. Some people with irritable bowel syndrome symptoms do not consult the profession (Thompson and Heaton 1976) others present, their major symptom complex in one of the three major subdivisions outlined. That enquiry reveals either symptoms or signs to demonstrate that abnormality exists in either or both of the other two subgroups as well. Moreover, it is not uncommon for a patient to convert from major symptomatology in one subgroup to major symptomatology in another (those interested in music will be aware that Verdi demonstrated all three subgroups of the irritable bowel syndrome at varying stages of his life). Thus when a patient has the irritable colon syndrome, no more than the irritable bowel syndrome patients symptoms and signs are predominantly related to abnormal colonic function at that moment in time. The major symptoms of this variant of the irritable bowel syndrome are abdominal pain and altered bowel habit (Fielding 1978). A subgroup are said to suffer from painless diarrhoea but it is my view that detailed enquiry and examination reveals that the vast majority of these have episodes of pain and interludes of constipation. The pain may occur anywhere in the abdomen but most frequently occurs in the left iliac fossa or hypogastrium. It varies in intensity through all shades of severity and in nature from colicky to spasmodic to persistent. In three quarters of patients it occurs after the ingestion of food and in those in a constipated phase roughly half get relief following defaecation. The clinical hallmark of the pain is the inappropriate attitude of the patient as the symptom is described. The patient is well dressed, usually has liberally applied make up and nail varnish, looks well and is smiling whilst poorly describing and with difficulty detailing what is for her severe pain.

Abnormal bowel habit occurs in over four fifths of patients. It may be constipation or diarrhoea yet detailed enquiry would reveal that both often coexist. I cannot over emphasise the need for detailed enquiry; patients who deny being constipated may admit to the intermittant daily passage of lumpy, firm to hard stools. Rarely patients complain of the passage of mucus per rectum yet over a quarter will admit to its passage on questioning. The patient may also complain of the passage of what they consider excess wind per rectum.

Symptoms relating to the other subgroups often coexist as do similar symptoms relating to the genito-urinary system. Over one third will give a history of previous abdominal surgery most frequently an appendicectomy or cholecystectomy, but there is also evidence that these patients have excess extra abdominal surgery (Fielding 1983). One half have a cancer phobia (Fielding 1977).

It has been suggested that a symptom complex of distention, relief of pain with bowel movement, loose bowel movements with onset of pain and more frequent bowel movements with pain, especially when combined with a history of the passage of mucus per rectum and a feeling of incomplete emptying after defaecation enable a confident positive prediction that the patient has the irritable bowel syndrome (Manning et al., 1978). In that study irritable colon syndrome patients were only separated from controls not matched for age and sex, who mainly suffered from such disorders as duodenal ulceration, gastroesophageal reflux, gastric ulceration and gall stones (only a minority had either inflammatory bowel disease or carcinoma of the colon). These disorders do not constitute a challenge in differential diagnosis. The symptom complex appears, is somewhat artificial and moreover takes no account of the description of the symptom so important a facet of clinical medicine for the alert physician. One fifth of the patients give a history of weight loss but this is due to being afraid of ingesting certain foods rather than to loss of appetite (Keeling and Fielding, 1975).

On examination these patients look well. Their examination is often described as normal but there are generalised, abdominal, perianal and rectal signs which considerably aid in diagnosis. The general and perianal signs are also present in patients with stress related non gastro-intestinal disorders. They include a sudden slowing of the pulse by some twenty beats per minute whilst it is being measured (Fielding, 1977), cool sweaty palms, neurodermatitis, brisk neurological jerks and increased clinically assessed anal sphincteric tone (Fielding, 1981). The abdominal and rectal signs are far more specific for irritable colon syndrome and are an excessively palpable and excessively tender colon, a right iliac fossa squelch sign, the experiencing of pain during digital insertion for rectal examination, an empty or nearly empty rectum, the presence of scybala in the rectum and a positive mucosal tap sign (Fielding 1978,1981)

Sigmoidoscopy must always be regarded as part of clinical medicine. As well as excluding tumour or inflammatory bowel disease it also positively helps in the diagnosis of irritable colon syndrome. First it shows indirect evidence of excess mucus production and or secretion by a combination of a glairy appearance of the mucosa, globules of mucus within the lumen and strands of mucus across the lumen. Secondly, it shows hypercontractility of the sigmoid colon following the insufflation of air; the speed of contraction following distention is characteristic, with the mucosa thrown into folds akin to small bowel mucosa . Finally this distention and contraction is nearly always accompanied by pain and in over half the cases this pain is identical to the patients own pain both in nature and site, even if that site is elsewhere in the abdomen (Keeling and Fielding, 1975).

Thus from our limited understanding of the pathophysiology of the irritable bowel syndrome we can apply that knowledge to history taking and examination and in the large majority of sufferers make a positive clinical diagnosis both of the disorder itself and as to which of its major subdivisions is or are present. From this knowledge we should be able to devise screening programmes to determine the extent of the problem in the community.

One final point: the irritable bowel syndrome may coexist with other gastrointestinal disorders: thus the findings of gallstones or even inflammatory bowel syndrome symptoms (Fielding, 1981 and Isgar et al., 1983) should not detract out attention from the true origin of such symptoms. Moreover, whilst one must thoroughly investigate elderly patients with, recent onset of or change in, irritable bowel syndrome symptoms the finding of serious disease must raise the question from which entity did the symptoms arise rather than necessarily attributing the symptoms to the discovered cancer. Our symptoms, signs, diagnoses correlates in gastroenterology are still in their infancy.

REFERENCES

Chaudhary, N. and Truelove, S. 1982.
  The irritable Colon Syndrome. Quart J Med; 31: 307-322.

Fielding, J. 1977.
  A year in Out Patients with the irritable bowel Syndrome.
  Irish Jour Med Sci; 146: 162-166.

Fielding, J. 1978.
  Clinical and Radiological Manifestations of the Irritable Bowel
  Syndrome.
  Jour Irish Colls Phys Surgs; 8: 11-15.

Fielding, J. 1979.
  The butterfly effects of stress. Gen Pract; Jan 19: p. 30.

Fielding, J. 1981.
  The diagnostic sensitivity of physical signs in the irritable
  bowel syndrome.
  Irish Med Jour; 74: 143-144.

Fielding, J. 1981.
  Activity and its assessment in Crohn's disease. In 'Recent advances
  in Crohn's disease'. Eds A. Pena, et al Martinus Nijoff Publishers,
  The Hague.

Fielding, J. 1983.
  Surgery and the irritable bowel syndrome: The singer as well as
  the song.
  Irish Med Jour; 76: 33-34.

Isgar, B., Harman M., Kaye, Md., Whorwell PJ., 1983.
  Symptoms of irritable bowel syndrome in ulcerative colitis in
  remission.
  GUT 24: 190-192.

Keeling, P. and Fielding, J., 1975.
  The irritable bowel syndrome. Irish Jour Colls Phys Surgs 4: 91-94.

Manning A., Thompson W., Heaton K., Morris A., 1978.
  Towards positive diagnosis of the irritable bowel.
  Brit. Med Jour; 2: 653-654.

Thompson W. and Heaton K., 1978.
  Functional bowel disorder: A new perspective.
  GUT; 19: 975(a).

STRESS AND INFLAMMATORY BOWEL DISEASE (IBD)

Gisela Huse-Kleinstoll, Th. Küchler, A. Raedler, K.-H.Schulz

Department of Medical Psychology, Dir. Prof.Dr. Margit von Kerekjarto

Medical Clinic, Dir. Prof. Dr. H. Greten

University Hospital Hamburg-Eppendorf - FRG

Abstract

Inflammatory bowel diseases (IBD) are chronic disorders of the intestinal mucosa. As their etiology and pathogenesis are unknown, the influence of psychological strain in IBD remains hypothetical. In this paper a psychoimmunological model of the pathogenesis is presented. Stress in IBD is defined as a result of a disturbance in individuation in early childhood with a subsequent failure in forming mature personal relationships. A lack of independence results in chronic stress in IBD patients and events like loss of a relative, demands for perfection or a strong threat from parental figures leed to a breakdown in adaptive psychic processing. This corresponds to a breakdown in the immune regulation of the intestinal mucosa. Recent studies on the influence of psychological stressors in animals and humans have demonstrated the impact of cortical and subcortical stimulation as well as enviromental stressors on the immune system. As both immune system and the personality show a life long adaption to environment a psychoimmunological approach seems to be appropriate for investigating stress in the pathogenesis of IBD.

Ulcerative Colitis and Crohn's disease are chronic reoccurring disorders of the intestinal mucosa whose etiology and pathogenesis are unknown (WEINER, 1977). While ulcerative colitis occurs almost exclusively in the large intestine, only rarely involving the terminal ileum, Crohn's disease may affect any part of the intestinal tract. It is observed about 20 % of the time in the colon and only sporadically in the stomach, esophagus and mouth. In some cases it is not possible to clearly distinguish between the two diseases. Both men and women are affected with an almost equal degree of frequency. The disease may occur for the first time at any age, but peaks occur at 18 and 50 years of age.(WEINER, 1977; SCHULTHEIS and v. UEXKÜLL, 1981) Clinical symptoms such as severe diarrhea, acute abdominal pains and intestinal bleeding are extremely detremental to the patient's condition in the acute stage of the disease. Heavy bleeding and severe attacks of diarrhea may place the patient's life in **danger** Patients with colitis face an increased risk of cancer which is intensified with the duration of the illness. For these reasons, in the course of the illness surgery is often unavoidable. Most patients experience relief of symptoms after an ileostomy.

Crohn's disease, on the other hand, reoccurs in about 50 % of the patients who have had the affected intestinal segment surgically removed. (LUKE et al., 1982)

Etiology and Pathogenesis

At present there exist no more than a series of hypotheses about the etiology and pathogenesis of inflammatory bowel diseases. Despite their characterization as "psychosomatic illnesses" the role of psychological factors in the pathogenesis of these disorders has not gone unchallenged. Genetic factors are often assumed because of frequent occurrence in individual families and because of differences between ethnic groups, but here it is quite conceivable that family communication style and cultural differences are responsible for these findings.

Among other factors which have been held responsible for etiology and pathogenesis are: viruses and bacteria, toxic-allergic factors, vascular-neurological factors as well as a disturbance of the body's immunological system.

A number of findings support the view that both diseases might involve similar pathological mechanisms (THAYER et al., 1969; TRUELOVE, 1971; ENGEL, 1973, SCHULTHEIS and v. UEXKÜLL, 1981). This is true as well from the psychodiagnostic point of view, where differences correlate more with age than with basic differences in personality structure. Among the factors mentioned as having been held responsible for the pathogenesis of inflammatory bowel disease, the hypothesis of a disturbance in the body's immunological system is the most promising both with regard to its plausibility and its relationship to psychological stress.

Personality and Personal Relationships

Information about psychological characteristics of patients with ulcerative colitis was presented by ENGEL (1955) in a survey of the results of the first 25 years of research in this area. Since that time they have been confirmed by numerous clinical observations and through the use of psychological tests.

According to these studies, personality structure of ulcerative colitis patients is characterized by attributes such as compulsiveness, neatness, conscientiousness, punctuality, obstinacy, indecisiveness, and overadaptivity (WEINER, 1977). The intellectual capabilities of ulcerative colitis patients are described as good to above average. The personality traits of Crohn's disease patients are described similarly ,although these patients are perhaps on the whole somewhat less obsessive (FREYBERGER et al., 1982).

In addition to the personality traits outlined above, patients with chronic inflammatory bowel disease are marked by a close symbiotic relationship to their mothers which points to a disturbance in the development of the mother-child relationship. While it is normal for children to separate themselves from their mothers during the period of toilet training from age 1 to 3, to begin to walk and learn to distance themselves from their mothers by means of outbursts of anger and defiance, the mother of patients with chronic inflammatory bowel disease reacts to the child's attempts to establish his independence with particularly strong anxiety, refuse to accept them and rebuff them.

Such mothers are as a rule unsure of themselves, timid, impulsive and overburdoned and have unrealistic values and moral standards (ENGEL, 1955; MOHR et al.,1958). In order to retain the affection of their unpredictable mothers, particularly sensitive children often react with an especially strong form of adaptation. Since this emotional learning process takes place at a very early age, the patient is not aware of it nor is he able to articulate it. The mother reacts in a similar way. In the close mother-child relationship each partner needs the other for his or her emotional stability.

The undissolved symbiotic ties of patients with chronic-inflammatory bowel disease to the key person of their early childhood or a substitute has the following effects:

1. They continually seek partners with whom they can establish similar close relationships (key person; substitutes).
2. They establish only superficial relationships at best with other persons.
3. These patients remain dependent and immature.
4. Their psychological boundaries are unstable and they tend towards paranoid psychosis.
5. Due to their lack of ability to define themselves, their **percep**tions of self and others are distorted (ARAJÄRVI et al., 1961; WIJSENBEER et al., 1968; FINCH and HESS, 1962), and
6. They tend towards compulsive domineering behavior as an attempt to compensate for their weaknesses in perception and orientation.

The specific stress of IBD-patients

The specific stress that patients with chronic-inflammatory bowel disease are subjected to has its roots in the extreme dependence of the patient on his parents or parent surrogates. Because of this dependency, they believe that they must subordinate themselves to the wishes of others up to the point of self-denial when involved in interpersonal conflicts, particularly when threatened with rejection by key persons, and frequently, in addition, subordinate themselves in order to avoid the anger of such persons or possible new conflict situations.

The patients' adaptiveness, while satisfying their need for support and security, at the same time intensifies their feelings of dependency. The patients' own feelings of anger and antipathy are experienced by them as a threat to their relationships and are, as a consequence, suppressed. If this suppression is not successful, guilt feelings easily arise in these patients. This constant internal pressure to adapt in these patients is the equivalent of a chronic stressor.

In addition there are changes in the patients' lives, so-called life events, which demand additional adapative energy until they are finally overtaxed and give up in resignation. In such situations the first manifestations of the disease or of a relapse often appear. These situations which are specific for each individual patient can only be identified with great patience on the part of the interviewer. GRACE (1953) compiled psychosomatic retrospective histories for 4 patients with Crohn's disease. His interviews with these patients lasted a minimum of 14 hours. Similar results are, however, reported by authors who have treated patients with chronic-inflammatory bowel disease over a period of many years (ENGEL,1955; PAULLEY, 1971).

The life events outlined here act on the patient as acute or prolonged stressors and must in addition be reckoned to the chronic stress the patient is submitted to. Acute stressors are, for example, the death of a parent or a child's operation. Prolonged stressors are, for example, the necessity of caring for a mother-in-law or a husband's temporary unemployment.

Under the pressure of prolonged stressors the strain on the patient slowly increases to the point where it is unbearable. The last mentioned stressors occur more frequently and are consistantly characterized by patients with phrases like "Will I ever get this all over with", "I never thought I would be done with it or get it finished" (GRACE, 1953).

Immunological and Psychoimmunological Findings

The correlations between psychological stress and the onset of chronic-inflammatory bowel disease which we have described here from the psychological point of view are, in our opinion, related to the pathogenetic model of inflammatory bowel disease as a distrubance of the immunological system. The connection is established by the following reasoning:

1. Inflammatory bowel disease is the result of a disturbance in the regulation of the immunological system associated with the mucous membrane.

2. In addition to autoregulation the immunological system is controlled through neuroendocrine and central nervous system circuits.

3. Moreover, the direct influence of psychological stress on immune reactions can be empirically established.

The theoretical and empirical basis of each of these points is briefly outlined below:

Point 1:

In the intestinal mucosa, immune reactions against intestinal antigens are characterized by at least two features:

1) The immune response is restricted to the mucosa exclusively, that means immunoreactivity in the mucosa often coexists with systemic tolerance and thus intestinal antigens do not necessarily provoke generalized immune reactions.

2) The prevalent effector molecule is found to be an IgA isotype antibody. The advantages of a predominant intestinal IgA production consist in a stability of these very antibodies against enzymatic activities, in the absence of complement fixing and in the inability of the IgA antibodies to arm killer lymphocytes. Thus IgA antibodies neutralize, but do not kill symbiotic bacteria and do not damage mucosal epithelium that carries cross-reacting cell surface structures (DOBBINS, 1982).

Both characteristics of mucosal immune response are governed by regulator T cells, as is shown in figure 1. B cells of IgM, G and A isotypes in the spleen as well as in the intestinal mucosa are suppressed by appropriate T cells with the exception of B cell clones in the mucosa which produce IgA antibodies with the help of a contrasuppressor T cell that inhibits the corresponding suppressor cell. These findings, however, result from various experiments using murine intestine (ELSON et al., 1979;

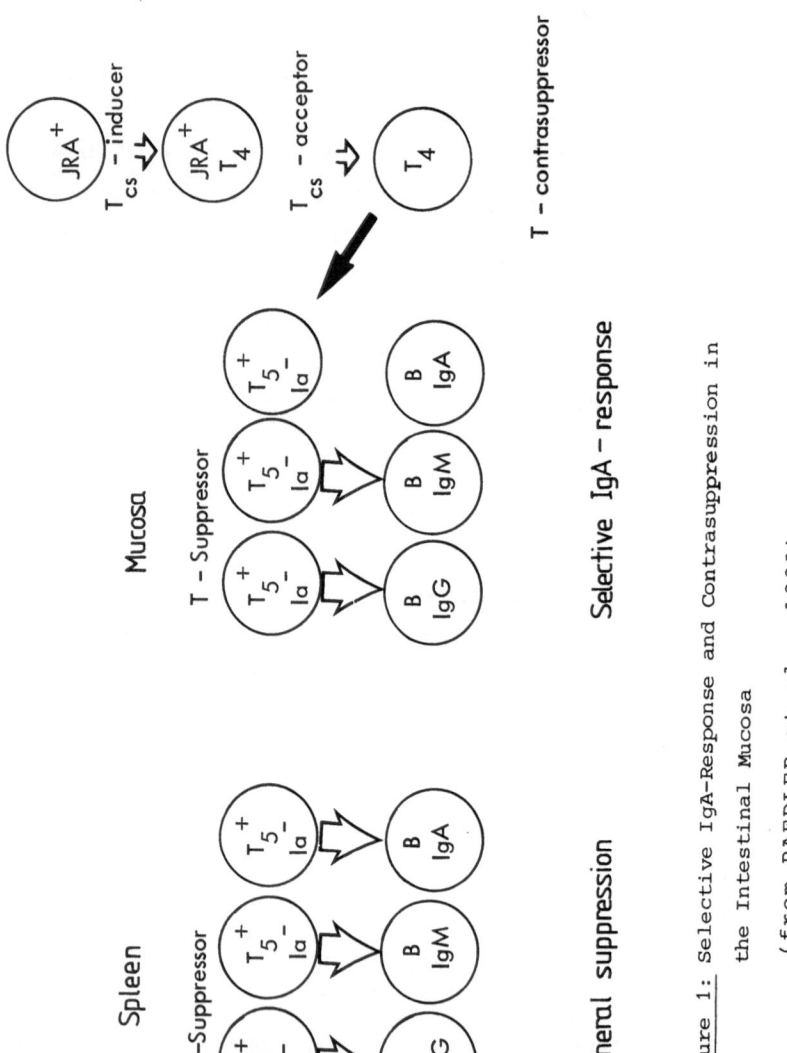

Figure 1: Selective IgA-Response and Contrasuppression in
the Intestinal Mucosa

(from RAEDLER et al., 1983)

GREEN et al., 1982).

Among the manifold data concerning the role of immune phenomena in
Crohn's disease there is one observation that, in our opinion, deserves
special attention. That is the predominance of IgG and IgM antibody production as compared to that of IgA in the altered mucosal lining (BAKLIEN
et al., 1975; GREEN et al., 1975; ROSEKRANS et al., 1980). Thus it seems
tempting to suggest that the pathogenesis of Crohn's disease is rooted in
an immunoregulatory disturbance resulting in an inhibition of IgA production and thus a depression of mucosal protection. Mucosa-damaging IgG
and IgM antibodies are produced instead. This pathogenetic sequence is
shown in figure 2. Recently we have been able to confirm this potential
ability of T regulatory cells isolated from the mucosa of patients
suffering from Crohn's disease to depress IgA- and favour IgG-production.
This was done in experiments using a pokeweed mitogen-driven B cell assay.
In conclusion, independent of the underlying etiological stimulus , the
pathogenetic events of Crohn's disease seem to involve a dysregulation of
the immune response within the altered intestinal mucosa.

Point 2:

The immune response is subjected to external regulatory operations.
Investigation of the cellular and subcellular immunological mechanisms with
respect to neuroendocrine signals and analysis of the afferent, central and
efferent pathways within the network of immune-neuroendocrine interactions
have led to the following experimental findings, described by BESEDOVSKY
and SORKIN (1981):
- Lymphocytes express receptors for hormones and neurotransmitters.
- Manipulation of neuroendocrine function affects the immune response.
- The activated immune system elicits changes in the neuroendocrine system.
- Afferent signals from the activated immune system elicit a response in
  neurons of the hypothalamus.
  Here we can add some of the most recent findings reported in RENOUX et
  al. (1982, 1983) and BOVBJERG et al. (1982).
- The production of T-cell producing factors and the NK activity in mice
  is found to be controlled by the brain neocortex.
- Classical conditioning procedures could be used to suppress immunologic
  reactivity.

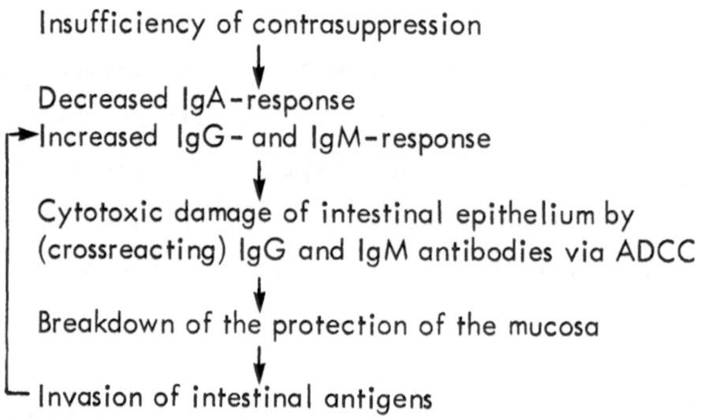

Insufficiency of contrasuppression

↓

Decreased IgA-response

Increased IgG- and IgM-response

↓

Cytotoxic damage of intestinal epithelium by
(crossreacting) IgG and IgM antibodies via ADCC

↓

Breakdown of the protection of the mucosa

↓

Invasion of intestinal antigens

Figure 2:   Pathogenesis of the immunoregulation
in IBD

(from RAEDLER et al., 1983)

Point 3:

The influence of various stressors on the hypothalamo-pituitary axis and the autonomous nervous system has been sufficiently documented. The interaction of these systems with the immunological system has been described above. Although it has long been common wisdom to note that excessive stress increases susceptibility to illness, it is only recently that the influence of experiential factors on immune response has been established, not only in animal studies but in human studies as well.

These experiments have demonstrated that:

- Environmental stressors change the susceptibility to pathogenic agents in murine systems.
- Environmental stressors modulate the antibody response and the reactivity of lymphocytes, the latter depending on the interval between stress and testing.
  A review of this work can be found in ADER (1981).
- In corresponding experiments with human volunteers PALMBLAD et al. (1976, 1979) showed that sleepdeprivation and exposure to noise effect the cellular immune response.
- In addition BARTROP et al. (1977) were able to demonstrate that prolonged exposure to the stresses of life, e.g. bereavement, provokes a depression of lymphocytic function, obviously without involvement of the endocrine system.

Conclusion

In conclusion we would like to present our hypothesis on the role of acute and prolonged stress in the etiology and pathogenesis of IBD. This is illustrated in figure 3.

Acute and prolonged stressors in the psychosocial environment influence the conscious (cortex) and unconscious memory (subcortical areas) of the person. Immune regulation in the intestinal mucosa is influenced by the central nervous system through two pathways: through the neuroendocrine pathway and through direct neural axis. We believe that genetic determination plays an independent role in immune regulation. Infectious agents and alimentory constituents play a more secondary role. This can be said as well of other metabolic factors.

Personal and immunological experience is stored in memory systems.

816

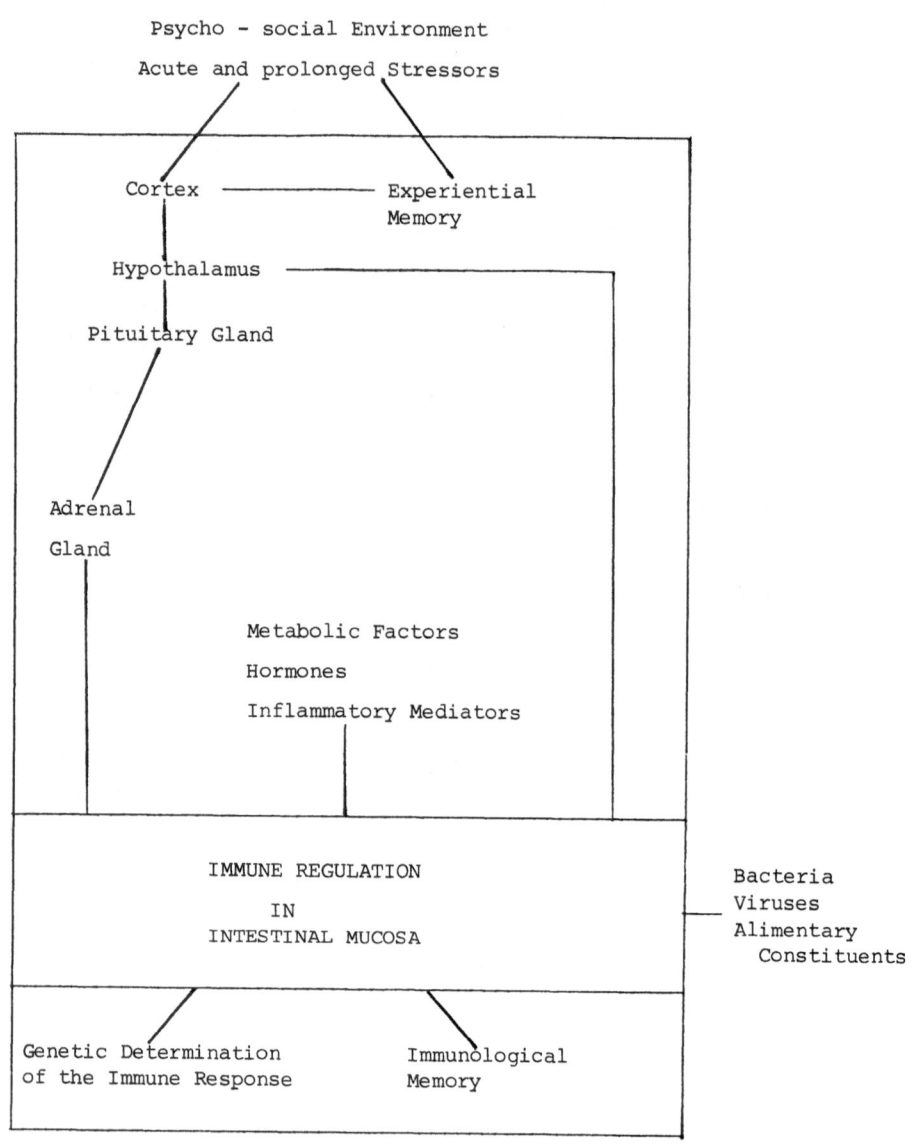

Figure 3: Hypothetical Model of Etiology and Pathogenesis
          of Inflammatory Bowel Disease (IBD).

Both personality and immune system are rooted in the experiences of early childhood and undergo a constant and parallel development. Both involve memory and the ability to recognize familiar and foreign elements they are confronted with. Further study of the affects of psychological stressors on immune reaction as an etiological factor in IBD are clearly needed. We feel that this line of investigation offers much promise both for clinical treatment of IBD and for deepening our general understanding of the interaction between somatic and psychosocial factors in human illness.

References

Ader, R., 1981. Behavioral Influences on Immune Responses. In: Weiss, S.M., Herd, J. A. and Fox, B.H. (eds.) Perspectives on Behavioral Medicine. Academic Press, New York

Arajärvi, T., Pentti, R. and Aukee, M., 1961. Ulcerative colitis in children. Psychological Study,Ann.Clin.Res. 7, 1

Baklien, K. and Brandtzaeg, P., 1975. Comparative mapping of the local distribution of immunoglobulin containing cells in ulcerative colitis and Crohn's disease of the colon. Clin.Exp.-Immunol. 22, 197

Bartrop, R.W., Lockhurst, E., Lazarus, L., Kiloh, L.G. and Penny, R.,1977. Depressed Lymphocyte Function After Bereavement. Lancet 1, 834-836

Besedovsky, H.O. and Sorkin, E., 1981. Immunologic-Neuroendocrine Circuits: Physiological Approaches. In: Ader, R. (ed.), Psychoneuroimmunology. p. 545, Academic Press, New York

Bovbjerg, D., Cohen, N. and Ader, R., 1982. The central nervous system and learning: A strategy for immune regulation. Immunology Today 3,287-291

Dobbins, W.O., 1982. Gut immunophysiology: A gastroenterologist's view with emphasis on pathophysiology. Am.J.Phys., 242, G1

Elson, C.O., Heck, J.A. and Strober, W., 1979. T-cell Regulation of Murine IgA Synthesis. J.Exp.Med. 149, 632

Engel, G.L. 1955. Studies of Ulcerative Colitis III. The nature of the psychologic processes. Am.J.Med. 19, 231

Engel, G.L. 1973. Ulcerative Colitis. In: A.E.Lindner (ed.) Emotional Factors in Gastrointestinal Illness. Amsterdam, Excerpta Medica.

Finch, S.M. and Hess, J.H., 1962. Ulcerative Colitis in Children. Am.J. Psychiatry 118,819

Freyberger, H., Wellmann, W., Ziegler, H., Nordmeyer, J., Künsebeck, H.-W. and Lempa, W., 1982. Psychosomatischer Aspekt der chronisch-entzünd-lichen Darmerkrankungen. Paper at the 37th Meeting of the Dtsch.Ges. für Verdauungs- und Stoffwechselkrankheiten, Stuttgart, Sept. 1982

Grace, William J., 1953. Life Stress and Regional Enteritis. Gastro-enterology 23, 542-553

Green, F.H.Y. and Fox, H., 1975. The distribution of mucosal antibodies in the bowel of patients with Crohn's disease. GUT 16, 125

Green, D.R., Gold, J., St.Martin, S., Gershon, R. and Gershon, R.K.,1982. Microenvironmental immunoregulation: Possible role of contrasuppressor cells in maintaining immune responses in gut-associated lymphoid tissues. Proc.Nat.Acad.Sci. USA 79, 889

Luke, M., P. Kirkegaard, and Christiansen, J. 1982. Long Term Prognosis After Resektion For Ileocolic Crohn's Disease. Br. J. Surg. 69,429-438

Mohr, G.J., Josselyn, I.M., Spurlock, J. and Barron, S.H., 1958. Studies in Ulcerative Colitis in Children. Am. J. Psychiatry 114,1067

Palmblad, J., Cantell, K., Strander, H., Fröberg, J., Karlsson, C.G., Levi, L., Granström, M. and Unger, P., 1976. Stressor Exposure and Immunological Response in Man: Interferon-producing Capacity and Phagocytosis. J. of Psychosom. Res. 20, 193-199

Palmblad, J., Petrini, B., Wasserman, J. and Åkerstedt, T., 1979. Lymphocyte and Granulocyte Reactions during Sleep Deprivation. Psychsosom. Med. 41, 273-278

Paulley, I.W., 1971. Crohn's Disease. Psychother. Psychosom. 19,111-117

Raedler, A., Raedler, E., Seyfarth, K., Scholz, K.U. and Klose, G., 1983. Immune Regulation in Crohn's Disease. In preparation

Renoux, G., Bizière, K., Bardos, P., Degenne, D., Renoux, M., 1982. NK Activity in mice is controlled by the brain neocortex. In: Herberman, R.B. (ed.) NK Cells and other Natural Effector Cells. Academic Press, New York

Renoux, G., Bizière, K., Renoux, M. and Guillaumin, J.M., 1983. The Production of T-cell-inducing factors in mice is controlled by the brain neocortex. Scand. J. Immunol. 17, 45-50

Rosekrans, P.C.M., Meijer, C.J.L.M., van der Wal, A.M., Cornelisse, C.J. and Lindeman, J., 1980. Immunoglobulin containing cells in inflammatory bowel disease of the colon: A morphometric and immunohistochemical study. GUT 21, 941

Schultheis, K.H. and Uexküll, v. T., 1981. Psychosomatische Aspekte des Morbus Crohn. In: Uexküll, v.T. (Ed.) Lehrbuch der Psychosomatischen Medizin. 2. Aufl., Urban & Schwarzenberg, München, Wien, Baltimore

Thayer, W.R., Brown, M., Sangree, M.H., Katz, J. and Hersh, T., 1969. Escherichia coli O:14 and colon hemagglutinating antibodies in inflammatory bowel disease. Gastroenterology 57, 311

Truelove, S.C., 1971. Course and prognosis. In. Engel, A. and Larsson, T. (eds.) Regional Enteritis. Nordiska, Stockholm

Weiner, H. 1977. Psychobiology and Human Disease.Elsevier, New York, Oxford, Amsterdam, p. 499

Wijsenbeek, H., Maoz, B., Nitzan, I. and Gill, R. 1968. Ulcerative Colitis Psychiatric and psychological study of 22 patients. Neurol. Neurochir. 71, 409

UPPER GI BLEEDING LESIONS RELATED TO- OR ASSOCIATED WITH-STRESS

M. Deltenre, A. Burette and M. De Reuck
GI Department Univ. Hosp. Brugmann
Free University Brussels (ULB)
BELGIUM

DEFINITION, INCIDENCE AND NATURAL HISTORY

Since Curling (1842) reported, one century ago, the occurence of acute gastroduodenal hemorrhage in patients with severe burns, diffuse or focal acute ulcerations of digestive mucosae have been described in many different clinical circumstances: shock, renal or respiratory failure, physical trauma, post operative course (Hastings et al, 1978; Lucas, 1981, and Van den Berg et al, 1980). The association between various stress conditions and gastrointestinal ulcerative lesions may be defined as a syndrome, observed in critically ill patients and revealed by hemorrhage in 4 to 15% of the cases (Kamada et al, 1980 and Pruitt et al. 1970).

The precise incidence of stress gastrointestinal lesions remains unknown but one may consider that these lesions are virtually present in all patients in stress condition (Czaja et al, 1974; Anon. 1978; Lucas et al, 1971 and Mc Elwee et al, 1979) . The incidence of symptomatic, that's to say bleeding, lesions seems to be reduced for the last 10 years (Ritchie, 1981), perhaps because of preventive management and improvement of medical care (Mc Elwee et al, 1979; Gordon, 1980; Halloran et al, 1980; Lorenz et al, 1980; Mc Dougall et al, 1977; Pescina, 1981; Priebe and Skillman, 1981; Stothert et al, 1980 and Zinner et al, 1981). Acute GI bleeding occurs in 5 to 20% of patients admitted in intensive care unit (Van der Berg et al, 1980; Halloran et al, 1980; Cotton et al, 1973; Hubert et al, 1980 and Morgan et al, 1977). The risk of acute hemorrhage seems to be proportional to the severity and duration of stress.

In our unit, the proportion of stressed patients among the population of acute upper GI bleeding is slightly diminishing since 1977, from 28% to 22%. During this five years period, the number of admissions in intensive care units of our hospital and the number of upper GI bleeding cases remained unchanged.

As described by Lucas et al (1971), who performed repeated endoscopies in patients with polytrauma, first lesions may appear on the sixth hour: diffuse, well-circumscribed whitish spots are observed in gastric fundus. After 36 hours, lesions become red, and round, oval or linear erosions appear. Afterwards, trans mucosal ulcer, with bleeding may develop.

Theoretically, we may define 3 stages: erosion which is superficial and limited to the mucosa, the "exulceratio simplex of Dieulafoy" which is deeper and involves the sub mucosa with possible vascular effraction and finally, acute ulcer with muscular layer's involvement. This third type of so-called "acute" lesion may be sometimes difficult to distinguish from a chronic ulcer. That's the reason why it is frequently impossible to assume that such an ulcer, when observed in a stressed patient with acute hemorrhage, is unquestionably related to the stress condition. It may be a pre existing, chronic lesion.

The acute Dieulafoy bleeding ulcer may be recognized almost certainly as a stress-related acute lesion: no sloughs nor real ulcer is visible, only a continuous arterial spurting of blood through the mucosa. In other examples of erosions or exulcerations observed in stressed patients with hemorrhage, endoscopy alone cannot distinguish the lesion from ulcerations commonly observed in out patients complaining of mild epigastric pain or dyspepsia. Moreover, some bleeding ulcers, observed in stressed patients, seems to be, endoscopically, chronic ulcers.

In stressed patients, acute upper GI bleeding occurs usually without any prodrome. Endoscopy, despite a remarkable efficiency for determination of the bleeding source (Cotton et al, 1973) is unable to demonstrate unquestionably the direct relationship between the lesion and the stress situation, in many cases. That's why, all critically ill patients who developped upper GI bleeding were included in this study. So, the following report will deal with bleeding lesions related to or associated with stress.

RESULTS

202 cases from a consecutive serie of 710 acute upper GI bleedings were studied over a five-year period (1977-1981). Mean age (65y 5 m), sex ratio, frequency of chronic alcoholism or anti inflammatory drugs intake were not significantly different from the whole population of hemorrhages. A previous history of peptic disease or gastro-duodenal surgery was reported respectively in 18 and 3.5% of the patients. 60% of the patients had minimal hemoglobinemia under 10 g% and endoscopy was performed within 24 hours in all cases.

Various stress conditions are reported in table I.

TABLE I: Main stress conditions in stressed patients with upper GI bleeding (n = 202)

A.	Cardio-Respiratory failure :	19%
B.	Acute Renal failure :	16%
C.	Liver failure :	7%
D.	Toxi-Infection, Severe metabolic disorders:	41%
E.	Polytrauma :	5%
F.	Post-operative course	12%

(Multiple stress factors in 24% cases)

Multiple stress factors were present in 24%. The predominant problem was cardio respiratory failure in 19%, including acute respiratory distress with at least 48 hours of artificial ventilation or chronic respiratory failure with O_2 pressure under 60 mmHg and CO_2 pressure above

50 mmHg. Acute renal failure with creatininemia above 3 mg% was the main problem in 16% and liver failure with ascites, jaundice, encephalopathy in 7% of the cases. In 41%, the main stress factor was related to toxi-infection with shock or prolonged metabolic coma. Polytrauma was considered as the main stress condition in 5% and post operative course in 12%.

Three hundred and twenty two lesions were observed, among 197 patients reported in table II below.

TABLE II: Upper GI bleeding lesions in stressed patients (n = 320 lesions in 197 patients)

DIFFUSE EROSO-EXULC. LESIONS

Esophagus :	58	18%)	
Stomach :	97	30%)	57%
Duodenum :	29	9%)	

FOCAL ULCER(S)

Stomach :	48	15%)	
Duodenum:	51	16%)	31%

OTHERS · 12%

Multiple lesions in 48% cases, no diagnosis in 5 cases.

Lesions were multiple in almost half of the cases and in 5 patients, diagnosis of the bleeding source was not done. 57% of the lesions were diffuse erosions and exulcerations mainly observed in the stomach and 31% were focal whether chronic-like ulcer or Dieulafoy acute ulcer (around 1/3 of focal lesions).

Conservative management, including endoscopic coagulation, was or had to be applied in most cases (190/202). 12 patients underwent surgery. Whatever the treatment, the overall mortality is high: 40%, but bleeding itself could be considered as the direct cause of death in 7% of the cases. It must be stressed that 13 patients out of 14 who died from hemorrhage had a focal bleeding lesion.

There is no relationship between the type of stress and the respective proportion of focal and diffuse lesions as shown in table III.

TABLE III : Relationship type of stress-type of lesion.

	n	Diffuse Lesions	Focal ulcer(s)	p
Cardio Resp.	44	54%	42%	NS
Renal fail.	39	66%	28%	NS
Liver fail.	17	54%	39%	NS
Toxi-infect.	98	61%	32%	NS
Polytrauma	13	75%	25%	NS
Post operative	28	77%	18%	NS
Total		57%	31%	

PROGNOSIS FACTORS

Have the type of stress situation a significant predictive value about the outcome? Table IV indicates the prognosis. Total mortality is significantly lower in patients who developped acute bleeding in post operative course but they were usually younger and had no other stress factor. There is a trend for higher mortality in patients with cardio respiratory failure. If we look at the group of patients in whom bleeding was considered as the direct cause for death, there is, and it is logical, no significant difference.

TABLE IV : Type of stress and prognosis.

	Death from bleeding		Total Mortality	
Cardio Resp.	11%	NS	57%	p < 0.1
Renal fail.	6%	NS	36%	NS
Liver fail.	11%	NS	55%	NS
Toxi-infect.	4%	NS	46%	NS
Polytrauma	11%	NS	33%	NS
Post operative	4%	NS	26%	p < 0.05
Total	7%		40%	

As previously underlined, the type of lesion is a reliable prognosis factor (Table V). There is a significant difference in the mortality rate

TABLE V : Type of lesion and prognosis

	Death from bleeding	Total Mortality
Diffuse eroso-ulcerative lesions	1.5%	34%
	p < 0.01	p < 0.02
Focal Ulcer(s) + or - diffuse lesions	13%	50%

of patients with focal lesions versus patients with diffuse lesions. Bleeding from focal ulcer is usually more severe and this is confirmed when we consider the significantly higher mortality rate in the group of patients who died from hemorrhage.

Another significant point, in terms of prognosis, is the presence of multiple stress factors. Overall mortality is significantly higher when 2 or more stress factors are present (54% vs 37%, p 0.05).

As in many diseases, old age is a bad prognosis factor: the mean age of patients who died, in our serie, was 6 years higher than mean age of patients who healed (69 y. vs 63 y. 4 m.) but the number of stress factors seems to be the predominant point: mean age of cases with fatal outcome was 73 years 7 months in the group with one stress factor and 66 years 11 months in the group with 2 or more stress factors.

Finally, alcoholism, anti inflammatory drugs intake, or previous history of peptic ulcer has no prognosis value in our serie. As attempted, the mortality rate is proportional to the severity of shock induced by acute hemorrhage (25% deaths if no shock, 31% with reversible shock, 60% if recurrent shock).

To summarize, we could say that upper GI lesions responsible for bleeding in stress related (or associated) acute hemorrhage are diffuse ulcerative lesions in 60% and focal ulcer(s) in 1 case out of 3. Overall mortality is high: 40% but fatal outcome related to hemorrhage is observed in 7%. There is no relationship between the type of stress and the type of bleeding lesion. Bad prognosis conditions are: multiple stress factors, focal lesions(s) responsible for bleeding and as suspected, severe shock and old age. Patients with acute UGIB in the post operative course have a significantly better prognosis than patients with other stress conditions.

MULTIPOLAR ELECTROCOAGULATION

In order to improve the prognosis of stress-related or -associated bleeding lesions, the management of the most dangerous lesions (focal bleding ulcer, frequently responsible for severe shock) must be considered.

For almost two years, we used for endoscopic management of acute upper GI bleeding, the multipolar probe designed by Auth and Silverstein from the Seattle group (1980). 44 patients have been treated so far: 15 cases of esophageal varices, 13 so-called acute GD ulcers, and 16 chronic ulcers. Conditions for endoscopic therapy were: arterial or variceal active bleeding at the time of endoscopy, or rebleeding lesion under conservative management whether bleeding was artesial or venous. The probe is composed of six equally spaced longitudinal micro electrodes and allows tangential approach of the lesion, a water jet can be delivered through a hole in the center of the tip to wash the lesion before treatment. With this system, no ground-plate is needed and tissue injury is minimal since the electric field between the electrodes falls off rapidly.

The control module is a low-voltage source with maximal output of 25 watts. There is a water pump for washing. The system can easily be taken to the bedside and requires minimal maintenance. This dial with ten graduations allows to control the amount of delivered energy. The relation between the dial graduations and energy delivered is almost linear: 2.5 watts by graduation.

Definite hemostasis has been obtained in 35 cases out of 44 and the best results were observed in the acute ulcer group (acute "Dieulafoy" ulcer). Out of 13 patients with severe bleeding from acute ulcer, primary hemostasis was obtained in all cases and only one patient rebled within 1 week. 12 patients definitely healed. In case of so-called chronic ulcer, definite hemostasis was obtained in 12 cases out of 16.

Multipolar electrocoagulation seems to be an interesting approach for the management of acute bleeding ulcer of stressed patients.

CONCLUSION

It has been demonstrated that endoscopy, despite an excellent diagnosis performance, does not improve the prognosis of acute upper GI bleeding, in general (Eastwood, 1981). Peterson et al (1981) showed that endoscopy was of no benefit for patients whose hemorrhage ceases spontaneously. But, what about the others? As stated by Conn (1981), many sub groups exist where therapeutic endoscopy might be useful: stressed patients might be a population where endoscopic management would be of some benefit.

The prognosis of acute upper GI bleeding is conditioned by 3 factors, at least: the nature of the lesion (determined by endoscopy), the general condition of the patient (clinical assessment) and the nature of bleeding itself. To precise this last point beyond the phenomenom of "visible vessel", a Doppler probe adapted to fiberscopes is now tested (Deltenre et al, 1983). This system allows to recognize venous or arterial flow in the bottom of ulcerations. To progress, homogenous groups must be clearly defined according to these 3 criteria. The efficiency of various management techniques must be compared group by group, in a randomized way. To study a significant number of patients, multicentric, european trials would be undertaken.

REFERENCES

Anonymous (1978).
Editorial: Gastrointestinal bleeding in acute respiratory failure.
Brit. Med. J., I, 531.

Auth, D.C., Gilbert, D.A., Opie, E.A., Silverstein, F.E. 1980.
The multipolar probe-a new endoscopic technique to control gastrointestinal bleeding. (abstr.) Gastrointest. Endosc., 26, 63.

Van den Berg, B., Ong, G.L., Bruining, H.A. 1980.
Risk factors for upper gastrointestinal bleeding in critically ill
patients (abstr.) Intens. Care Med., 6, 54.

Conn, H.O. 1981.
To scope or not to scope. (Editorial) New Engl. J. Med., 304, 967-969.

Cotton, P.B., Rosenberg, M.T., Waldram, R.P.L., Axon, A.T.R. 1973.
Early endoscopy of oesophagus, stomach and duodenal bulb in patients
with hematemesis and melena. Brit. Med. J., 2, 505-509.

Curling, T.B. 1842.
On acute ulceration of the duodenum in cases of burn. Med.-chir. Trans.
25, 1.

Czaja, A.J., McAlhany, J.C., Pruitt, B.A. 1974.
Acute gastroduodenal disease after thermal injury. New Engl. J. Med.,
291, 925-929.

Deltenre, M., De Reuck, M., Silverstein, F.E., Martin, R.W., Burette, A.,
Gilbert, D.A. 1983.
Système Doppler adapté à l'endoscopie digestive: Etude pilote. Acta
Gastro-ent. Belg., (in press).

Eastwood, G.L. 1981.
Does the patient with upper gastrointestinal bleeding benefit from
endoscopy ? Dig. Dis. Sc., 26, (suppl.), 22s-26s.

Gordon, D.L. 1980.
Prophylaxis against acute stress erosions. Sth. Med. J., 73, 424-426.

Halloran, L.G., Zlass, A.M., Gayle, W.E., Wheeler, C.B., Miller, J.D. 1980.
Prevention of acute gastrointestinal complications after severe head
injury: a controlled trial of cimetidine prophylaxis. Amer. J. Surg.,
139, 44-48.

Hastings, P.R., Skillman, J.J., Bushnell, L.S., Silen, W. 1978.
Antiacid titration in the prevention of acute gastrointestinal bleeding.
New Engl. J. Med., 298, 1041-1045.

Hubert, J.P., Kiernan, P.D., Welch, J.S. Remine, W.H., Beahrs, O.H. 1980.
    The surgical management of bleeding stress ulcers. Ann. Surg., 191,
    672-679.

Kamada, T., Fujamoto, H., Kawano, S., Noguchi, M., Hiramatsu, K., Masuzawa,
    M., Sato, N. 1977.
    Acute gastroduodenal lesions in head injury. Amer. J. Gastroenterol.,
    68, 249-253.

Lorenz, W., Fischer, M., Rohde, M., Troidl, H., Reimann, H.J., Ohmann, C.
    1980.
    Histamine and stress ulcer: new components in organizing a sequential
    trial on Cimetidine prophylaxis in seriously ill patients and defi-
    nition of a special group at risk. Klin. Wschr., 58, 653-665.

Lucas, C.E., Sugawa, C., Riddle, J., Rector, F., Rosenberg, B., Walt, A.J.
    1971.
    Natural history and surgical dilemma of "stress" gastric bleeding.
    Arch. Surg., 102, 266-273.

Lucas, C.E. 1981.
    Stress ulceration: the clinical problem. Wld J. Surg., 5, 139-149.

McDougall, B.R.D., Bailey, R.J., Williams, R. 1977.
    H2-receptor antagonist and antiacids in the prevention of acute gastro-
    intestinal hemorrhage in fulminant hepatic failure. Lancet, I, 617-619.

McElwee, H.P., Sirinek, K.R., Levine, B.A. 1979.
    Cimetidine affords protection equal to antacid in prevention of stress
    ulceration following thermal injury. Surgery, 86, 620-626.

Morgan, A.G., McAdam, W.A.F., Walmsley, G.L., Jessop, A., Horrocks, J.C.,
    De Dombal, F.T. 1977.
    Clinical findings, early endoscopy and multivariate analysis in
    patients bleeding from the upper gastrointestinal tract. Brit. Med. J.,
    II, 237-240.

Pescina, J. 1981.
    Use of Cimetidine in the prevention of stress ulcers in intensive care
    units. Clin. Therap., 3, 453-455.

Peterson, W.L., Barnett, C.C., Smith, H.J., Allen, M.H., Corbett, D.B.1981.
    Routine early endoscopy in upper-gastrointestinal-tract bleeding. A
    randomized, controlled trial. New Engl. J. Med., 304, 925-929.

Priebe, H.J. and Skillman, J.J. 1981.
    Methods of prophylaxis in stress ulcer disease. Wld. J. Surg., 5, 223-
    233.

Pruitt, B.A., Foley, E.D., Moncrief, J.A. 1970.
    Curling's ulcer: a clinico-pathological study of 323 cases. Ann. Surg.,
    172, 523-536.

Ritchie, W.P. jr. 1981.
   Stress ulcer and erosive gastritis. Wld J. Surg., 5, 135–137.

Stothert, J.C., Simonowitz, D.A., Dellinger, E.P., Farley, M., Edwards,
   W.A., Blair, A.D., Cutler, R., Carrico, J. 1980.
   Randomized prospective evaluation of Cimetidine and antacid control of
   gastric pH in the critically ill. Ann. Surg., 192, 169–174.

Zinner, M.J., Zuidema, G.D., Smith, P.L., Mignosa, M. 1981.
   The prevention of upper gastrointestinal tract bleeding in patients in
   an intensive care unit. Surg. Gyn. Obst., 153, 214–220.

PART 5

Acute effect of psychological stress on the cardiovascular system: models and clinical assessment

edited by

A. L'Abbate

NEED FOR CLINICAL MODELS: PHYSIOLOGICAL VERSUS EPIDEMIOLO-GICAL STUDY

L. DONATO

C.N.R. Clinical Physiology Institute and Institute of Medical Pathology, University of Pisa, Pisa, Italy.

Circumstantial evidence on the role of psychological stress in cardiovascular disorders is part of the daily experience of every cardiologist. This evidence concerns at least three of the main areas of cardiovascular diseases, namely arterial hypertension, cardiac arrhythmias and ischemic heart disease.

The physiological frame within which the above evidence should be considered was originated by the pioneering studies of Cannon (1) on the behavioural response of cats to barking dogs (the "fight or flight" reaction), by those of Selye (2, 3) on the General Adaptation Syndrome to different noxious agents, by the studies of Moruzzi and Magoun (4) on the reticular formation and the EEG arousal induced by its stimulation, by those of Lacey (5) on the autonomic component of the arousal response, and many others who contributed to clarify the neuro-endocrine mechanisms involved and to develop techniques for their assays.

Sixty years after Cannon, we still have to resort to their model do deal with the problem of psychological stress and of the resulting somatic changes. Figs. 1 and 2 (which should be read in sequence) are an attempt to summarize existing knowledge and reasonable hypotheses.

834

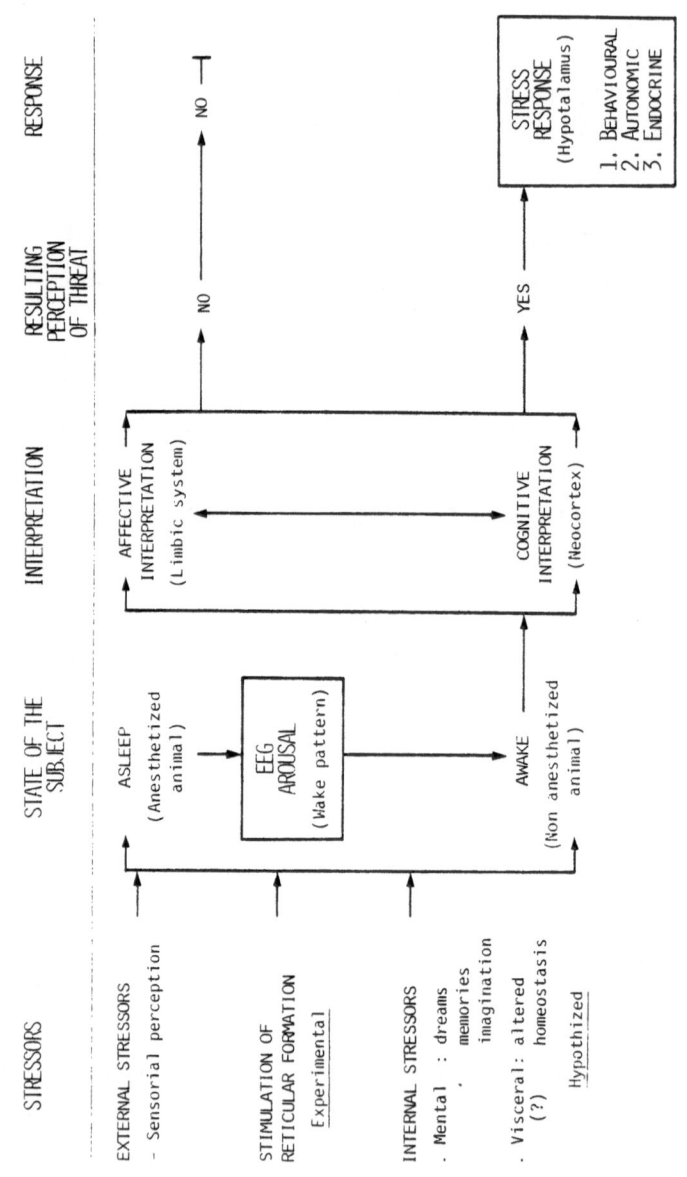

FIG. 1 - STRESS RESPONSE IN RELATION TO TYPE OF STRESSOR, STATE OF THE SUBJECT, AFFECTIVE AND COGNITIVE INTERPRETATION AND PERCEPTION OF THREAT.

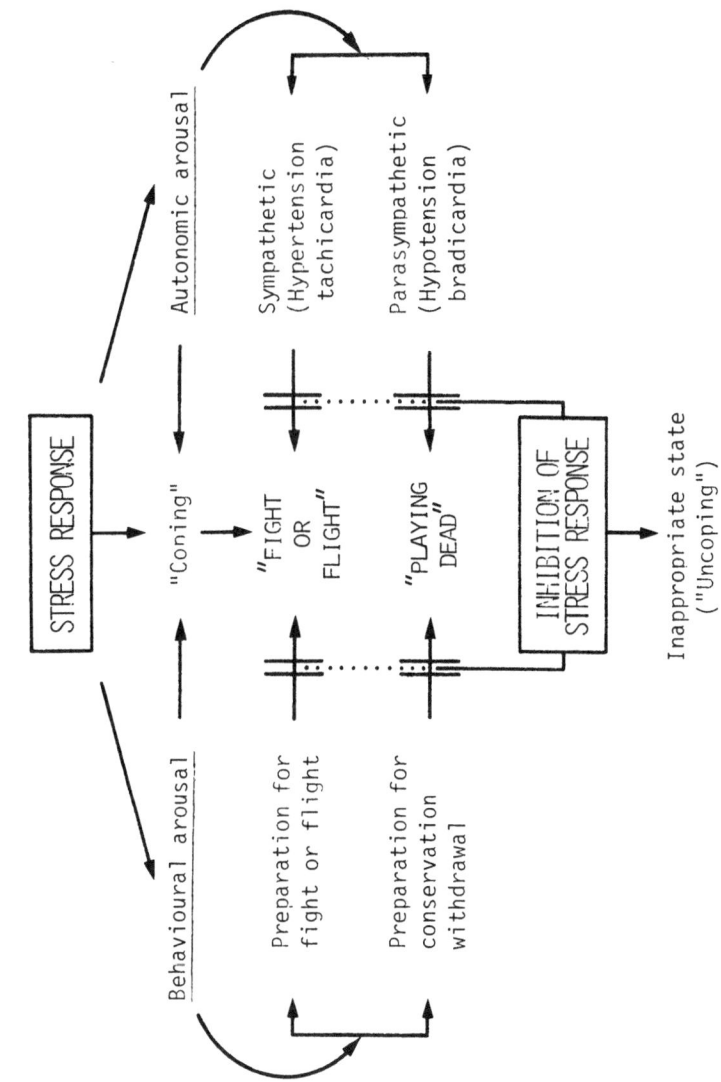

FIG. 2 – BEHAVIOURAL AND AUTONOMIC COMPONENTS OF STRESS RESPONSE (ENDOCRINE IS NOT SHOWN)

Progressive increase in prevalence of cardiovascular disorders has been related to the increased confrontation of the human individual with an increasingly difficult environment full of "barking dogs", and consequently to greater need for adaptive changes and more chances for inappropriate response and adaptive breakdown. The socially integrated modern man, can seldom respond with fight, flight or "playing dead" in front of his everyday stressors.

It is mostly on the basis of such circumstantial considerations and of limited epidemiological studies that the concept of cardiovascular disorders as life style related diseases has evolved and progressively gained credit. However, as to practical outcome of this trend, we must recognize that so far it has produced little specific clues and effective guidance to better prevention, treatment or rehabilitation of patients. I do not wish to deny that regular physical activity and "jogging" might be adequate substitute for "fight and flight" in our daily bereavement; however, efforts to transfer neurophysiological and endocrinological knowledge into an active approach for better understanding the mechanisms of cardiovascular disorders have been very limited in number and success, and even more so in terms of practical implications. This is even more surprising considering the large availability of drugs capable of affecting the various components of the arousal phenomena.

The pattern is different, as it will certainly come out from this meeting, in the areas of arterial hypertension, cardiac arrhythmias and ischemic heart disease.

Arterial hypertension. In the case of arterial hypertension, while attempts to demonstrate the occurrence of
purely psychogenic disease have not provided conclusive
evidence, social conflict has been shown capable of playing
a pathogenic role in animals with genetic predisposition to
hypertension (6); moreover some kind of stress, such as
sound-withdrawal, may lead to self-sustained hypertension
in rats (7).

In humans Wheathley (8) has produced evidence that
approssimately 90% of the hypertensive patients exhibit
anxiety symptoms of varying degrees, and that relief of
anxiety accompanies reduction in blood pressure: however,
it is not possible at present to determine whether this is
cause, effect or coincidence.

In the area of arterial hypertension increasing ability
to pick up the involved pathophysiological mechanisms and
availability of effective drugs not only have drastically
reduced mortality, but have also reduced the enphasis on
the role of psychogenic stress. In fact most of the
preventive effort is now addressed to identifying subjects
at risk, and markers for predispositions to hypertension or
to severe organ damage in hypertension. If we were able to
sort out these subjects it would indeed be of importance to
minimize their exposure to psychosocial stress. For
instance, we could define the most appropriate job profile
for the predisposed individual if we had practical and
objective ways to measure individual specificities in
response to different stressors. Such a clinical model
would probably be of great value also in understanding the
role of psychogenic stress in this disease.

Cardiac arrhythmias. The area of cardiac arrhythmias
seems to be one in which increased knowledge of the in-

volved neural and electrophysiological mechanisms, combined
with increased ability to study ecg pattern in ambulant
conditions by Holter monitoring, has produced marked
progress in the approach to the condition.

Beta-blockers administration to normal individual prior
to performance of stressful tasks has been advocated as a
preventive measure (9). We feel that such an approach
should not be implemented in an indiscriminate way, without
a better knowledge of the individual physiological state.

The same applies to indiscriminate beta-blockade to
reduce sudden death after myocardial infarction. Sudden
death may follow emotional stress both via excess
sympathetic or parasympatethic activation. In extrapolating
to the individual patient the results of population studies
it is worth stressing that reduced mortality in the overall
population does not rule out the possibility (due to pathophysiological inhomogeneities in the population) of a
combined effect of a true reduction (beyond the overall
population average) in a susceptible subset, with an actual
increase in mortality in a different subset of the
population (10).

This is an area in which an adequate comparative trial
on the basis of a well defined clinical and clinical physiological protocol may be of great value.

Ischemic heart disease. The most complex case is indeed
that of ischemic heart disease, and it is the one in which
the severity of the disease, the multiplicity of the
syndromes, the pathogenetic uncertainties, and the therapeutic limitations have encouraged formulation of theories and
sometimes of simplistic conclusions, as to the role of both
somatic and psychogenic factors.

In 1959 Friedman and Rosenmann (11) suggested the association of the so called type-A personality with higher prevalence of atherosclerosis and coronary disease. In 1969 Carruthers (12) put forward a hypothesis linking aggressive behaviour and atheroma, that was recently updated (13) taking into account biochemical correlated of emotion as distinct from those of anxiety. However, attractive as they may be from the speculative point of view, most of these schematizations have difficulties in coping with the highly diversified universe of ischemic heart disease, and with the complexity of pathogenetic mechanisms (14). In the approach to this problem the following points should be given full consideration:

- clinically homogeneous patients and syndromes are not necessarily homogeneous in their pathogenetic mechanisms nor in the state of their coronary vessels;

- only a small fraction of patients with ischemic heart disease appear to respond to emotional stressors;

- responders to emotional stressors often present ischemic episodes under circumstances in which an emotional conponent cannot be identified;

- spontaneous vasospastic angina, which under many respect could be the natural candidate for "emotionally induced angina" does not respond to cathecolamines stimulation (15), and is less sensitive to beta-blockers treatment than secondary or "effort angina" (16).

On the other hand preliminary observations by Biagini et al. (17) in our Institute suggest that sleep deprivation, a measure which has been applied successfully to the relief of depression, reduces the frequency of ischemic episodes in spontaneous angina, thus confirming the importance of psychogenic mechanisms.

This critical condition, certainly multifactorial in its pathogenic mechanisms, with periods of spontaneous exacerbation and amelioration, should be approached on the basis of an adequate selection of patients, preferably with documented vasospastic angina and a specific and well defined protocol including evaluation of the coronary, myocardial, neural and endocrine response to a set of acute physical, pharmacological and emotional stressors.

As working hypotheses, one could test the possibility that autonomic arousal evoked by emotional stressors acts as a modulator for the patency response of the vessel to local factor. It might also be of interest to test whether the arousal phenomenon could be evoked not only by external or endogenous mental stressors, but also by visceral stimulation: may regional ischemia, or contractile changes in the myocardium through appropriate afferent signals to the reticular formation (known to be accessible to both visceral and somatic "impression", 18) evoke the arousal phenomenon in its various components?

In conclusion, what is needed is to take inspiration from daily clinical experience and existing epidemiological evidence in the attempt to bridge the gap between neuro-endocrine physiology and cardiology on one hand, epidemiology and clinical science on the other. Advances in methodology, possibility of assaying hormones and neurotransmitters, of continuous recording resting and evoked EEG potentials in clinical conditions, combined with the availability of adequate methods to assess and follow the state of the coronary circulation and of the myocardium should now make possible this objective approach.

REFERENCES

1) Cannon WB. 1929. Bodily changes in pain, hunger, fear and rage, 2nd ed., Appleton, New York.

2) Selye H. 1936. A syndrome produced by diverse nocuons agents. Nature (Lond.), 138, 32.

3) Selye H. 1976. Forty years of stress research: the principal remaining problems and misconceptions. Can. Med. Ass. J., 115, 53.

4) Moruzzi G, and Magoun HW. 1969.: EEG. Clinical Neurophysiol., 1, 455.

5) Lacey JI. 1967. In "Psychological Stress" (Eds. Appley M.H., and Trumbull R.). 14 Appleton-Century- Grafts, New York).

6) Friedman R, and Dahl LK. 1975. The effect of chronic conflict on the blood pressure of rats with a genetic susceptibility to experimental hypertension. Psychosom. Med., 37, 402.

7) Marwood JF, and Lockett MF. 1981. Stress induced hypertension in rats. In "Stress and the Heart" (Ed. D. Wheathley), Raven Press, New York, p. 229.

8) Wheathley D. 1981. Anxiety and hypertension. In "Stress and the Heart" (Ed. D. Wheatley) Raven Press, New York, p. 273.

9) Griffith D, Pearson R, and James IM. 1981. Stressful situations. In "Stress and the Heart" (Ed. D. Wheathley) Raven Press, New York, p. 65.

10) Donato L. 1983. Technology assessment in cardiology. In "Frontiers in Cardiology for the 80s" (Eds. L. Donato and A. L'Abbate). Academic Press Inc. (London), Ltd, p. 321.

11) Friedman M, and Rosenman RH. 1959. Association of a specific overt behaviour pattern with increases in blood cholesterol, blood clotting time, incidence of arcus senilis and clinical coronary artery disease, J.A.M.A., 169, 1286.

12) Carruthers MA. 1969. Aggression and atheroma. Lancet, 11, 170.

13) Taggart P, and Carruthers M. 1981. Behaviour patterns and emotional stress in the etiology of coronary artery disease: cardiological and biochemical correlates. In. "Stress and the Heart" (Ed. D. Wheathley) Raven press, New York, p. 25.

14) Maseri A, Chierchia S, and L'Abbate A. 1980. Pathogenetic mechanisms underlying the clinical events associated with atherosclerotic heart disease. Circulation, 62 (Suppl. V): V-3.

15) Chierchia S, Crea F, Davies GJ, Berkenboom G, Crean P, and Maseri A. 1982. Coronary spasm: any role for alpha receptors? Circulation, 66 (Suppl. II) 247 (Abstract).

16) Parodi O, Simonetti I, L'Abbate A, and Maseri A. 1982. Verapamil versus propranolol for angina at rest. Am. J. Cardiol., 50, 923.

17) Biagini A, Emdin M, L'Abbate A, Guazzelli M, Maggini P, and Donato L. 1983. Sleep deprivation in spontaneous angina. Unpublished observations.

18) Papez JW. 1937. Arch. Neurol. Psychiatr. Chicago, 38, 725.

PSYCHOSOCIAL STRESS: ENDOCRINE AND BRAIN INTERACTIONS AND THEIR RELEVANCE
FOR CARDIOVASCULAR PROCESSES

B. BOHUS and J.M. KOOLHAAS
Department of Animal Physiology, State University of Groningen,
P.O.Box 14, 9750 AA HAREN (The Netherlands)

ABSTRACT

Organization of endocrine and cardiovascular responses to psychosocial stimuli occurs at four levels: in the limbic-midbrain system, the
hypothalamus, the pituitary gland and in the target organs (including
besides peripheral organs the brain). Alternations at all four levels may
be the cause of maladaptive response to environmental events. Experiments
with various rat models exposed to psychosocial stimuli suggest the importance of endocrine mechanism in the central organization (or desorga-
nization) of the acute cardiovascular responses.

INTRODUCTION

In order to respond adequately and thereby to adapt to the physical or

psychosocial environment requires a chain of behavioural, autonomic, endo-

crine and metabolic responses. These homeostatic responses may be con-

sidered in the tradition of Cannon (1915) and Selye (1936) as essential

physiological "stress" responses which are ultimately integrated by complex

brain mechanisms. In the recent years an increasing interest has been fo-

cussed upon mechanisms that assure homeostasis when the organism is exposed

to adverse psychological or social stimuli. It appears that these

stimuli are among the strongest to activate the release of pituitary and

adrenal hormones (e.g.,Mason,1968;Levine et al.,1972). That aversive psycho-

logical stressors affect the function of the cardiovascular system is also

well documented (e.g.,Cohen and Obrist., 1975).

The notion that psychosocially aversive events may be important in

the etiology of cardiovascular diseases is a long standing one but despite

of supporting psychological, physiological and demographic findings (Henry

and Cassel, 1969; Folkow and Neil, 1971; Cohen and Obrist, 1975), the sig-

nificance of stress for the development of cardiovascular diseases is

still controversial (Galosy and Gaebelin, 1977). The knowledge on the me-

chanisms of acute cardiovascular pathology that may develop as the consequence of aversive psychosocial stimuli is also far from being complete.

This paper describes the proposed organization of endocrine and cardiovascular responses to stressful stimuli. Subsequently, a few models to investigate the organization of the stress-response are presented. Finally, a short review of our work on the influences of neuropeptides and steroid hormones on autonomic responses related to psychological stress is described. These findings suggest that endocrine-brain interaction may contribute to the organization or derangement of cardiovascular responses to stress.

ORGANIZATION OF ENDOCRINE AND CARDIOVASCULAR RESPONSES TO ENVIRONMENTAL STRESSFUL STIMULI

While a wide range of responses is elicited by diverse stressful conditions in animal and man, surprisingly little is known about the precise organization of these stress responses. That the primary knowledge on this subject stems from studies on the pituitary-adrenocortical system is rather obvious because of the classical coupling between stress and the adrenal cortex. The overview as presented here, presupposes a comparable organization of the various endocrine and physiological responses at four levels.

The limbic-midbrain system in conjunction with certain cerebral cortical areas can be considered as the first level of organization. This system is composed of limbic areas such as the amygdaloid complex, the septal area and the hippocampus on one hand, and the ascending reticular activating system including the ascending projections of the aminergic nuclei in the brainstem on the other hand. This system integrates sensory informations (input from the environment or eventually imagination of certain psychosocial environment in man), and visceral and endocrine signals (inputs from the milieu interieur). These structures play a role in the organization of primary adaptive responses such as learning, memory, motivation, etc., and the "emotional" background of the behaviour (see Isaacson, 1974; Routtenberg, 1972; Gray, 1982). Neuroendocrine studies suggested that the limbic system may modulate the response to stress (both inhibition, particularly by hippocampus and septum, and activation, e.g. amygdala). The reticular activating system is primarily activating while the ascending serotonergic system seems to be of inhibitory nature (see Bohus, 1975a; Smelik and Vermes, 1980).

The involvement of limbic system in cardiovascular regulation has been demonstrated by a few studies (e.g. Holdstock, 1969; Gebber and Klevans, 1972). The involvement of the mesencephalic reticular activating system in the induction of an acute, neurogenic hypertensive response has been shown in both anesthetized and non-anesthetized animals (Sharpless and Rothballer, 1961; Bohus, 1974b, Bohus et al., 1983a; Kawasaki et al., 1980). Observations with surgical transections in the brainstem have suggested that a part of the pressor response evoked by electrical stimulation of the mesencephalic reticular formation is mediated through higher, probably limbic structures (Bohus et al., 1983a). Cardiovascular responses can be induced by the stimulation of the cell body areas of the ascending serotoninergic system in the medullary raphe area (Adair et al., 1977) and of the dorsal noradrenergic bundle system in the locus coeruleus (Kawamura et al., 1978).

The hypothalamus can be regarded as the second level of organization. Extensive anatomical connections with both limbic and brainstem structures provide the morphological basis of the communication between the first and second level of organization. The function of the hypothalamus in the organization of endocrine responses has more than one aspect. First of all, the neural informations provided by the limbic-midbrain system in the form of neurotransmission is "translated"into hormonal messengers: the various neurotransmitters activate or inhibit the synthesis and/or release of releasing or inhibiting factors/hormones such as the corticotrophin releasing factor (CRF; see Smelik and Vermes, 1980). The releasing/inhibiting factors on one hand reach the pituitary gland through the portal circulation thereby regulating the release of stress-related hormones. On the other hand, releasing hormones may have effects that are not mediated through the pituitary gland. Behavioural actions of TRH and LHRH have already been demonstrated (see Witter and De Wied, 1980). Recently it has been shown that CRF may cause behavioural effects that mimic the action of "stressful" stimuli (Sutton et al., 1982). A further endocrine function of the hypothalamus is the synthesis of neurosecretory hormones that are transported to the posterior pituitary.

A novel aspect of the endocrine function of the hypothalamus is that it contains the cell bodies of extensive peptidergic neuronal system that terminate in the limbic-midbrain structures. Vasopressinergic and oxytocinergic cell bodies are located in the supraoptic and paraventricular

nuclei while the suprachiasmatic nuclei contain vasopressin only. The arcuate nucleus contains the cell bodies of the opiomelanocortin system
(see Watson and Akil, 1980; Swaab, 1980). How stressful stimuli affect
the central release of these brain-born hormones is not known yet. Rossier
et al. (1977) have reported a depletion of hypothalamic β-endorphin following stressful stimulus of foot shock suggesting a release of stored
peptide.

The role of the hypothalamus in the organization of acute cardiovascular responses that resemble stress responses of fight or flight is well
established (e.g. Hilton, 1975). Posterior hypothalamic stimulations also
activate the release of adrenomedullary catecholamines which have been
considered for a long time as important stress hormones (Eferakeya and
Bunag, 1974). The release of catecholamines from the adrenal medulla
following hypothalamic manipulation may also contribute to cardiovascular
pathology. Nathan and Reis (1975) reported that fulminating arterial hypertension with pulmonary edema after damaging the anterior hypothalamus is
due to the release of adrenomedullary catecholamines.

The third level of organization is in the pituitary gland from where
the stress-hormones for the peripheral endocrine glands and other hormone-
sensitive organs are released. One may distinguish between first and
second generation of stress hormones. The first generation is represented
by the classically known stress hormones such as ACTH, corticosteroids and
the adrenomedullary catecholamines. The second generation consists of pituitary hormones such as prolactin and vasopressin. These hormones are
released by psychosocial stimulation into the bloodstream (Brown and
Martin, 1974; Thompson and De Wied, 1973). Their function as stress hormones has been recognized recently (Bohus, 1983). In addition, a new group
of peptide hormones has been discovered - i.e. endorphins and enkephalins.
It appears that ACTH and endorphins and also the rather recently discovered
γ-MSH originate from the common precursor molecule opiomelanocortin. Stimuli
that release ACTH from the anterior pituitary also lead to the release of
β-LPH and γ-MSH and a lesser degree to that of β-endorphin. The secretory
activity of the intermediate lobe of the pituitary seems to be regulated
by separate mechanisms. Besides β-endorphin, α-MSH is released among others
following stressor stimuli (see O'Donohue and Dorsa, 1982). Enkephalins,
the other group of opiate-like peptides originate from precursors that
are different from opiomelanocortin. Enkephalins may be released from the

pituitary gland, but the adrenal medulla is also a likely source of these peptides.

The fourth level of organization is at the level of the target sites. The peripheral organs or organ systems such as the cardiovascular, immune and neuromuscular systems, the liver, the kidney, etc. (Target Organs I) have been long considered as the only targets of stress hormones. The recognition that the brain is a major target organ (Target Organ II) for these hormones has opened new vistas in understanding the adaptive funct- ions of the neuroendocrine systems. Stress hormones, particularly neuro- peptides of second generation, affect physiological functions of the brain thereby basic behavioural processes such as learning, memory, moti- vation etc. (see Bohus, 1981). It was suggested that dysfunctions in the central effects of neuropeptides may be of etiological importance in adap- tive diseases (De Wied, 1979).

Considering the brain as a major target of the stress hormones closes the circuit of the brain-neuroendocrine interactions. Studies involving the stress hormones of both first and second generation suggest that the limbic-midbrain system may be considered as their major site of action. For example, the hippocampus contains specific receptors through which corticosteroids may affect adaptive brain functions (Bohus et al., 1982a). The hippocampus, the amygdala, the dorsal septum, the dorsal raphe nucleus in the brainstem have been considered as brain sites through which vaso- pressin may affect such adaptive processes as memory (Bohus et al., 1982b). Considering the organizational function of the limbic-midbrain system (first level) in physiological stress responses such as the cardiovascular changes, neuroendocrine influences on these physiological responses can be assumed. Before discussing this aspect of neuroendocrine-brain interactions, the question of suitable behavioural models would be discussed in short.

PSYCHOSOCIAL STRESS, BEHAVIOUR, NEUROENDOCRINE AND CARDIOVASCULAR RESPONSES

Many attempts have been made during the last two decades to develop suitable animal models to study the consequence of stressful events in acute and chronic cardiovascular pathology (Galosy and Gaebelein, 1977). Terms like fear, anxiety, rage, disappointment on one hand, uncertainty, conflict, controllability, anticipation, helplessness on the other hand, were usually used to describe the chain of events and circumstances under which

sustained endocrine responses to stress could be observed and the likely-
hood of pathological alterations in the cardiovascular system occurred.
Practically in all these studies electric shock was employed as punishment with or without avoidance possibilities. These techniques have a
couple of drawbacks. First of all, it is very difficult if not impossible
to separate the direct consequences of electric shock (painful stimulus)
from that of the psychological attributes of punishment. Secondly, somatic
responses that accompany escape and/or avoidance may require different
circulatory states than the psychic ones (Obrist et al., 1976).

In order to avoid these problems, cardiac changes of the rat were investigated in a passive avoidance paradigm (Ader et al., 1970) where the
stressful effect of punishment is well separated from that of the fear of
punishment. Briefly, passive (inhibitory) avoidance behaviour (avoiding
to enter a preferred dark compartment from an extensively lit elevated
platform) and accompanying cardiac changes were studied 24 h after experiencing the punishment. Measurement of cardiac rhythm by recording electrocardiogram was performed by means of radiotelemtry in the free-moving
rat and determining the duration of R-R intervals (interbeat intervals)
with the aid of a computer (Bohus, 1974a). It appeared that inhibitory
avoidance behaviour is accompanied by tonic and phasic changes in cardiac
rhythm. Tonic changes were represented by a slow (bradycardiac) heart rate
during the entire avoidance period. Phasic changes occurred when the rat
approached or partially entered the dark environment where punishment had
been received 24 h earlier. The phasic changes consisted of abrupt decrease
in heart rate and occasionally arrhytmia was observed. The magnitude of
the tonic bradycardia was correlated with the intensity of punishment,
while such a correlation was absent for the phasic response (Bohus, 1977).
The stressful character of the situation was also indicated by an increase
in plasma corticosterone level which reached its maximum 15 min after the
onset of the test (Bohus, 1975b).

In some experiments the rats were forcedly exposed to the former shock-
compartment by placing them directly into the dark. Heart rate decreased
markedly also in this paradigm (De Loos et al., 1979). Elevation of plasma
corticosterone level reached its maximum 5 min after the onset of forced
exposure indicating strong stressful character of the situation (Bohus,
1975b).

Another animal model which seems to provide an excellent mean to study

the influence of acute or eventually chronic stress on the cardiovascular
system has been introduced by Koolhaas et al. (1983) in this department.
The model combines ethological analysis of the behaviour of the rat in
seminatural social conditions and the direct, continual measurement of
blood pressure as the index of cardiavascular responsiveness to psychosocial stress situations. The pattern of plasma corticosterone levels
increases in the victors and loosers of the social interactions and
the increase of blood pressure suggests that the model could be differentially used to investigate the psychophysiology of hypertension.

NEUROENDOCRINE INFLUENCES ON CARDIAC RESPONSES DURING EMOTIONAL BEHAVIOUR

Neuropeptides that markedly affect adaptive behavioural responses
but also oestrogens, appeared to influence the cardiac response to emotional stressor of free passive avoidance or forced exposure procedure
of free-moving rats (Bohus et al., 1983b). Briefly, ACTH 4-10, a behaviour-
ally active fragment of the stress hormone ACTH/α-MSH induces a marked
tachycardia in the avoidance situation. It was suggested that increased
heart rate of ACTH 4-10 treated rats was probably due to an increased
sympathetic influence on heart rate that is normally minimal during these
kinds of emotional behaviour. That neonatal chemical sympathectomy
prevented the appearance of tachycardia in ACTH 4-10 treated rats supported
this view. Sympathetic influences are evoked by more intense stressors in
which the organism is actively engaged in the preparation or execution of
activities that will cope with stress (Obrist et al., 1970). Tachycardia
in ACTH 4-10 treated rats may occur because the peptide signals intense
stress to the appropriate brain centers. This signalization may result
in a facilitated arousal state that increases the probability of appropriate coping behaviour (Bohus et al., 1983b). This hypothesis seems to
be valid for man as well (see Brunia and Van Boxtel, 1978; Branconnier
et al., 1979; Breier et al., 1979).

Vasopressin or fragments of this peptide enhanced the magnitude of
the bradycardiac response of the rat during a facilitated avoidance behaviour. Hereditary diabetes insipidus due to the genetic absence of vasopressin was accompanied by behavioural deficit and the absence of the
phasic but not of the tonic bradycardiac response during passive avoidance
behaviour (Bohus, 1977).

- Another aspect of the role of vasopressin and related peptides in central cardiovascular regulation has been observed in experiments where electrical stimulation of the mesencephalic reticular formation of urethane-anaesthetized rats was used as a model to induce acute neurogenic hypertensive (pressor) responses. It appeared that vasopressin and related peptides administered intracerebrally diminish the magnitude of the pressor responses. It was considered that neurohypophyseal hormones acting in the brain may serve as a protective mechanism against the overresponse of the cardiovascular system to stressful stimuli (Versteeg et al., 1982; Bohus et al., 1983a).

Finally, female sex hormones may also serve a similar protective function. In agreement with former observations by Von Eiff et al. (1971) it was found that bradycardiac response to emotional stress (forced exposure paradigm) is absent in female rats during oestrus while males and females show comparable reactions when the latter are in dioestrus. Administration of oestradiol to ovariectomized female rats mimicked the influence of natural oestrus on the cardiac response (De Loos et al., 1979).

Taken together, neuroendocrine mechanisms may profoundly alter the organization of acute cardiovascular responses to psychological stressor stimuli. This modulatory actions seem to be of physiological significance in integrated behavioural, autonomic and endocrine coping mechanisms.

CONCLUDING REMARKS

The complex organization of the physiological stress responses suggests that whenever the function in one of the four levels is disturbed malfunctioning in the environment-body interaction may occur. Recent observations suggest the importance of endocrine function in the organization of various adaptive responses. Accordingly, dysfunctions in the neuroendocrine systems, either hyper- or hypoactivity, may cause desintegration of adaptive brain mechanisms and thereby may contribute to autonomic pathology. The cause of the neuroendocrine dysbalance may be hereditary (e.g. hereditary hypothalamic diabetes insipidus) or acquired. Alterations in the endocrine function as the consequence of aging should also be considered. The availability of proper models to investigate the impact of psychosocial stimuli upon the brain, behaviour and physiology allows us to explore further the mechanism of organization at the four levels. The use of rats with hereditary autonomic disturbances - e.g. the various strains of the spontaneously hypertensive rats, etc. - further widens the possibilities to get more knowledge on the pathomechanisms of autonomic disorders in relation to psychosocial stressors. Accordingly, an integrated view in which recent environment, genetic, developmental and aging as possible contributing factors are considered may help to find proper animal models to predict the risk of certain individuals for a general breakdown of adaptation.

ACKNOWLEDGEMENTS

Part of the authors studies that are reported here has been supported by the Saal van Zwanenberg Stichting, the Stichting Farmacologisch Studie-fonds and the Nederlandse Hartstichting. The secreterial aid by Mrs. Joke Poelstra-Hiddinga is greatly acknowledged.

REFERENCES

Adair JR, Hamilton BL, Scappaticci KA, Helke DJ, Gillis RA. 1977. Cardio-vascular responses to electrical stimulation of the medullary raphe area of the cat. Brain Res. 128: 141-145.

Ader R, Weijnen JAWM, Moleman P. 1972. Retention of a passive avoidance response as a function of the intensity and duration of electric shock. Psychon.Sci. 26: 125-128.

Bohus B. 1974a. Telemetered heart rate responses of the rat during free and learning behavior. Biotelemetry 1: 193-201.

Bohus B. 1974b. The influence of pituitary peptides on brain centers controlling autonomic responses. In: Integrative Hypothalamic Activity. Eds. D.F.Swaab and J.P.Schadé. Progr.Brain Res. vol. 41, Elsevier Amsterdam. pp. 175-183.

Bohus B. 1975a. The hippocampus and the pituitary-adrenal system hormones. In: The Hippocampus, vol. 1: structure and development. Eds. R.L.Isaacson and K.H.Pribram. Plenum Press, New York and London. pp. 323-353.

Bohus B. 1975b. Environmental influences on pituitary-adrenal system functions. In: Les Endocrines et la Milieu. Ed. H.-P.Klotz. Problèmes Actuels D'Endocrinologie et de Nutrition. Série No. 19. Expansion Scientifique Française, Paris. pp. 55-62.

Bohus B. 1977. Pituitary neuropeptides, emotional behavior and cardiac responses. In: Hypertension and Brain Mechanisms. Eds. W.de Jong, A.P. Provoost and A.P.Shapiro. Progr.Brain Res. vol. 47. Elsevier Amsterdam. pp. 277-288.

Bohus B. 1981. Neuropeptides in brain functions and dysfunction. Int.J. Ment.Health 9: 6-44.

Bohus B, de Kloet ER, Veldhuis HD. 1982a. Adrenal Steroids and Behavioral adaptation: relationship to brain cortocoid receptors. In: Adrenal Action on Brain. Eds. D.W.Pfaff and D.Ganten.Springer, Berlin, pp.108-140.

Bohus B, Conti L, Kovacs GL, Versteeg DHG. 1982b. Modulation of memory processes by neuropeptides: interaction with neurotransmitters systems. In: Neuronal Plasticity and Memory Formation. Eds. C.Ajmone Marsan and H. Matties. Raven Press, New York. pp.75-87.

Bohus B, Versteeg CAM, de Jong W, Cransberg K, Kooy JG. 1983a. Neurohypophyseal hormones and central cardiovascular control in the neurohypophyssis. Eds. B.A.Cross and S.Leng. Progr.Brain Res. Vol. 60. Elsevier Amsterdam. pp.463-375.

Bohus B, de Jong W, Hagan JJ, de Loos W, Maas CM, Versteeg CAM. 1983b. Neuropeptides and steroid hormones in adaptive autonomic processes: implications for psychosomatic disorders. In: Integrative Neurohumoral Mechanisms. Eds. E.Endröczi, D.de Wied, L.Angelucci and U.Scapagnini. Elsevier Biomedical Press, Amsterdam. pp. 35-49.

Branconnier RJ, Cole JO, Gardos G. 1979. ACTH 4-10 in the amelioration of neuropsychological symptomatology associated with senile organic brain syndrome. Psychopharmacology 61: 161-165.

Breier C, Kain H, Konzett H. 1979. Personality dependent effects of the ACTH 4.10 fragment on test performance and on concomitant autonomic reactions. Psychopharmacology 65: 239-245.

Brown GP, Martin, JBM. 1974. Corticosterone, prolactin, and growth hormone responses to handling and new environment in the rat. Psychosom. Med. Vol. 36, no. 3, 241-247.

Brunia CHM, van Boxtel A. 1978. MSH-ACTH 4-10 and task induced increase in tendon reflexes and heart rate. Pharmacol.Biochem.Behav. 9: 615-618.

Cannon WB. 1915. Bodily changes in pain, hunger, fear and rage. Appleton New York.

Cohen D, Obrist PA. 1975. Interactions between behaviour and the cardiovascular system. Circulation Res. 37: 693-706.

O'Donohue TL, Dorsa, DM. 1982. The opiomelanotropinergic neuronal and endocrine systems. Peptides 3: 353-395.

Eferakeya A, Bunag RD. 1974. Adrenomedullary pressor responses during posterior hypothalamic stimulation. Am.J.Physiol. 227-1: 114-118.

Von Eiff AW, Plotz EJ, Beck KJ, Czernik A. 1971. The effect of estrogens and progestins on blood pressure regulation of normotensive women. Am.J. Obstet.Gynec. 109: 887-892.

Folkow B, Neil E. 1971. Circulation. New York: Oxford University Press.

Galosy RA, Gaebelein CJ. 1977. Cardiovascular Adaptation to Environmental Stress: Its Role in the Development of Hypertension, Responsible Mechanisms, and Hypotheses. Biobeh.Rev., Vol.1: 165-175.

Gebber GL, Klevans LR. 1972. Central nervous system modulation of cardiovascular reflexes. Fed.Proc. 31(4): 1245-1252.

Gray JA. 1982. The Neuropsychology of Anxiety: An Enquiry into the Functions of the Septo-Hippocampal System. Oxford Univ.Press, New York.

Hilton SN. 1975. Ways of viewing the central nervous control of the circulation - old and new. Brain Res. 87: 213-219.

Holdstock TL. 1969. Autonomic reactivity following septal and amygdaloid Lesions in white rats. Physiol.Behav. 4: 603-607.

Isaacson RL. 1974. The Limbic System. Plenum Press, New York.

Kawamura H, Chesterfield GG, Frohlich ED. 1978. Cardiovascular alteration by nucleus locus coeruleus in spontaneously hypertensive rat. Brain Res. 140: 137-147.

Kawasaki H, Watanabe S, Ueki S. 1980. Brain Res.Bull. 5: 711-718.

Koolhaas JM, Schuurman T, Fokkema DS. 1983. Biobehavioural basis of coronary prone behaviour. Dembrosky and Schmitt, eds. in press.

Levine S, Goldman L, Coover GD. 1972. Expectancy and the pituitary-adrenal system. In: Physiology, Emotion and Psychosomatic Illness. CIBA Foundation Symposium 8 (new series). Elsevier, Excerpta Medica, North Oîland, Amsterdam, pp. 281-291.

Loos de WS, Bohus B, de Jong W, de Wied D. 1979. Reduction of heart-rate reactions to emotional stress by ovarian hormones in rats. J.Endocrinol. 81: 138P-139P.

Mason JW. 1968. A review of psychoneuroendocrine research on the pituita-ry-adrenal cortical system. Psychosom.Med. 30: 576-607.

Nathan MA, Reis DJ. 1975. Fulminating arterial hypertension with pulmona-ry edema from release of adrenomedullary catecholamines after lesions of the anterior hypothalamus in the rat. Circulation Res. 37: 226-235.

Obrist PA, Webb RP, Sutterer JR. 1970. The cardiac-somatic relationship: some reformulation. Psychophysiol. 6: 569-587.

Obrist PA. 1976.The cardiovascular-behavioral interaction - as it appears today. Psychophysiology 13: 95-107.

Rossier J, Greuch ED, Rivier C, Long N, Guillemin R, Bloom FE. 1977. Foot-shock induces stress and increases β-endorphin levels in blood but not in the brain. Nature 270: 618-620.

Routtenberg A. 1972. Memory as input-output reciprocity: an integrative neurobiological theory. Ann.N.Y.Acad.Sci. 193: 159-174.

Selye H. 1936. A syndrome produced by diverse nocuous agents. Nature 138: 32-33.

Sharpless SK, Rothballer AB. 1961. Humoral factors released from intra-cranal sources during stimulation of reticular formation. Am.J.Physiol. 200(5): 909-915.

Smelik PG, Vermes I. 1980. The regulation of the pituitary-adrenal system in mammals. In: General, Comparative and Clinical Endocrinology of the Adrenal Cortex. Ed. I.Chester Jones and I.W.Henderson. Academic Press, New York. Vol.3. pp.1-55.

Sutton RE, Koob GF, Le Moal M, Rivier J, Vale W. 1982. Corticotropin releasing factor produces behavioural activation in rats.Nature 297:331-333.

Swaab DF. 1980. Neurohypophysial hormones and their distribution in the brain. In: Hormones and the Brain. Eds. D.de Wied and P.A.van Keep. MTP Press Limited, London, pp. 87-113.

Thompson EA, de Wied D. 1973. The relationship between the antidiuretic activity of rat eye plexus, blood and passive avoidance behaviour.. Physiol.Behav. 11: 377-380.

Versteeg CAM, Bohus B, de Jong W. 1982. Attenuation by Arginine- and Desglycinamide-lysine-vasopressin of a centrally evoked pressor res-ponse. J.Auton.Nerv.Syst., 6: 253-262.

Watson SJ, Akil H. 1980. On the multiplicity of active substances in single neurons: β-endorphin and α-melanocyte stimulating hormone as model system.In: Hormones and the Brain. Eds. D.de Wied and P.A.van Keep. MTP Press Limited, London, pp. 73-86.

de Wied D. 1979. Schizophrenia as an inborn error in the degradation of β-endorphin - a hypothesis. Trends Neurosci. 2: 79-82.

Witter A, de Wied D. 1980. Hypothalamic-pituitary oligo peptides and behavior. In: Handbook of the Hypothalamus, Vol. 2: Physiology of the Hypothalamus. Eds. P.J.Morgane and J.Panksepp. Marcel Dekker, New York. pp. 307-451.

HORMONAL RESPONSE TO ACUTE STRESS: FOCUS ON OPIOID PEPTIDES

R.E. Lang, K. Kraft, Th. Unger and D. Ganten
German Institute for high blood pressure research and Department of Pharmacology, University of Heidelberg, Im Neuenheimer Feld 366, D-6900 Heidelberg, F.R.G.

There are several ways to characterize a persons' reaction to stressful situations. Most commonly, the response of the autonomic nervous system as reflected in heart frequency, blood pressure, or respiratory rate is used as a simple measure of the emotional response. Another possibility is to determine the endocrine reaction pattern by measuring the plasma levels of those hormones known to be increased during stress.

In addition to catecholamines and the glucocorticoids of the adrenal cortex there are a number of pituitary hormones which have been reported to be elevated following physical exercise and also mental stress. Regarding the anterior pituitary these are ACTH -one of the longest known so called "stress hormone"- and β-endorphin which both derive from the same biosynthetic precursor, furthermore growth hormone and prolactin. Among the neurohypophysial hormones vasopressin has long been considered as a hormone to be released in response to stressful stimuli. Experiments supporting this view however have often been based on indirect observations or on bioassay measurements. Using specific radioimmunoassays, at least in rats, no changes in plasma vasopressin concentrations were observed following immobilization, noise, or exposure to cold (Keil and Severs, 1977, Husain et al. 1979). In contrast, we recently found that plasma oxytocin levels are highly responsive to such stimuli (Lang et al., 1981)

With the discovery of the enkephalins in the adrenal medulla, which are secreted concomitantly with catecholamines, a further hormonal system probably involved in the organisms' response to stress has emerged. In the following, the question will be discussed whether an analysis of blood hormones could be of help in evaluating the effect of stress on the functional cardiovascular system and in particular which hormones would be of interest in this respect.

1. In general, measurements of each of the above mentioned so called "stress hormones" indipendently of whether they are related to the cardiovascular system or not, may improve the characterization of a persons' reaction in a stressful situation. The assessment of both the cardiovascular answer and the neuroendocrine response might allow us to investigate whether a dissociation between both is possible. In this case it might be of interest to see if there are some people reacting quite normally with their hormonal systems to a certain stress, which however, show an overshooting response in heart rate and blood pressure. This would mean that the failure to adapt adequately to a stimulus comprises not the whole personality but consists only in a defect in the cardiovascular system. Moreover, such an observation would tell us that the cause of maladaptation must be found beneath the level of psychological stress perception and its processing.

2. If one of the hormones released during stress was directly involved in cardiovascular processes, it would be of particular value to know its response in persons with cardiovascular diseases as compared to healthy controls. However, apart from catecholamines, none of the above mentioned hormones has been reported so far to induce cardiovascular effects with one exception: the opioid peptides.

The opioid peptides β-endorphin (β-END) and enkephalins, originally discovered in the brain and pituitary gland, are found in a large number of peripheral organs and occur also in the blood where they may function as hormones. The major origin of circulating β-END is considered the anterior pituitary gland. Plasma enkephalins may mainly derive from the adrenal medulla, but other parts of the sympathetic system may be an additional source since we recently detected considerable amounts of leucine-enkephalin (leu-ENK) and methionine-enkephalin (met-ENK) in the guinea pig heart, which almost completely disappeared following chemical sympathectomy with 6-OH-dopamine (Lang et al. 1983a). In the chromaffine cells of the adrenal gland enkephalins are stored and released together with catecholamines (Viveros et al. 1979). Separation of adrenargic from nonadrenargic chromaffine granules by density centrifugation and subsequent analysis for enkephalin content revealed, that at least in bovine adrenals enkephalins are probably co-stored with adrenaline in the same granules but not with noradrenaline (Lang et al. 1983b). This suggests that enkephalins by acting as hormones might have the same target organs as adrenaline, e.g. the vascular smooth muscle cells.

In a number of animal experiments both β-END and enkephalins have been shown to affect blood pressure. Systemic administration of β-END to urethane anaesthetized rats has been reported to produce an immediate fall in blood pressure lasting up to 90 min. which was abolished by the opiate antagonist naloxone (Lemaire et al. 1978). A fall in blood pressure has also been observed following i.v. injection of enkephalins in anesthetized rats (Moore et al. 1980, Wei et al. 1980). Controversially, in conscious dogs and rats enkephalins have been shown to increase blood pressure and heart rate (Sander et al. 1981, Simon et al. 1978). This discrepancy might be explained by a direct action of systematically injected enkephalins at baroreceptor reflexes which may be prohibited by anesthesia due to the blunting effect of narcotics on central cardiovascular reflexes.

In men, a stable enkephalin analogue (FK 33824) was shown to lower erect blood pressure three hours after intravenous injection. When baroreceptor reflex activity was tested using the slope of the mean pressure: heart period relationship after infusion of sodium nitroprusside a significant reduction in baroreflex sensitivity was found after administration of this compound. Under the same circumstances, the opiate antagonist naloxone increased baroreflex sensitivity (Reid et al. 1982).

A number of animal experiments concerned with the influence of naloxone on the development of shock indicated that endogenous opioid peptides may be involved in the development of hypotension in various forms of shock (Holaday and Faden 1978, Faden and Holaday 1979). These encouraging results achieved in animals prompted several investigators to try the antagonist in patients with shock refractory to conventional therapy (Dirksen et al. 1980, Peters et al. 1981 and Wright et al. 1980). In numerous cases including cardiogenic, hemorrhagic and septic shock, a significant improvement of the circulatory indices was reported to occur immediately after administration of naloxone. This effect was mostly attributed to a blockade of the depressor effects of β-END, but no attempts were made in experimental or clinical studies to identify the nature of the opioid peptides which supposedly were inhibited by naloxone and to measure their plasma levels during shock.

We have therefore recently examined the effect of hemorrhagic shock on the concentrations of β-END, met-ENK and leu-ENK in the plasma of pentobarbital anesthetized dogs (Lang et al. 1982 a). Blood samples were collected from the femoral vein and adrenal blood was obtained from the vena phrenica. The peptides were measured by specific radioimmunoassays recently developed in our laboratory after extraction from plasma using small columns packed with octadecasilyl silica. Cross reactivity of β-END antibody with β - Lipotropine was about 4%, the met-ENK and leu-ENK antibodies crossreacted less than 5% with leu-ENK and met ENK, respectively. Lowering of the mean arterial blood pressure (MAP) to 40mmHg by bleeding cause a rise in β-END, met-ENK and leu-ENK. Simultaneous femoral and adrenal vein sampling revealed tenfold higher concentrations of leu-ENK in the adrenal effluent indicating that the adrenal gland was the main source of plasma enkephalins. The ratio of met-ENK to leu-ENK in adrenal blood was approximately 3:1. β-END levels remained elevated during the three hours shock period whereas the enkephalin levels tended to decrease probably due to exhaustion of the adrenal gland. Volume repletion caused a significant reduction of the β-END and enkephalin levels in the plasma (Table 1)

Table 1: β-END and M-ENK and L-ENK concentrations (pg/ml plasma) in femoral vein (f.v.) and adrenal vein (a.v.) blood during helorrhage induced hypovolemia (means of three studies). Note the vast increases of M-ENK and L-ENK during hypovolemia and their return to below control levels after volume repletion while -END levels persist at higher levels

	MAP (mmHg)	β-END f.v.	M-ENK a.v.	L-ENK a.v.	f.v.
30 min after laparatomy	117	40.7	422.5	164.3	
hypovolemia					
0	40	145.6	1060.5	293.5	23.0
1h	40	161.6	2849.0	892.4	124.7
2h	40	172.6	715.0	468.6	42.9
3h	40	134.4	394.5	324.8	
volume repletion					
0	94	48.2	166.5	21.6	
1h	92	43.2	164.5	32.8	
2h	84	33.0	171.5		

These data demonstrate that hemorrhagic shock leads no only to a mobilization of β-END from the pituitary but moreover stimulates the adrenal gland to release enkephalins which are costored with catecholamines in the adrenal medulla. Calculated on a molar basis, the plasma concentrations of leu-ENK substantially exceeded those of β -END. Enkephalins have therefore to be taken into consideration in addition to β -END when a possible role of circulation opiod in the pathogenesis of shock is discussed.

Enkephalins may also be involved in the adaptation of the organism to physical exercise. Recently we measured the leu-ENK concentrations in plasma of volunteers before and immediately after a 100m sprint (Lang and Bieger, in preparation). Physical exercise appears to be a strong stimulus for the release of enkephalins into the blood. We found in this study Leu-ENK resting levels of 14.5 $\pm$ 3.8 pg/ml (mean $\pm$ SD, n=15), which increased to 39.1 $\pm$ 8.23 pg/ml (n=13) after sprint. Apart from metabolic functions one possible role of enkephalins under these circumstances may be the control of blood flow by a direct action at the blood vessels. Experiments in isolated arteries suggest a vasodilatatory effect of opioid peptides, which may be brought about by an inhibition of the adrenargic innervation of blood vessels. Using isolated perfused femoral artery preparations in pentobarbital anesthetized cats intraarterial administration of leu-ENK or met-ENK has been found to produce dose-dependent decreases in perfusion pressure (Moore et al. 1982). Both β -END and enkephalins were reported to dilate in vivo the arterioles of the hamster cheek pouch, whereby β -END showed a higher potency (Wong et al. 1981). The pressure changes elicited by electrical field stimulation of the isolated perfused rabbit ear artery are markedly attenuated by the addition of enkephalins and their analogues (Ronai et al. 1982). Finally, when the isolated perfused cat spleen is prelabeled with ^{3}H-norepinephrine, perfusion with met-ENK results in a dose dependent inhibition of nerve stimulation-mediated release of dopamine- β -hydroxylase and ^{3}H-norepinephrine overflow (Gaddis and Dixon 1982). This effect can be blocked by the opiate antagonist naloxone. Perfusion of the spleen with naloxone alone has no effect on the basal or nerve-stimulated release of dopamine- β -hydroxylase and ^{3}H-norepinephrine. These findings provide evidence for the existence of a presynaptic opiate receptor on peripheral adrenergic neurons which affects the catecholamine release in an inhibitory way. The observation that naloxone does not influence the nerve stimulated release of norepinephrine may indicate that blood borne enkephalins may function as physiological ligands at this receptors.

In conclusion, the strong morphological and functional interrelationship between opioid peptides and the sympathetic nervous system suggests a possible role of enkephalins in the pathogenesis of cardiovascular diseases. Indeed, recent findings in spontaneously hypertensive rats (SHR) support such an idea. It was reported that in this strain the enkephalin contents in the adrenal gland and sympathetic ganglia were reduced by more than 50% as compared to normotensive controls (Di Giulio et al. 1979). Since SHR are considered a model of hypertension which in many aspects corresponds to essential hypertension in men, the determination of enkephalin plasma levels in human essential hypertension would be a task of particular importance. In view of the close association of opioid peptides with the sympathetic nervous system and blood pressure control an evaluation of enkephalins and β-END should be included into studies on the influence of stress on the cardiovascular system.

REFERENCES

Dirksen R, Otten MH, Wood GJ, Verbaan CJ, Haalebos MMP, Verdouw PV, Nijhuis GMM. 1980. Naloxone in shock. Lancet 1360-1361.

Faden AI, Holaday JW. 1979. Opiate antagonists: a role in the treatment of hypovolemic shock. Science 205: 317-318.

Di Giulio AM, Yang H-YT, Fratta W, Costa E. 1979. Decreased content of immunoreactive enkephalin-like peptide in peripheral tissues of spontaneously hypertensive rats. Nature 278: 646-647.

Gaddis DR, Dixon WR. 1982. Presynaptic opiate receptor-mediated inhibition of endogenous noroponephrine and dopamine- -hydroxylase release in the cat spleen, independent of the presynaptic alpha adrenoceptors. J Pharm and Exper Therapeutics 233, I: 77.

Husain K, Manger WM, Rock TW, Weiss RJ, Frantz AG. 1979. Vasopressin release due to manual restraint in the rat: role of body compression and comparison with other stressful stimuli. Endocrinology 104: 641-644.

Holaday JW, Faden AI. 1978. Naloxone reversal of endotoxin hypotension suggests role of endorphins in shock. Nature 275: 450-451.

Keil LC, Severs WB. 1977. Reduction in plasma vasopressin levels of dehydrated rats following acute stress. Endocrinology 100: 30-38.

Lang RE, Rascher W, Hermann K, Heil J, Unger Th, Ganten D. 1981. Stimulation of oxytocin but not vasopressin secretion in response to stress. Pfluger's Archiv, Europ J Physiol, Suppl to Vol 391, abstract 99: 25.

Lang RE, Bruckner UB, Kempf B, Rascher W, Sturm V, Unger Th, Speck G Ganten D. 1982a. Opioid peptides and blood pressure regulation. Clin and Exp Hypertension-Theory and Practice A4 (1-2): 249-269.

Lang RE, Hermann K, Dietz R, Gaida W, Ganten D, Kraft K, Unger Th. 1983a. Evidence for the presence of enkephalins in the guinea pig heart. Life Sci 32: 399-406.

Lang RE, Taugner G, Gaida W, Gante D, Kraft K, Unger Th, Wunderlich I. 1983b. Evidence against co-storage of enkephalins with noradrenaline in bovine medullary granules. Europ J Pharmacol 86: 117-120.

Lemaire I, Tseng R, Vincent M, Remond G. 1978. Systemic administration of -endorphin: potent hypotensive effect involving a serotonergic pathway. Proc Natl Acad Sci 75: 6240-6242.

Moore RH, Dowling DA. 1980. Effects of intravenously administered Leu-or Met-enkephalin on arterial blood pressure. Peptides 1: 77-87.

Moore RH, Dowling DA. 1982. Effects of enkephalins on perfusion pressure in isolated hindlimb preparations. Life Sci 31: 1559-1566.

Peters WP, Friedman PA, Johnson MW, Mitch WE. 1981. Pressor effect of naloxone in septic shock. Lancet 529-532.

Reid JL, Rubin PC, Elliott HL. 1982. Clinical studies with centrally acting drugs: implications for central blood pressure regulation. Neurosc Lett 10: 26.

Ronai AZ, Harsing LG, Berzetei IP, Bajusz S, Vizi ES. 1982. Met^5-enkephalin-Arg-Phe acts on vascular opiate receptor. Europ J Pharmacol 79: 337-338.

Sander GE, Giles TD, Kastin AJ, Quiroz AC, Kaneish A, Coy DH. 1981. Cardiopulmonary pharmacology of enkephalins in the conscious dog. Peptides 2: 403-407.

Simon W, Schaz K, Ganten U, Stock G, Schlor KH, Ganten D. 1982. Effects of enkephalins on arterial blood pressure are reduced by propanolol. Clin Sci Mol Med 55: 237s-241s.

Viveros OH, Diliberto EJ, Hazum E, Chang KJ. 1979. Opiate-like materials in the adrenal medulla: evidence for storage and secretion with catecholamines. Mol Pharmacol 16: 1101-1108.

Wei ET, Lee A, Chang JK. 1980. Cardiovascular effects of peptides related to the enkephalins and -casomorphin. Life Sci 26: 1517-1522.

Worn TM, Koo A, Choh Hao Li. 1981. Beta-endorphin: vasodilating effect on the microcirculatory system of hamster cheek pouch. Int J Peptide Protein Res 18: 420-422.

Wright DJM, Phillips M, Weller MPI. 1980. Naloxone in shock. Lancet 1361.

EMOTIONAL STRESS AND HEART DISEASE : CLINICAL RECOGNITION AND ASSESSMENT

K. McIntyre,
Harvard University School of Public Health
Boston, Massachusetts 02115 USA

and **C. Jenkins,**
University of Texas Medical Branch
Galveston, Texas 77550 USA

SUMMARY

The authors review recognition of the stressed individual, the stressors and, to the extent possible, the individual with cardiovascular responses to stress which raise the risks of acute CHD endpoints.

1. ACUTE STRESS REACTIONS

Research Evidence and Probable Mechanisms

Ventricular arrhythmias, including cardiac arrest, appear to be provokable in predisposed individuals by psychological stress. Thus, a 39 years old man who suffered cardiac arrest (ventricular fibrillation) on two occasions and exhibited numerous premature ventricular beats had normal coronary arteries and appeared to have these arrhythmias on a psycho-emotional basis (Lown, NEJM 294:623, 1976) and also improved as a result of relaxation techniques and was controlled by beta-adrenergic receptor blockade. This relationship between emotions and dysrhythmia has been further documented in clinical and epidemiologic studies (Reich et al. 1981 : Jenkins, 1981) Experimental models of myocardial ischemia have demonstrated that stress increases the severity of ventricular dysrhythmias (Hoffman, (1978)).

The body usually responds to emotional arousal in the same way it responds to physical stress, in as much as the effects of all are mediated through the sympathetic nervous system. The response usually includes increases in heart rate, blood pressure, blood flow to muscles, and a decrease in blood flow to kidneys, stomach and intestines.

Where emotion, physical stress or exercise cause these physiological changes, the oxygen demand of the heart is proportionately increased sometimes to a great degree. Electrical irritability of the heart may increase. Each of these factors becomes more dangerous in patients in whom heart disease is already present.

Cardiac rhythm abnormalities, from minor to life-threatening, have been found to be associated with emotional stress and anxiety. (Regestein, (1975), Bove, (1977) and Lown et al., (1976)).

One example of the effect of stress on heart rhythm in day-today living conditions was demonstrated by a study by Taggart et al., (1969) who monitored individuals with and without diagnosed coronary artery disease while they drove in busy city traffic. Both groups showed increased heart rates some in excess of 140 beats per minute. One clearly defined index of inadequate oxygenation of the myocardium, electrocardiograhic ST-segment changes, developed in three out of 32 healthy drivers. In 13 patients with known coronary artery disease, the ST-segment and T-wave abnormalities were more pronounced, five developed multiple ventricular ectopic beats.

Another example comes from a study of two groups, 23 normal subjects and a second group of seven with coronary artery disease. Both groups were monitored by ECG while performing a task uncomfortable to them, delivering a speech in public. (Taggart, P., (1973)).
Heart rates up to 180 beats per minute, ectopic beats and elevations of plasma catecholamine and free fatty acids were observed in both groups. Ischemic ST-segment changes by ECG occured in six of the seven coronary subjects, and five had multiple or multifocal ventricular ectopic beats. The coincidence of ischemia and arrhythmia may be a trigger for ventricular fibrillation and sudden cardiac death. A beta-blocking agent, which interferes with the impact of stress upon the heart, supressed the tachycardia and ECG changes in both groups.

Clinical Signs and Symptoms

Major life changes raise the risk of acute stress reactions, but many people who encounter losses or crises which are culturally-defined as major do not manifest the disorganization of behaviour or cognition, nor the burst of painful emotions which mark acute stress reactions. Conversely, for some persons reacting with great disorganization or distress, it is difficult to identify a major external precipitant. Often in such cases an event or realization which would be considered trivial by most observers, is given profound meaning by the subject and is judged to be the precipitant of the Acute Stress Reaction.

Although anecdotal evidence abounds on both sides of the issue, controlled prospective research suggests that merely encountering life changes, with the probable exception of an extreme loss (e.g. death of a spouse), is not significantly predictive of near future CHD.

Clinically, acute stress reactions are marked by a sudden onset of strong emotions, particularly anxiety or depression, but also anger or withdrawal. The ability of patients to carry out usual role functions is impaired -- they cannot work effectively, cannot concentrate on difficult tasks or those involving long sequences of adaptive responses. They may even appear to be in a panic or a "state of shock". During such reactions, persons often withdraw, neglect their usual responsibilities, and become temporarily dependent. The patient may consult a health professional, counsellor or clergyman either via self-referral or referral by family or friends. Usually the patient or the referring person has already made the diagnosis of acute stress reaction and linked it to a precipitant.

Acute stress reactions are easily noted by their sudden onset. Help is usually sought because it is one of the more socially acceptable risk indicators. For these reasons it is less necessary to conduct screening for acute than for chronic stress reactions.

Test and Measures

For research purposes measurement of the magnitude of stressor or the intensity of the reaction may be desired. The magnitude of the external crisis may be estimated by use of any of the several life crisis inventories now published. The first and most commonly used is the Schedule of Recent Experience (SRE) described by Holmes and Rahe (1967). Subsequently, scales to achieve similar purposes have been published by Paykel, Prusoff, and Uhlenhuth (1971), Dohrenwend (1973), Hurst, Jenkins and Rose (1978) and Horowitz et al (1977).

Assessment of the behavioral and affective reaction to acute stress may be performed by scales developed to measure anxiety, depression, hostility, withdrawal and confusion. In mildly affected patients, self-report scales may be used, but when the reaction is severe, the patient may not be able to respond reliably (or at all) to a self-administered scale. This observation in itself can be used as a rough measure of the intensity of the reaction in an otherwise literate and capable person.

In reactions of greater severity, check lists of overt affect and behaviour completed by trained health workers may be used effectively. A number of such scales have been developed for use by nurses, psychologists, physicians and research technicians.

2. SLEEP DISTURBANCE AND DIFFICULTIES

Research Evidence and Probable Mechanisms

Trouble in obtaining adequate restful sleep is a widespread problem. It is sufficiently important in the USA that the Institute of Medicine (IOM) of the National Academy of Sciences recently conducted a major study which was published under the title "Sleeping Pills, Insomnia and Medical Practice". (Anon, 1979). The IOM found that the majority of patients seeking medical attention for sleep problems are actually suffering from physical or psychological disorders.

Often the worsening sleep problem is the presenting symptom which leads to discovery and treatment of the underlying problem. As discussed in recent reviews of medical literature, sleep difficulties appear to be a risk factor for CHD and sudden cardiac death. Sleep disturbance may also reduce the threshhold for exertional angina and increase the frequency of emotionally precipitated angina (Jenkins et al. 1982), as well as being prospectively associated with future angina pectoris, myocardial infarction, and total mortality (Floderus, 1974; Friedman et al., 1974; Thomas and Greenstreet, 1973; Thiel et al., 1973 and Bengtsson et al., 1973).

Clinical Signs and Symptoms

Many patients accept inadequate sleep as part of their life circumstances and fail to report it spontaneously to the physician. For this reason, specific questions about sleep should be included in the medical history. Different forms of sleep problems imply different causes and different remedies, therefore specific questions should be asked about trouble falling asleep, trouble staying asleep (e.g. awaking far too early and not being able to get back to sleep), waking up several times per night (one awakening for the purpose of micturition is common in adults, perticularly with increasing age), and awakening after the usual amount of sleep feeling tired and worn out.

Better data is obtained by asking a question specifically such as : "How often do you have trouble failling asleep? : -- or "How many times last month. . .?" If one asks : "Are you bothered by insomnia?", a negative answer could mean either failure to understand the term "insomnia", or no sleep difficulty or a sleep difficulty which, though severe, has been accepted as tolerable. Asking for specific estimates permits comparisons on subsequent occasions to determine if hygienic changes in activity patterns are leading to improvement in sleep.

Test and Measures

For research and screening purposes, self-administered forms have been found useful to assess type and frequency of sleep disturbance. A form used successfully in a longitudinal study of heart surgery patients is as follows :

HOW OFTEN in the past month
DID YOU

	(0) Not at All	(1) 1-3 days	(2) 4-7 days	(3) 8-14 days	(4) 15-31 days
Have trouble falling asleep	----	----	----	----	----
Have trouble staying asleep	----	----	----	----	----
Wake up after your usual amount of sleep feeling tired and worn out	----	----	----	----	----

A similar form was used in a study of stress and health change in a professional group whose work involves heavy responsibility and stress and frequent shift changes. In this group of 400 adult men (under age 50) about 10% reported problems on 15 or more days of the last month, another 25% reported problems on 4 to 14 days, about a third had problems 1 to 3 days, and the remaining third had no problem nights at all for that symptom on each of the above sleep disturbance questions.

In a group of 250 men and women awaiting coronary artery bypass or cardiac valve surgery, 20-25% reported each of three sleep problems 15 or more days, and about 50% reported not having the problem at all in the most recent month.

3. FATIGUE AND EMOTIONAL DRAIN

Research Evidence and Probable Mechanisms

Among patients who suffer myocardial infarction, a state of physical and mental exhaustion has been described as a common prodrome (Bending, 1956). One study reported increased tiredness and generally poor health in over 100 patients who survived myocardial infarction in the one week prior to the event (Hollingshead, 1975). The suggestion has been made that such manifestations of "emotional drain" may be manifestations of sub-clinical cardiovascular disease and as such may represent prodromes rather than risk factors for this particular clinical endpoint of coronary heart disease.

Wolf (1967) described a Sisyohean pattern, (or a "Sisyphus reaction") in which the individual "strives without joy". Hackett goes on to say "rarely is this individual rewarded by the satisfaction of accomplishment". Instead he doggedly toils on to a state of "psychic exhaustion" or "emotional drain" described by Wolf and Bruhn (1974) as a "precursor to myocardial infarction and sudden death". Their evidence, although retrospectively collected nevertheless merits further controlled research and clinical caution.

One of the barriers to clarification of the possible role of fatigue, emotional drain, or vital exhaustion as a risk factor for cardiovascular crises is that data have not been systematically and uniformly collected in research studies.

Fatigue and exhaustion are common symptoms indeed, no less now in our age of labor-saving devices than in eras and places where long hours of heavy exertion provided ample reason for such states. Perhaps therein lies a key to the distinction of healthy fatigue, from that form having serious implications. Studies of prodromata of sudden coronary death suggest that what is clinically significant is a sudden exacerbation of fatigue or exhaustion in the absence of adequate physical reasons such as prolonged exertion or inadequate sleep (Feinleb et al., 1975).

The combination of unusual fatigue with emotional drain has recently been studied as the concept of "vital exhaustion". It is marked by a sense of the loss of the energy to think, to plan, to strive --- even a feeling that one has lost the energy to go on living with one's usual effectiveness (Appels et al. (1980). The presence of this syndrome has been reported to be more common in persons prior to myocardial infarction than in health, control groups. Research in earlier years (Bruhn, et al., 1969) supports the possibility that this general concept may represent a risk indicator for CHD deserving further study.

Clinical Signs and Symptoms

Clinical inquiry regarding fatigue and exhaustion should be careful to :-Rule out sleep difficulties and sustained heavy exertion as causes; Rule out the presence of other systemic diseases as causes;-Establish that the fatigue has recently exacerbated and is not merely the continuation of a chronic condition;-Establish that a change in function, observable to others, has taken place, and not just a change in subjective feelings. This can be done by questioning family members of the patient or by eliciting specific examples (such as activities given up, increased use of naps).

Tests and Measures

Several relevant scales have been widely used. The Fatigue Scale of the Profile of Mood States (McNair et al. 1971) has been used in psychological studies and psychopharmacoligic experiments but not, to our knowledge, in cardiovascular research. The Maastricht Questionnaire (Appels et al., 1979), has evidenced validity in discriminating groups of myocardial infarction patients. Preliminary evidence suggests a high correlation with depression scales. It has been translated into English and is beginning to be used in research in the USA.

There may be other scales available to measure this concept with which the authors are not familiar.

CHRONIC EFFECTS

1. CORONARY-PRONE BEHAVIOUR -- TYPE A

Research Evidence and Probable Mechanisms

The Type A behavior pattern is not the same as "stress"; it represents neither a stressful stimulus nor a distressed response, but rather is a style of behavior with which some persons habitually respond to circumstances that arouse them (Jenkins, 1979). The pattern can be reliably rated and is a deeply ingrained, enduring trait.

Friedman (1969) defined Type A behavior as : "A characteristic action-emotion complex" found in people who are constantly struggling to reach poorly defined goals in the shortest time possible.

Some reviewers now conclude that there is a strong link between Type A behavior and the clinical emergence of myocardial infarction and that Type A behavior has "about the same strength of associations with coronary artery disease prevalence and incidence as do other standard risks factors". (Jenkins, C.D. (1978)). On the other hand, Hackett and Rosenbaum state : "While no body of data yet exists that can refute Jenkins statement, to accept it without serious reservations is a step few clinicians are prepared to take". Nevertheless, in 1970-1980 an expert review panel· of biomedical and behavioral scientists convened by the National Heart, Lung & Blood Institute concluded that "available scientific evidence demonstrates an association between Type A behavior and an increased risk of clinically apparent CHD...".

Laboratory research into possible mechanisms by which the Type A pattern might raise risk of CHD has progressed to the point where some promising hypotheses can be offered. It is quite clear that the Type A pattern, like other psychosocial risk indicators for CHD discussed in this chapter, exerts its influence through the central nervous system. Subjected to a common challenge (eg., biochemical, physiological, psychological) extreme Type A and Type B men have been shown to react quantitatively differently in terms of pituitary hormones, serum insulin, adrenalin, and blood platelet aggregation : (Friedman, et al., 1975; Williams, et al., 1978; Simpson, et al., 1975). In addition, Type A men show greater increases in systolic blood pressure than Type B men when confronted with competitive challenge. (Manuck, et al., 1978).

Clinical Signs and Symptoms

The Type A behavior pattern is marked by exaggerated hurry, impatience, hostility, competitiveness, and achievement striving. Type A persons typically eat fast, talk fast, walk and drive fast, and become singularly upset when others slow them down. The Type A person looks very alert and seems prepared to move quickly. His/Her movements are abrupt, gestures emphatic, and speech staccato. They have a "short fuse" for becoming angry. Not all Type A persons have all of these behavioral features. The strong presence of several of them or the moderate presence of somewhat more, are enough to classify a person as Type A.

Much more complete clinical descriptions of the Type A pattern and guides for its recognition by physicians and health workers have been published. No one should attempt clinical detection of the Type A pattern without first thoroughly studying these resources and, if possible, working with a person who has received specific training in Type A assessment. The first method to be developed for the assessment of the Type A pattern was the Structured Interview (SI) of Friedman and Rosenman (1978). This consists of 25 to 30 questions asked in specific manners, the responses to which are judged more by their nonverbal characteristics than by the specific logical content of the reply. The SI must be conducted by an interviewer trained both in the techniques of timing and challenge in administering the questions and in the modes of rating the vocal and motor behavior of the respondent as well as the content of replies. (The authors of this contribution can provide information regarding where formal training in Type A assessment may be obtained. This picture changes from year to year).

Tests and Measurements

The following references are prime sources for more complete description of the Type A pattern and its detection.

For screening and research purposes, particularly when large number of persons are to be assessed, the structured interview method becomes impractical to administer. Self-administered machine-scoreable tests are the most economic method of measuring Type A. The most widely used of these tests at present is the Jenkins Activity Survey (JAS), which is a multiple choice questionnaire, computer-scored, which yields separate scores for the overall Type A pattern, and for its component dimensions: speed and impatience, job involvement, and hard-driving competitive (Jenkins et al., 1979).

Other self-administered tests having published evidence of validity include the Bortner Short Rating Scale (Bortner, 1969), the Framingham Brief Type A Scale (Haynes et al., 1978). These three tests have all been developed for use with adults. Variations of this testing approach have been developed for use with adolescents, for example, the Mathews (1968), Youth Test for Health (MYTH) and still other work is underway for measuring the construct in younger children. (Butensky et al., 1976).

Much work is underway in this field and additional methods for assessing the Type A pattern in different age and occupational groups may soon be offered. As for the measurement of any other risk factor --- biological or psychosocial --- it is essential that the prospective user insist upon adequate validation of methods before assuming that they necessarily measure what is claimed for them.

2. ANXIETY AND DEPRESSION

Research Evidence and Probable Mechanisms

Although chronic anxiety and depression have not characteristically been found to be precursors of myocardial infarction, nevertheless they have demonstrated a truly predictive relationship to the future development of angina pectoris and cardiac death (Jenkins, 1980). This may be due to an association with increased deposition of atherosclerosis which has been found associated with chronic anxiety and depression in cross-sectional studies involving coronary angiography (Zyzanski, et al., 1976), or may be due to other features of emotionally labile persons, such as greater autonomic and endocrine response or a lower threshold for perceiving or reporting pain.

Clinical Sign and Symptoms

The distinction between acute and chronic anxiety is important to underscore. These have been studied in depth by Spielberger (1975) and others under the labels of state anxiety and trait anxiety. Brief episodes of anxiety have been discussed earlier. Chronic anxiety is associated with a variety of physiologic changes mediated by the autonomic nervous system and endocrine glands. The mechanisms linking these affects to the cardiovascular system are probably similar to those discussed under Acute Stress Reactions, but are less well studied. Chronic anxiety is manifested by feelings of vague fear (the sources of which are usually unknown), feelings of tension, worry, increased alertness and activation, increased defensiveness, and decreased feelings of relaxation, contentment or calmness. Anxiety differs from fear in that it does not have a conscious, well defined object.

A reason for acute anxiety can often be discerned (for example, in students, approaching exams or uncertainty about a personal relationship). The reasons behind chronic anxiety are often less obvious and may require a person with psychological insights or skills to detect. Clinical recognition of anxiety is fairly straightforward and can be approached by a confidential interview, after rapport is achieved, focusing on feelings, psychophysiologic signs, interference with appetite, sleep, and other bodily functions, the interviewer being alert to detect increased defensiveness or "jumpiness".

Depression can also express itself acutely, such as a brief episode of depression after a defeat, or as chronic depressive feelings which continue at varying levels for long periods of time after a major loss. Clinically, depression is marked by sadness, discouragement about the future, diminished interest in everyday activities, lowered self-esteem, a slowing of reaction time, feeling unworthy, hopeless, and discouraged, interference with sleep (particularly early waking), loss of appetite, loss of sexual interest, and in severe cases a slowing down of physiological functioning. The observant clinician will discern depression in the "body language" of facial expression, posture, speech, and motor movement (including walking and gesturing). The psychological reasons for depressive episodes may either be circumstances of "real" loss or of symbolic losses or threats of loss.

Often the etiology of anxiety and depression is psychodynamically much more complex than this, and, particularly in the case of depression, physiological and endocrine factors including subtleties of brain enzyme balance may be involved. If this complexity is suspected, a specialist in the field should be consulted.

Test and Measures

For screening and research purposes a number of well validated measures of anxiety and depression are available. Among the more common anxiety scales are Spielberger's State-Trait Anxiety Inventory (Spielberger et al., 1970), the Zung Anxiety Scale (Zung 1971), the Bending

874

short form of the Taylor Manifest Anxiety Scale (MMPI) (Bending, 1956). There are a variety of similar scales in English as well as in other languages.

For depression among the more commonly used scales are the Zung (1965) Depression Scale, the Dempsey (1969) short form of the MMPI Depression Scale (D30), and the Beck and Beamesderfer (1974) Depression Inventory. There are other suitable scales as well, but the key requirement for all psychosocial research is that these measures be well validated and have a sufficient number of items to provide a scale score with a high reliability.

3. LIFE PROBLEMS AND DISSATISFACTIONS

Research Evidence and Mechanisms

Long sustained, chronically troubling life problems, dissatisfactions, and interpersonal conflicts have been reported in the scientific literature on numerous occasions to be precursors to angina pectoris and, with less frequency, precursors to myocardial infarction. The prospective studies using this class of variables have been conducted in Israel and Sweden, cross-sectional studies have been reported from other countries. This literature has been reviewed upon occasion, but not recently.

The overall implications of the findings are that the intensity and duration of the problem are more important in determining risk to CHD than the specific content of the difficulty. Problems and conflicts in the areas of family, work, coworkers, superiors, and finance all have been implicated at least once in various studies. Parallel to the research on anxiety, depression and neuroticism, the association of life problems and dissatisfactions seems stronger for angina pectoris than for myocardial infarction.

The central nervous sysem is probably the primary conduit by which sustained disturbing life problems affect the cardiovascular system, perhaps by way of chronic states of anxiety, depression and sleep loss, or by driving the person to escape or cope by immersing in Type A behavior. These problems may also lead to efforts to cope by over-eating, alcohol abuse, or heavy smoking.

Clinical Signs and Symptoms

Clinical recognition of these problems can be achieved by a comprehensive probing interview touching upon each of the areas of emotional investment listed above. This may be a psychosocial indicator which can be determined as well or better by interviews with family members and co-workers than it can be by interviews with the index person.

Test and Measures

To our knowledge there are no interview schedules or questionnaires for life problems and dissatisfactions which have been commonly used in cardiovascular research in the United States. The reader is referred to the English translations of the interview items used in the prospective Israeli Heart Study (Medalie, et al., 1973) and the psychosocial discord index used by Theorell and colleagues in Sweden (Theorell, et al., 1976) and Floderus-Myrhed, 1974). A number of other life satisfaction inventories have been used in sociological and psychological research, and these can be located through abstract compendia in these respective disciplines.

4. SOCIOECONIMIC DISADVANTAGE

Research Evidence and Probable Mechanisms

The relation of socioeconomic status (SES) to CHD risk has changed over the last 40 years in industrialized nations. In the earlier era in industrialized nations and currently in developing countries, persons in higher socioeconomic status have been at higher risk to CHD than their more disadvantaged neighbors. A reversal in this risk gradient has taken place, however, so that in most Western European countries and the U.S. and Canada, persons at lower levels of occupation, education, and income have now come to be at far higher risk for CHD and coronary death than their advantaged fellow citizens (Kitagawa and Hauser, 1973, Jenkins and Zyzanski, 1980, Marmot 1981). Areas with high unemployment rates, high percentages of unskilled laborers, low median education, and high frequencies of substandard or overcrowded housing, tend to be the ecologic areas with the highest rates of cardiovascular death and disability.

The mechanism of action is unproven, but there are probably several. Population surveys have shown people of lower SES in the U.S.A. to have higher average blood pressures, higher relative weight, and to smoke more cigarettes. One study has shown CHD patients with lower education to have more malignant ECG patterns on monitoring than similar patients having average or higher education (Weinblatt et al., 1978).

Clinical Signs and Symptoms

Socioeconomic level can be determined rather simply either in the clinical or research setting by questions regarding descriptions of regular occupation (or principal occupation during working life if the subject has recently changed jobs or is retired), level of schooling completed, type of housing (single family dwelling, flat, small apartment, rooming house), whether access to bathroom and kitchen facilities is exclusive to the subject's family or is shared with other households, ratio of number of persons to number of rooms in household. Asking the amount of family income is sometimes a sensitive question and has been found to be a much weaker correlate of coronary risk in groups and neighborhoods than are level of occupation and of education.

Tests and Measures

There are a variety of well-established research measures for determining socioeconomic level in sociological literature. The scales by
A.B. Hollingshead (1957) and by W.L. Warner et al., (1949), are examples.
The U.S. Bureau of the Census has a number of standard data collection
questions and a system for coding occupational titles which has proved
valuable in area studies of cardiovascular mortality, such as those by
Kitagawa and Hauser (1973).

5. SOCIAL MOBILITY AND STATUS INCONGRUITY

Research Evidence and Probable Mechanisms

After a series of strong studies supporting the validity of social
mobility (Syme, 1947) and status incongruity (Shekelle et al., 1969) as
risk indicatos for CHD in large populations in the 1960's, subsequent
research reports have been less consistent in their findings. The authors
are uncertain whether these variables are related in a primary sense to
CHD risk or whether earlier findings were based on a secondary association through a more proximal psychosocial risk factor such as the coronary-prone behavior pattern, anxiety and depression, or life problems
and dissatisfactions.

Clinical Signs and Symptoms

The clinical approach to determining whether social mobility and
status incongruity present a problem to a subject would be by means of
a series of questions which would determine whether the subject's early
life experiences had adequately prepared him/her for adjusting effectively to one's current social and occupation circumstances. Persons who have
undergone a number of major dislocations over the course of their lifetime and who therefore may have a continuing (though perhaps subliminal)

sense of "not belonging" where they are would be presumably the most likely to suffer ill effects to the cardiovascular system.

Status incongruity is the condition of simultaneously possessing conflicting characteristics of different social or cultural groups. This problem can be easily seen where there are discrepancies among levels of eduction, occupation, income, quality of housing, and organizational membership. Status incongruity is evidence that certain segments of a person's life situation (but not all) have changed during his lifetime. It also signifies that conflicting expectations or demands may be placed on that person, either by himself or others in his heterogeneous social space.

Tests and Measures

The measures of social mobility most often used in cardiovascular research are those developed by Syme et al., (1965), Wardwell and Bahnson, (1973). The simplest and apparently most promising index of status incongruity is that developed by Shekelle (Shekelle, et al., 1969). In essence, this index rates different markers of social level on scales having an equal number of intervals and congruent frequency distributions. Individuals whose several markers take on a wide range of values are considered to have status incongruity, whereas those whose ratings are more homogeneous are hypothesized to be less exposed to this source of inconsistency.

REFERENCES

Appels, A. (1980). Psychological prodromata of Myocardial infarction and sudden death. Psychother Psychosom 34:187-195.

Beck, A.T., Beamesderfer, A. (1974). Assessment of depression: the depression inventory. In Pichot P (ed.) Psychological Measurement in Psychopharmacology: Modern Problems in Pharmacopsychiatry. Vol. 7, Basel (Switz.) Karger.

Bendig, A.W. (1956). The development of a short form of the Manifest Anxiety Scale. J Consulting Psych. 20:384.

Bengtsson, C., Hallstrom, T., Tibblin, G. (1973). Social factors, stress experience, and personality traits in women with ischaemic heart disease, compared to a population sample of women. Acta Med Scand (suppl) 549:82-92.

Bortner, R.W. (1969). A short rating scale as a potential measure of pattern A behavior. J Chronic Dis 22:87-91.

Bruhn, J.G., Chandler, B., Wolf, S. (1969). A psychological study of survivors and nonsurvivors of myocardial infarction. Psychosom Med 31: 8-19.

Butensky, A., Faralli, V., Heebner, D., Waldron, I. (1976). Elements of the coronary-prone behavior pattern in children and teenagers. J Psychosom Res 20:439-444.

Dempsey, P. (1964). A unidimensional depression scale for the MMPI. J Consult Psych 28:364-370.

Dohrenwend, B.S., Dohrenwend, B.P. (eds.) (1973). Stressful Life Events: Their Nature and Effects. New York, John Wiley & Sons Inc.

Feinleib, M., Simon, A.B., Gillum, R.F., Margolis, J.R. (1975). Prodromal symptoms and signs of sudden death. Circulation 52 (Suppl 3) : III 155-159.

Floderus, B. (1974). Psychosocial factors in relation to coronary heart disease and associated risk factors. Nord Hyg Tidskr Suppl 6.

Friedman, M. (1969). Pathogenesis of Coronary Artery Disease. New York, McGraw-Hill Book Company, p 75-135.

Friedman, G.D., Ury, H.K., Klatsky, A.L. et al. (1974). A psychological questionnaire predictive of myocardial infarction: results from the Kaiser-Permanente Epidemiologic Study of Myocardial Infarction. Psychosom Med 36:327-343.

Friedman, M., Byers, S.O., Diamant, J., Rosenman, R.H. (1975). Plasma catecholamine response of coronary-prone subjects (Type A) to a specific challenge. Metabolism 24:205-210.

Haynes, S.G., Levine, S., Scotch, N., Feinleib, M., Kannel, W.B. (1978). The relationship of psychosocial factors to coronary heart disease in the Framingham study. I. Methods and risk factors. Amer J Epidemiol 107:362-383.

Hoffman, F. (1978). In neural mechanisms in cardiac arrhytmias. B.J. Schwartz Ed., Raven Press, New York.

Hollingshead, A.B. (1957). Two-factor Index of Social Position. New Haven, (Conn.) A.B. Hollingshead.

Holmes, T.H., Rahe, R.M. (1967). The social readjustment rating scale. Journal Psychosom Res 11:213-218.

Horowitz, M., Schaefer, C., Hiroto, D., Wilner, N., Levin, B. (1977). Life event questionnaires for measuring presumptive stress. Psychosom Med 39:413-431.

Hurst, M.W., Jenkins, C.D., Rose, R.M. (1978). The assessment of life change stress: a comparative and methodological inquiry. Psychosom Med 40:126-141.

Institute of Medicine (1979). Report of a Study: Sleeping Pills, Insomnia, and Medical Practice. IOM Publication 79-04. Washington: National Academy of Sciences.

Jenkins, C.D. (1978). Behavioral risk factors in coronary artery disease. Ann. Rev. of Med. 29:543.

Jenkins, C.D. (1979). The coronary-prone personality. In "Psychological aspects of myocardial infarction and coronary care"; Gentry and Williams Ed., St. Louis, Mosby 1:5-30.

Jenkins, C.D. (1979). The coronary-prone personality, Chapter 1, p 5-30 in Gentry W.D., WIlliams, R.B., (eds.) Psychological aspects of myocardial infarction and coronary care. Second edition. St. Louis, Mosby.

Jenkins, C.D., Zyzanski, S.J., Rosenman, R.H. (1979). The Jenkins Activity Survey. New York: Psychological Corporation.

Jenkins, C.D., Zyzanski, S.J. (1980). Behavioral risk factors and coronary heart disease. Psychother Psychosom 34:149-177.

Jenkins, C.D. (1981). The epidemiology of sudden cardiac death and its behavioral antecedents, in Biobehavioral factors in sudden cardiac death. Solomon et al. Ed. Nat-Acad press 13-46.

Jenkins, C.D. (1982). Psychosocial risk factors for coronary heart disease. Acta Med Scand (Suppl.) 660:123-136.

Jenkins, C.D., Stanton, B.A., Klein, M.D., Savageau, J.A., Harken, D.E. (1982). Correlates of angina pectoris among men awaiting coronary bypass surgery. Psychosom Med (In press).

Kitagawa, E.M., Hauser, P.M. (1973). Differential mortality in the United States: A study in socioeconomic epidemiology. Cambridge: Harvard University Press.

Lown, B., et al. (1976). Basis for recurring ventricular fibrillation in absence of coronary heart disease, and its management NE Jour. Med. 294:623.

Manuck, S.B. Garland, F.N. (1979). Coronary-prone behavior pattern, task incentive and cardiovascular response. Psychophysiology 16:136-142.

Marmot, M. (1978). Changing social distribution of heart disease. Brit Med J 2:1109.

Mathews, K.A. (1978). Mathews youth test for health (MYTH), described in Mathews K.A., Assessment and developmental antecedents of the coronary-prone behavior pattern in children. p 207-217, in Dembroski T.M., Weiss S.M., Shields J.L., et al., Coronary-Prone Behavior. New York: Springer.

McNair, D.M., Lorr, M., Droppleman, L.F. (1971). Profile of mood-states: Manual. San Diego: Educational and Industrial Testing Service.

Medalie, J.H., Snyder, M., Groen, J.J., et al. (1973). Angina pectoris among 10,000 men: 5 year incidence and univariate analysis. Am J Med 55:583-594.

Paykel, E.S., Prusoff, B.A., Uhlenhuth, E.H. (1971). Scaling of life events. Arch Gen psychiatry 25:340-347.

Regestein, Q.R. (1975). Relationship between psychological factors in cardiac rythm and electrical disturbances. Compr. Psychiat. 16:137.

Reich, P., De Silva, R.A., Loun, B. and Murawski, B.J. (1981). Acute psychological disturbances preceding life-threatening ventricular arrhytmias. Sons. Am. Med. Ass. 233-235.

Rosenman, R.H. (1978). The interview method of assessment of the coronary-prone behavior pattern. p 55-69, in Dembroski T.M., Weiss S.M., Shields J.L., Haynes S.G., Feinleib M., (eds.) Coronary-Prone Behavior New York: Springer.

Rove, A.A. (1977). The cardiovascular response to stress psychosomatics 18:13.

Shekelle, R.B., Ostfeld, A.M., Paul, O. (1969). Social Status and incidence of coronary heart disease. J Chronic Dis 22:381-394.

Simpson, M.T., Olewine, D.A., Jenkins, C.D., Ramsay, F.H., Zyzanski, S.J., Thomas, G., Hames, C.G. (1974). Exercise-induced catecholamines and platelet aggregation in the coronary-prone behavior pattern. Psychosom Med 36:476-487.

Spielberger, C.D., Gorsuch, R.L., Lushene, R.E. (1970). Manual for the State-trait anxiety Inventory (self-evaluation questionnaire.) Palo Alto: Consulting Psychologists Press.

Spielberger, C.D. (1975). The measurement of state and trait anxiety: Conceptual and methodological issues. p 713-725, in Levi L;, Emotions- Their Parameters and Measurement. New York: Raven.

Syme, S.L., Borhani, N.O., Buechley, R.W. (1965). Cultural mobility and coronary heart disease in an urban area. Amer J Epidem 82:334-346.

Syme, S.L. (1975). Social and psychological risk factors in coronary heart disease. Mod Concepts Cardiovasc Dis 44: (4): 17-21.

Taggart, P., et al. (1969). Some effects of motor car driving on the normal and abnormal heart. British Med. Jour. 4:120.

Taggart, P. (1973). Electrocardiogram plasma catecholamines and lipids and their modification by oxyprenolol when speaking before an audience. Lancet 2:341.

Theorell, T., Askergren, A., Olsson, A., Akerstedt, T. (1976). On risk factors for premature myocardial infarction in middle-aged building construction workers. Scand J Soc Med 4:61-66.

Thiel, H.G., Parker, D., Bruce, T.A. (1973). Stress factors and the risk of myocardial infarction. J Psychosom Res 17:43-57.

Thomas, C.B., Greenstreet, R.L. (1973). Psychobiological characteristics in youth as predictors of five disease states: suicide, mental illness, hypertension, coronary heart disease and tumor. Johns Hopkins Med J 132:16-43.

Wardwell, W.I., Bahnson, C.B. (1973). Behavioral variables and myocardial infarction in the Southeastern Connecticut Heart Study. J Chronic Dis 26:447-461.

Warner, W.L., Meeker, M., Eells, K. (1949). Social class in America. Chicago: Science Research Associates.

Weinblatt, E., Ruberman, W;, Goldberg, J.D., Frank, C.W., Shapiro, B.S. and Chaudhary, B.S. (1978). Relation of education to sudden death after myocardial infarction. New Eng J Med 299:60-65.

Williams Jr. R.B., Friedman, M., Glass, C.D., Herd, J.A., Schneiderman, N. (1978). Section Summary: Mechanisms linking behavioral and patho- physiological processes. 119-128, in Dembroski, T.M., Weiss S.M., Shields J.L., Haynes S.G., Feinleib M., (eds.) Coronary-Prone Behavior. New York: Springer.

Wolf, S. (1967). The end of the rope. The role of the brain in cardiac death. Canad. med. Ass. Jour. 97:1022.

Wolf, S. and Bruhn, J.G. (1974). Journal of psychosomatic res. 18:187.

Zung, W.W.K. (1965). A self-rating depression scale. Arch Gen Psychiatry 12:63-70.

Zung, W.W.K. (1971). A rating instrument for anxiety disorders. Psychosomatics 12:371-379.

Zyzanski, S.J., Jenkins, C.D., Ryan, T.J., Flessas, A., Everist, E.M.(1976) Psychological correlates of coronary angiographic findings. Arch Inter Med 136:1234-7.

POSSIBILITIES AND LIMITATIONS OF LONGTERM STUDIES ON THE EFFECT OF PSYCHOLOGICAL STRESS ON CARDIOVASCULAR FUNCTION

A. Perski and T. Theorell
National Institute for Phsychosocial Factors in Health
Box 60210
S-104 01 STOCKHOLM - Sweden

In the clinical world most of the psychophysiological processes that are studied are short, minutes, hours or days. Rigorous longterm studies on the effect of psychological stress on cardiovascular function are more uncommon. As clinicians we quite often have distinct ideas about patients who continue to consult us, but we lose information about all the others. Epidemiological studies, on the other hand, often have longterm perspectives, but most of them have superficial information about the individual patient. Therefore an important goal for future research will be to limit longterm studies to small groups of subjects, preferably those who really run the risk of becoming ill. The idea would be that these subjects interact with dangerous environments in disease producing ways. Relationships would then stand out much more clearly and action would perhaps prove to be more fruitful.

Sources of limitation:

a. Complex interactions

    A numer of large-scale prospective studies have taken into account various kinds of psychosocial factors in the longterm pathogenesis of coronary heart disease. Examples are

    1. Behaviour
    2. Life events
    3. Work stress
    4. Lack of social support

    As we all know these different classes of variables interact with one another. For example, in a study of 7 000 middle-aged building-construction workers in Stockholm we recorded how self-reported life events during one year related to risk of myocardial infarction during the first year of follow-up. If we disregard the most obvious pitfall, that events could occur during the months after the start of follow-up before the infarction, there are still many difficulties in interpreting the results. Several authors have pointed out that personality, coping style and physiological make-up may decide whether an event is important or not. In fact, most of our findings were negative. Total life change score for the past year did not predict elevated risk for myocardial infarction, but those with elevated scores of psychosocial discord, a type A-related variable, and high life change scores reported more cardiovascular and other illness during the follow-up (1, 2). These observations are consistent with those of Glass (3) and Siegrist (4). We also observed that there may be a complex interaction between age, group piece wage, and physical work load which could result in psychological strain possibly accelerating the onset of a myocardial infarction (5).

b. Age

In the construction-worker study, the only individual life event that was significantly more frequently reported by those who were to develop myocardial infarctions in the near future than by others, "increased responsibility at work" was reported among subjects with or without myocardial infarction older than or younger than 51 years. Figure 1 divides the whole population into four five-year age strata. It is obvious from the table and the figures that the reported prevalence of "increased responsibility last year" decreased continuously from age 40 to age 60 among these men. It was only above age 50 when "increased responsibility at work" is relatively unexpected that we saw any relationship with risk of myocardial infarction.

Table 1
Relative risk (RR) of developing a myocardial infarction (MI) during the first year of follow-up when having reported "Increased responsibility at work" during the past 12 months. Building construction worker study. (n = 6728, excluding those with heart disease, hypertension or diabetes sick-leave during preceding year.)

Age	41 - 51 years			52 - 61 years		
	RR		MI noMI	RR		MI noMI
Increased respon-sibility	1.35	Yes No	2 390 11 2910	3.50	Yes[X] No	4 236 15 3131

[X] $p < 0.05$ two-tailed (Fisher test)

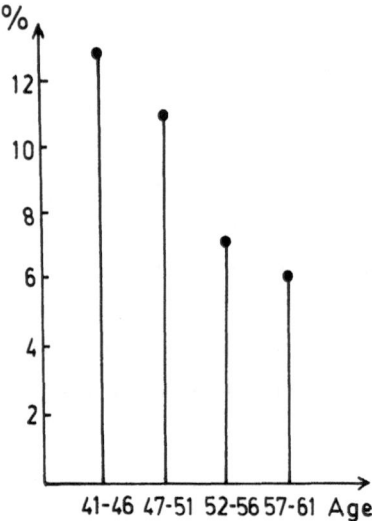

%

Perceived increased
responsibility at work

Fig.1 Frequency (%) of reported "Increased responsibility during the
       past year". Building-construction workers with no longlasting
       work absenteeism (no more than 30 consecutive days) last year.
       Total n = 5155.

c. Genetic factors

    In a study of 17 monozygotic and 13 dizygotic male middle-aged to elderly
    twin pairs we studied the acute cardiovascular effects of conversations
    about longterm difficulties and life events on hemodynamic parameters.
    Fig. 2 shows the intra-pair correlations for systolic blood pressure.
    Observations for diastolic blood pressure, peripheral vasoconstriction
    and serum growth hormone were quite similar. In the monozygotic pairs,
    but not in the dizygotic, the systolic blood pressure showed progressively
    more and more intra-pair similarity under this kind of stress. Thus, it
    would seem that under conditions that call forth our primitive defense
    reactions, the genes may be more important than at rest (6). Still
    slightly less than half of the variance was explained indicating that
    the environment is also quite important.

d. Sympatho-adrenomedullary activity at rest

    It has been pointed out that the basal sympatho-adrenomedullary activity
    may be one of the most important physiological links between psychosocial
    factors and risk of cardiovascular illness. In a study that we are just

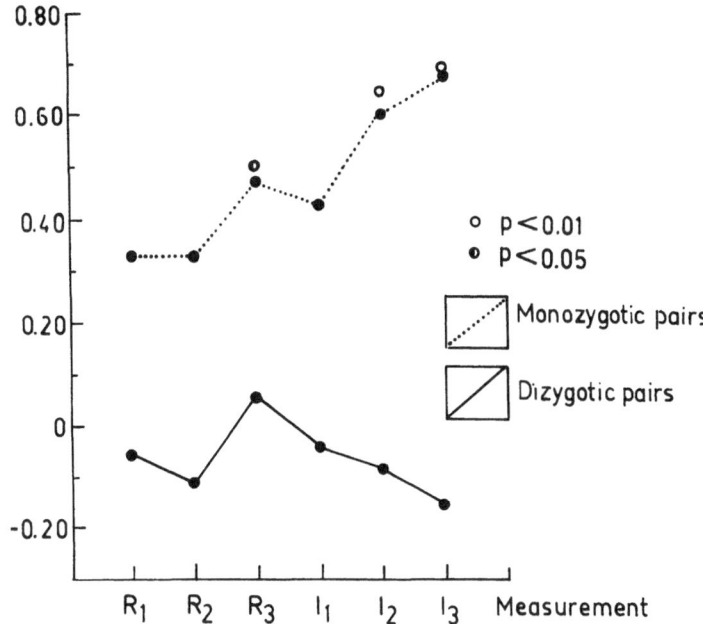

Fig. 2   Intrapair correlation coefficients for systolic blood pressure
in 17 monozygotic and 13 dizygotic middle-aged to old men with
varying degrees of coronary heart disease, during rest (R_1-R_3)
and conversation about life events and difficulties perceived
during childhood (I_1), working life (I_2) and marital life (I_3).

now finishing, a sample of 18-year-old men with <u>elevated</u> blood pressure
at rest (> 146 mm Hg <u>and</u> > 90 mm Hg diastolic) was followed up after 10
years of work experience. They were compared with men who had participated
in the same screening procedure but had "normal" (126-130 mm Hg syst.)
or "low" blood pressure (100-104 mm Hg syst.) at rest. The group with
initially elevated blood pressure turned out to have significantly
higher plasma adrenaline levels at rest ($p < 0.03$) than the other
groups. Furthermore, all subjects were divided into those who had been
working mostly in "strain jobs", in this case rushed tempo <u>and</u> low
opportunity to learn new things, according to a standard classification
system (7). Fig. 3 shows average blood pressures during work-related
activities ("work") and during other activities ("home"). The blood
pressure was measured every hour during an ordinary work day. It was
observed that the blood pressure mostly increased during work-related
activities in the initially blood pressure elevated (and "high adrenaline")
group in "strain" jobs but not in the other groups. The three-way analysis
of variance was significant for systolic blood pressure ($p < 0.01$) and

888

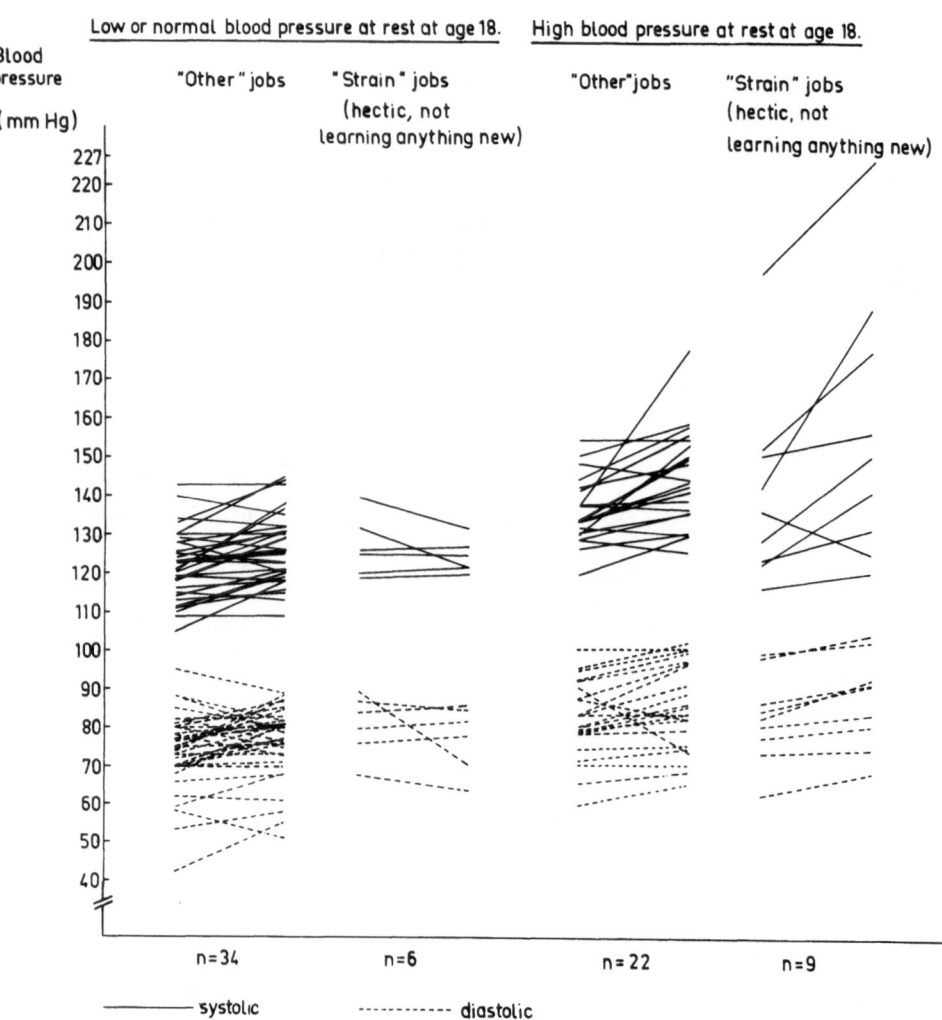

Fig. 3

for diastolic blood pressure ($p < 0.05$). These findings should be seen against the background of a significant but low-order relationship between work in hectic and dull occupations and elevated risk of myocardial infarction (7).

2. Sympatho-adrenomedullary activity to challenge

In research on physiological response to psychosocial challenges the prevalent concept has been activation theory. Cannon (8) envisioned that when we perceive a situation as threatening or dangerous, our central nervous system releases the inhibition in the sympathetic nervous system, thus pronouncing a massive, unspecific sympathetic discharge. That concept has assumed a "wired", uniform organization of autonomic nervous system and a static link to the cortex. However, in recent years the assumptions of this model have been challenged. Instead, a more dynamic model of specificity theory has been proposed (9). This more complex model proposes both stimulus-response (SR) specificity as well as individual-response(IR) specificity of autonomic responses. SR specificity is defined as a tendency for a stimulus to evoke characteristic responses from a group of patients. IR specificity is defined as a tendency for an individual to emit characteristic responses to a group of stimuli. Considerable evidence has been gathered in recent years that specificity theory is a valid concept and that autonomic response patterns are highly organized and "plastic". The most important line of evidence comes from the concept of "central command" in autonomic neurophysiology (10). Autonomic response patterns are organized in the brain and are based on functional considerations. They can be elicited by direct brain stimulation (11) or by voluntary intent (12). Those patterns also seem to be modifiable through conditioning or voluntary effort (13, 14). Recognition of the fact that the autonomic nervous system is both individually specific as well as subject to change poses another source of limitation in both long term and acute studies of stress on the cardiovascular system. An unified set of physiological responses to psychosocial stimuli cannot be expected. Emphasis should instead be concentrated on defining which stimuli and which individual autonomic patterns are specific to which cardiovascular diseases (hypertension, arrhythmias, coronary spasms, or atherosclerosis). Psychological stress testing could then be used to identify high risk individuals for specific disorders and contribute a new dimension to diagnosis.

A second consequence in addition to environmental interventions is that a _complementary_ approach to prevention of cardiovascular diseases can be used. A modification of environmental conditions (crisis situations job structure and social network) can be supplemented by modifications of physiological response patterns by means of pharmacological and physiological techniques (such as relaxation, biofeedback and autogenic training) whenever feasible and meaningful.

Possibilities

Several large-scale longterm studies have indeed demonstrated significant but mostly low-order associations between psychosocial factors and cardio-vascular disease risk. We have listed above the reasons for limitations of those studies. Improved focus on complex interactions between environmental factors and individual characteristics of patients, may provide a much clearer picture of factors relevant to breakdown of human adaptation.

References

1. Theorell T, Flodérus B and Lind E. Relationship of disturbing life-changes and emotions to the early development of myocardial infarction and other serious illnesses. 1975. Int. J. Epidemiol. 4:281.
2. Theorell T. Selected illnesses and somatic factors in relation to two psychosocial stress indices - a prospective study on middle-aged construction building workers. 1976. J. Psychosom. Res. 20:7-20.
3. Glass DC. Behavior patterns, stress and coronary disease. 1977. Wiley, New York.
4. Siegrist J, Dittman K, Rittner K and Weber I. 1980 Soziale Belastungen und Herz-infarkt. Eine Medizin-Soziologische Fall-Kontroll-Studie. 1980. Enke Verlag, Stuttgart.
5. Theorell T and Flodérus-Myrhed B. Workload and risk of myocardial infarction. A prospective psychosocial analysis. 1977. Int. J. Epidemiol. 6:17-21.
6. Theorell T, Schalling D, de Faire U and Askevold F. Personality traits and physiological reactions to a stressful interview in a group of twins with varying degrees of coronary heart disease. 1979. J. Psychosom. Res. 23:89.
7. Alfredsson L, Karasek R A and Theorell T. Myocardial infarction risk and psychosocial work environment: An analysis of the male Swedish working force. 1982. Soc. Sci. Med. 16:463.
8. Cannon W B. Bodily changes in pain, hunger, fear ang rage. 1929. Appleton and Co, New York.
9. Engel B T and Moos R M. The generality of specificity. 1967. Arch. Gen. Psychiat. 16:574-582.
10. Korner P I. Central nervous control of autonomic cardiovascular function. 1979. In Berne R M and Geiger S R (eds.). Handbook of Physiology, The Cardiovascular System I. Waverly Press, Baltimore, MD.
11. Eldridge F L, Millhorn D E and Waldrop T G. Exercise hyperpnea and loco-motion: Parallel activation from hypothalamus. 1980. Science, 211:844-846.

12. Goodwin W M, McCloskey D T and Mitchell J M. Cardiovascular and res-
    piratory responses to changes in central command during isometric
    exercise at constant muscle tension. 1972. J. Physiol. 226:173-190.
13. Joseph J A and Engel B T. Instrumental control of cardioacceleration
    induced by central electrical stimulation. 1981. Science, 214:341-343.
14. Perski A, Engel B T and McCroskery. The modification of elicited cardio-
    vascular responses by operant conditioning of heart rate. 1982. In Cacioppo
    J T and Petty R E (eds.) Perspectives in cardiovascular psychophysiology.
    1982. The Guilford Press, New York, N.Y.

INTERACTION BETWEEN SHORT-AND LONG-TERM STRESS
IN CARDIOVASCULAR DISEASE

J. Siegrist

Institute of Medical Sociology,Faculty of Medicine
University of Marburg,Marburg FRG

ABSTRACT

This paper examines the hypothesis that chronic experience of social
stress lowers the threshold for adaptive coping with acute stressors
such as negative life changes and thus increases the vulnerability for
cardiovascular breakdown.Results from a large retrospective case - control
study on 38o male patients with first acute myocardial infarction(AMI)(age
3o - 55) and a matched control group are presented which confirm the hypothesis. Additional support is given by a 18 month follow-up of 7o % of the
AMI patients.
Preliminary findings from an ongoing prospective study on 4oo initially CHD-free industrial workers reveal that indicators of nonadaptive coping
such as prolonged emotional upset and severe sleep disturbances are linked
to cardiovascular risks. It is also shown that nonadaptive coping emerges
from social contexts which generate long-term as well as short-term experiences of distress. We conclude that psychosociologic parameters should
be integrated into transdisciplinary studies of populations at cardiovascular risk,combining experimental and epidemiological approaches towards
analysis of cardiovascular pathology.

INTRODUCTION

Scientific approaches to the study of human disease usually favour
experimental manipulations which bear the advantage of testing causal
relationships and of controlling for intervening effects. It goes unsaid
that most important insights are generated by this approach. However,in
humans experimental manipulation is limited with regard to intensity and
quality of possibly noxious stimuli as well as with regard to time of
exposure. Thus the problem of "ecological validity" and predictive power
of experimental results in real human life conditions has been raised,and
this is the point where epidemiology as a distinct approach to the study
of human disease enters the scene. It may be that a strict organizational
and intellectual separation of basic experimental sciences on the one hand
and epidemiologic and socio-behavioural sciences on the other hand even
creates obstacles to innovative knowledge. A similar effect may occur if

we adhere too strictly to a separation between those who study short-term impact of stress and those who study its long-term effects. This paper presents some of the still marginal research on interaction between short- and longterm stress in cardiovascular disease from a biosocial perspective. A rather extensive review on the role of chronic social stressors in the development of cardiovascular disease is presented elsewhere in these proceedings(Siegrist et al.1983 a). And by now,several excellent reviews of the field are available(Dembroski et al.1978,Elliott et al.1982,Orth-Gomer et al.1983).

Our knowledge of how neuroendocrine reactions to acute or repetitive stressors influence patterned bodily regulations with the consequence of pathophysiologic outcome is still fragmentary. Perhaps the best exception to this statement is Folkows work on the role of structural adaptive changes in the vascular bed in the early development of essential hypertension(Folkow 1981). It seems that overrides of normal homeostatic feedback loops play an important role in the initiation of such processes and that hypothalamic activation caused by stress experience increases the excitability and responsiveness of involved systems rendering inhibition of this increase more and more difficult(Henry 1982). Cumulative stress even of a mild nature may lead to qualitative change in early stages of pathophysiology. Cumulative stress experience may also be crucial in triggering breakdown and onset of disease such as acute myocardial infarction or malignant cardiac arrhythmias(Verrier et al.1981).

Our hypothesis states that chronic experience of social stress lowers the threshold for adaptive coping in the presence of acute stressors and by this increases the vulnerability of a person at cardiovascular risk. Of course,an appropriate test of this hypothesis calls for prospective time-series analysis of selected populations,combining socio-behavioural, neurobiological and clinical information. Since this time of study is still rare,our presentation concentrates on findings from retrospective and prospective social epidemiologic studies.

MATERIALS AND METHODS

Own research on the topic includes a large retrospective case - control study on 38o mals patients with AMI(age 3o-55)and a healthy control group matched with the sample half by age,sex and occupational status.

70% of all AMI patients could be followed-up over a period of 18 months (for details see Siegrist et al., 1982). A second still ongoing study is of a prospective nature: 400 male blue-collar industrial workers (age 25-55) who are exposed to heavy workload (such as noise, shiftwork, piecework) and to considerable job insecurity are followed over a period of several years. Only CHD free subjects are included (base line ECG). Cardiovascular screening and extensive interview on psychosocial risks are executed approximately once a year. This study aims at linking changes in cardio-vascular risk factors and new cardiac events to chronic and subacute psychosocial stressors. In selected subgroups, neurohormonal reactions to a standardized stress test are being studied. Blood lipid, blood pressure, body weight and height are assessed, information about smoking, physical activity, family history of CVD, angina pectoris (Rose Questionnaire) and medication is gathered through interview. Chronic social stressors are explored in the following areas: Occupational biography, chronic workload, social support and interpersonal difficulties. Acute social stressors are registered by an inventory on negative life changes (see below). Psychological testing includes relevant dimensions of coronary-prone behaviour (Dittmann et al., 1983) and neuroticism. Information about time budget, extra job obligations, emotional upset and sleep disturbances is collected.

In study I, degree of adaptive coping is measured by a score of subjective impact of experienced negative life events (addition of weighted items on 8 dimensions of stress experience to a total score ranging from 1 to 44 points per event, assuming that high scores of subjective impact indicate low adaptive coping (Siegrist et al., 1982, Siegrist et al., 1983b). In study II, sustained emotional upset and severe recent sleep disturbances are included as indicators of non-adaptive coping.

RESULTS

We first present results from the retrospective study. According to our hypthesis a linear relation is expected between amount of chronic stress experience and amount of subjective impact by recently experienced negative life changes. Figure 1 shows results of an analysis of variance on 380 AMI patients where an index of chronic subjective workload is related to scores of subjective impact. Similar results are obtained by single dimensions of chronic workstress (e.g. time urgency, F=9.59;

p < 0.001; high responsibility F=21.59; p<0.001). All F-values remain statistically significant if we include only persons into analysis who experienced at least one negative life event during the last two years (N=295).
This is done in order to control for a possible bias created by the proportion of zero-values in the sample.

The same trends can be found in the area of chronic interpersonal
difficulties, lack of social support and unfavourable socioeconomic background. Table 1 presents results of analyses of variance in 380 AMI subjects where eleven indicators of chronic difficulties are related to the
amount of subjective impact by experienced life events. In nine of eleven
indicators F-values as well as t-tests are statistically significant.

Of course, subjective information gathered by persons who recently experienced a severe illness may be distorted. To some extent control of distortion is possible by measuring psychological characteristics such as denial and neuroticism.

Figure 1 . Analysis of variance between chronic workload and life event-
scores (subjective impact) in 380 subjects with myocardial infarction..

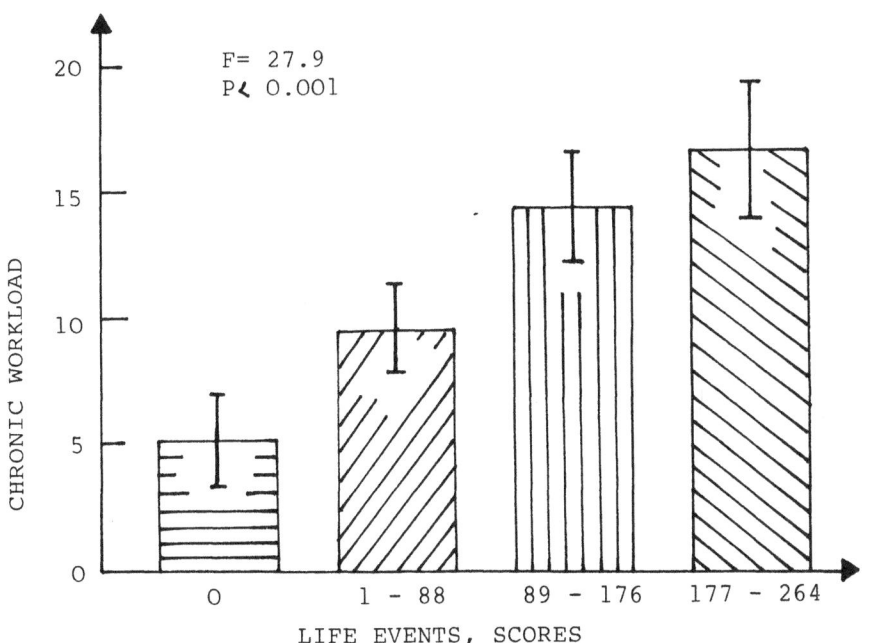

As expected, subjective impact of life event is somewhat lower in subjects with high denial ($F=3.56$, $p<0.05$) and higher in subjects with high neuroticism ($r=0.29$). However, if we calculate partial correlations between chronic workload and scores of subjective impact by life events, controlling for denial, coefficients of correlations show only minimal changes (from 0.25 (unadjusted) to 0.22 (adjusted)). The same holds true for neuroticism (from 0.39 (unadjusted) to 0.35 (adjusted)).

Table 1. Relationship between indicators of chronic interpersonal difficulties and life event scores (analysis of variance in 380 AMI subjects).

Chronic interpersonal difficulties	Life event-scores		
	F	P	t-test:P
marital conflicts	6.1	0.001	0.01
difficulties to trust spouse	8.2	0.001	0.01
problems with children	19.0	0.001	0.01
social isolation	2.1	ns	ns
bad relations to neighbours	2.9	0.05	0.01
troubles with people in general	37.3	0.001	0.01
bad housing condition	7.4	0.001	0.01
general job dissatisfaction	6.1	0.05	0.01
high occupational mobility	0.7	ns	ns
high turnover in job positions	2.8	0.05	0.001
high degree of overwork	3.1	0.05	0.001

We then can conclude that several separate measures of chronic stress experience in middle-aged males who recently survived their first acute myocardial infarction are statistically associated with an indicator of poor adaptive coping in the presence of acute negative life events. Interactions of long-term and short-term stress experience may create cumulative effects which overwhelm adaptive efforts of individuals and, by this, precipitate cardiovascular breakdown. It is of special interest in this context that "score of subjective impact by life events" was the best discriminator in a multiple stepwise discriminant analysis of 190 AMI subjects and 190 matched controls (standardized coefficient of discriminant function $=0.48$); the discriminant analysis included nine psychosocial and somatic risk factors). Number of severe acute life events (more events at all in the AMI group) as well as time pattern (more events during months preceeding illness onset in the AMI group) were significantly different between disease and control group (Siegrist et al., 1982).

Finally some very preliminary findings from the prospective study on industrial workers are reported. Currently only base line-values (first point of measurement) are available, as the second panel is being executed

right now. Within one year, two definite cardiac events occured in the
sample, one being fatal, the other non-fatal. Due to small number of hard
cardiac events at present time, the following is still highly tentative.

Table 2. Two indicators of non-adaptive coping with social stress (sustained emotional upset and severe recent sleep disturbances) in relation to
cardiovascular risks.

	% with RR > 160 and/or 105	% with ECG-abnor-malities (Minne-sota Code)	% with new AMI
Total sample (N=416)	11%	18%	0.5%
Subsample sustained emotional upset (N=45)	22%	13%	4%
Subsample severe sleep disturbances (N=11)	55%	36%	18%

Both indicators of non-adaptive coping, but especially severe sleep
disturbances, show an increased cardiovascular risk (one exception being
ECG-abnormalities in the subsample of emotional upset). It may be interesting that both cases of new AMI occured in the subsample of six persons
who showed both sustained emotional upset as well as severe recent sleep
disturbances. Finally it is important to notice that nine of eleven subjects with severe sleep disturbances suffer from excessive chronic workload, from irregular work schedules or job insecurity and that mean frequency of experienced severe negative life events during the last two
years was 2.3 per person . Severe sleep disturbances and emotional upset
thus emerge from social contexts which generate long-term as well as short-
term experiences of distress.

CONCLUSIONS

Until now only a few studies in the field of cardiovascular stress
research included time-series designs and explored possible impact of
chronic exposure to social stress on the threshold for adaptive coping in
the presence of acute stress. Friedman and co-workers in a pioneering study
showed that the blood cholesterol level of bank accountants with moderate
chronic workload increased when they worked under pressure to close their

books. Bank accountants going through periods of workload, have also been shown to respond with more vulnerability (heart rate acceleration) when exposed to a standardized physical test than otherwise (Taggart, personal communication, quoted in Theorell 1983)). In a Swedish prospective study on building construction workers, concrete workers over the age of 50 with marked psychosocial work strain one year before final data collection had an especially high risk of myocardial infarction. It is assumed that the older concrete workers at the same time suffered from chronic workload due to the fact that norms for group piecework as a basis for wages were increasingly difficult to fullfil with advanced age (Theorell et al., 1977, Orth-Gomer et al., 1983).

These preliminary findings together with the information presented in this paper call for transdisciplinary longitudinal studies combining neuro-endocrine and cardiovascular parameters of non-adaptive coping with socio-logical and psychological parameters which might increasingly be able to define populations at cardiovascular risk. Within our field many methodological and conceptual problems are still unsolved. For example, conceptual clarification of what is called quality of stress experience is needed. We have suggested elsewhere that experiences of active distress, i.e. of a synergistic activation of the defense reaction and of the conservation-whithdrawl reaction are closely related - at least in middle-aged male industrial populations - to efforts to maintain achieved social and occu-pational status which in turn assure oneself of a sense of social identity (Siegrist et al., 1982). Evidently, such an approach is just a first step towards an integration of biological and sociological concepts (Henry 1982).

REFERENCES

Dembroski,T.M.,Weiss,S.M.,Shields,J.L.,Haynes S.G.,Feinleib,M.(eds)1978.
    Coronary prone behavior.Springer,Berlin,Heidelberg,New York
Dittmann,K.H.,Matschinger,H.,Siegrist,J.1983 b.
    Fragebogen zur Messung von Kontrollambitionen.In press.
Elliott,G.R.,Eisdorfer,C.(eds)1982.
    Stress and human health.Springer New York,Heidelberg,Berlin
Folkow,B.1981.
    Central and peripheral mechanisms in spontaneous hypertension in rats.
    In:H.Weiner,M.A.Hofer,A.J.Stunkart(eds):Brain,behavior and bodily
    disease.Raven Press,New York

Friedman,M.,Rosenman,R.H.,Carroll,V.1958.
  Changes in serum cholesterol and blood clotting time in men sub-
  jected to cyclic variation of occupational stress.Circulation
  17:852
Henry,J.P.1982.
  The relation of social to biological processes in disease.Social
  Science and Medicine 16:369
Orth-Gomer,K.,Perski,A.,Theorell,T.1983
  Psychosocial factors and cardiovascular disease-a review of the
  current state of our knowledge.In press.
Siegrist,J.,Dittmann,K.H.,Rittner,K.,Weber,I.198o
  Soziale Belastungen und Herzinfarkt.Enke,Stuttgart
Siegrist,J.,Dittmann,K.H.,Rittner,K.,Weber,I.1982.
  The social context of active distress in patients with early
  myocardial infarction.Social science and medicine 16:443-453
Siegrist,J.,Dittmann,K.H.,Siegrist,K.,Weber,I.1983 a.
  Chronic social stress and cardiovascular disease-a selective
  review.In press.
Siegrist,J.,Dittmann,K.H.1983 b.
  Das Inventar zur Erfassung lebensverändernder Ereignisse.In press.
Theorell,T.,Floderus-Myrhed,B.1977.
  "Workload"and risk of myocardial infarction-a prospective psycho-
  social analysis.J.Epidemiol.6:17
Theorell,T.1983.
  Physiological issues in establishing links between psychosocial
  factors and cardiovascular illness.In press.
Verrier,R.L.,Lown,B.1981.
  Autonomic nervous system and malignant cardiac arrhythmias.In:
  H.Weiner,M.A.Hofer,A.J.Stunkard(eds):Brain,behavior and bodily
  disease.Raven Press,New York

CLINICAL CLUES OF NEURO-HUMORAL INTERPRETATION OF THE GENESIS OF CORONARY SPASM.

A. L'ABBATE
Fisiologia Clinica, Istituto del Consiglio Nazionale delle Ricerche presso l'Università degli Studi di Pisa, Via Savi, 8 Pisa, Italia

1. INTRODUCTION

In the last ten years the clinical angiographic documentation that large coronary vessels actively contract and relax and the evidence provided by hemodynamic monitoring that ischemic attack may not be preceded by any increase in myocardial oxygen demand, have challenged the traditional view of the mechanisms of angina pectoris, myocardial infarction and sudden death.

Although experimental investigators have long appreciated the capability of both large and small coronary arteries to contract and relax in responce to a large variety of stimuli, clinically coronary vascular dynamics have been confined to the adaptation of arteriolar tone to the metabolic state of the myocardium. Thus atherosclerotic lumen reduction has been considered the only factor capable of competing with metabolically mediated vasodilation, limiting oxygen delivery to the heart and decreasing myocardial oxygen tension. In addition, complications of coronary atherosclerosis such as thrombus and plaque rupture have been considered responsible for myocardial infarction. Acceptance of coronary vasospasm was reserved for Prinzmental's angina, a "variant" and rare form of myocardial ischemia.

Today the concept of large coronary artery vasoconstriction is attracting increasing clinical interest as a possible mechanism not only for Prinzmetal's angina but also for typical angina, myocardial infarction and sudden death. Perhaps most rilevant is the acquisition that major forms of heart disease, even if associated with atherosclerotic lesions, may be dependent on smooth muscle contraction of capacitance vessels and/or other factors primarily affecting coronary blood flow by alterating blood rheology.

Although all identified factors which affect coronary microcirculation (intrinsic and extrinsic) are undeniably critical in the genesis, severity and perpetuation of ischemia, spasm of conduit arteries and interference with myocardial blood flow, upstream from the potent compensatory dilator forces of anoxic metabolism in myocardium, is no longer a fact that can be ignored.

2. CLINICAL FEATURES OF CORONARY VASOSPASM

In this paper an attempt has been made to review the main clinical features of vasospastic angina and to identify facts more than theories. When possible, the results of the transposition of experimental models into the clinical setting have also been reported.

On a clinical basis, the state of the art on the patho-genetic mechanisms of myocardial ischemia can be summarized as follows:

- atherosclerotic coronary lesions are found in the great majority of patients with ischemic heart disease. However, carefull investigation of the pathogenesis of the ischemic attacks document that while severe atherosclerosis is responsible for the great majority of episodes during exercise, those at rest occur because of a primary reduction of blood flow, independent of increases in myocardial O_2 demand, in either the presence or absence of atheroscleroris.

- Angiograms obtained during ischemia at rest show coronary lumen reduction in association with a wide range of organic narrowing. The majority of coronary spasms occur in vessels with, and at the level of, severe atherosclerotic plaques.
- Indirect evidence suggests that spasm may occur for days, months or years in the same vascular segment, (same leads showing transient ischemia at repeated electrocardiograms), although in a few cases direct and indirect evidence of simultaneous or subsequent spasm in different vascular districts has been reported.
- Spasm can also occur during muscular exercise, often immediately after.
- Only few patients have the "pure" form of vasospastic angina or a "pure" exertional angina; the great majority (almost 70%) having a mixed form.
- Half or more of the ischemic episodes, regardless of their pathogenesis, are pain free. A certain correlation exists between presence of symptoms and severity and duration of the attack.
- The type of ecg alteration found during ischemia appears to be related to the transmural extention of the ischemia which is in turn related to the dignity of the narrowed vessel, the severity of narrowing, and the presence or absence of collateral flow. Thus ST-segment elevation, the characteristic ecg alteration of Prinzmetal's angina, is only an extreme of a spectrum of ecg changes associated with vasospastic angina, including ST-segment depression and T wave abnormalities.
- Prolonged ecg monitoring of patients with angina at rest reveals in many of them spontaneous rhythmicity in the incidence of the attacks (1) with a period varing from

minutes to hours or days (Fig. 1, 2 and 3).

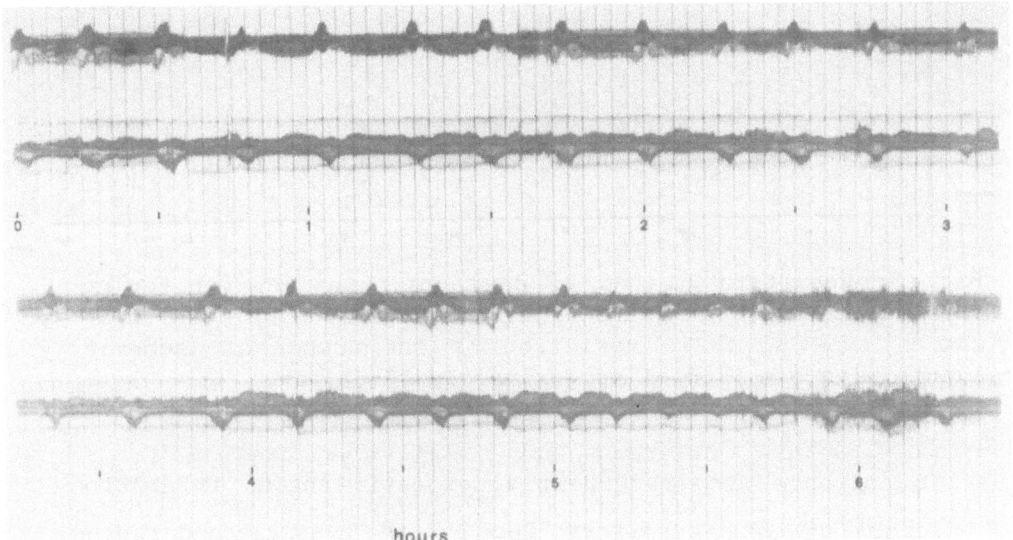

Fig. 1: Compact play-back of six-hour-two-lead-electro-cardiographic monitoring in a patient with angina at rest. Transient ischemia (positive and negative spikes respectively in the two leads) occurs with a period of approximately 15 min. All the episodes in this case were symptom free.

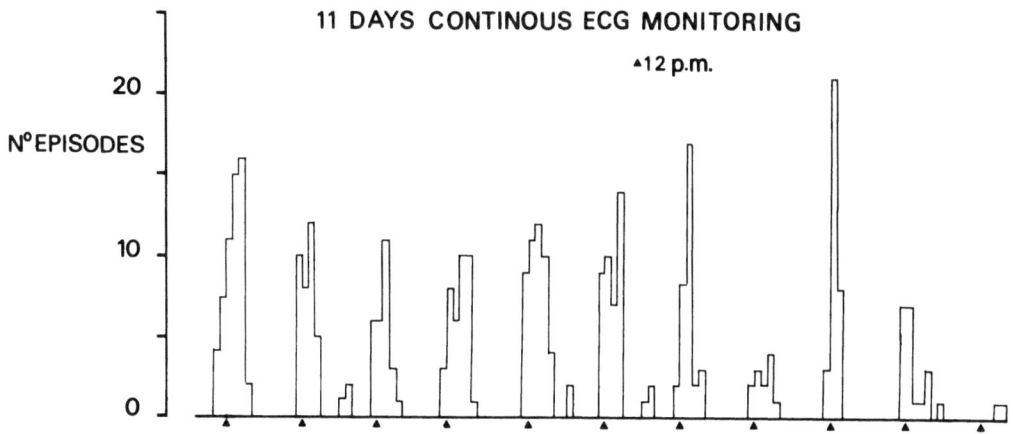

Fig. 2: Incidence of transient ischemic episode at rest as detected by 11 day continuous ecg monitoring in a patient with angina at rest. Episodes occur mostly during the night and a reproducible circadian distribution of ischemia is evident. Triangles on the abscissa indicate midnight time.

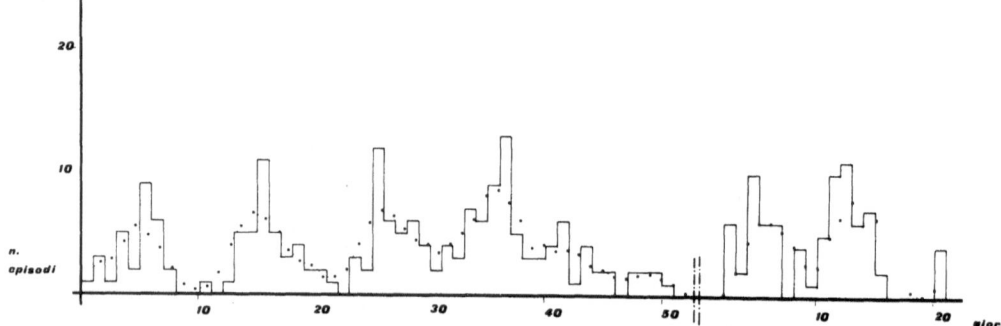

Fig. 3: Result of 51 and 21 day continuous ecg monitoring in a patient with angina at rest. Both periods of hospitalization show a cyclic variation of the number of ischemic events with a period of approximately 10 days.

- In patients with recurrent vasospastic angina infarction occurs in the territory fed by the vessel undergoing spasm.
- Coronary vasospasm can provoke rhythm disturbances and sudden death.
- Physical, psychological and pharmacological stimuli can provoke coronary vasospasm, ergonovine maleate (2) being the most effective and specific stimulus. More detailed informations on this topic is reported elsewhere in this workshop (3).
- The most effective drugs in releving and preventing spasm are nitrates and calcium-antagonists.

3. MECHANISM OF CORONARY SPASM

The mechanism of spasm is still unknown.

The results of experimental studies leave little doubt that the conduit coronary has the capacity to constrict and dilate in response to neuro-humoral, ionic, metabolic and pharmacological stimuli. Under appropriate conditions, large coronary vessels can be made to constrict independently of the response of small coronary arteries and of the nutritional state and the needs of the myocardium.

In considering the hypothetical mechanisms leading to spasm one should consider first the possibility of spasm being 1) primarily related to a coronary local defect including an abnormal response to physiological remote influence or 2) the response to an abnormal remote factor. As discussed below, a third possibility exists that in presence of a severe organic stenosis "physiological" changes in coronary tone may affect the degree of the stenosis ("dynamic stenosis") (4).

3.1 Local factors

The recurrence of spasm for long period of time in the same vessel, or even more in the same vascular segment, strongly suggests that a local abnormality is needed to trigger the exaggerated smooth muscle contraction. This consideration is quite obvious in the case of focal spasm occurring in normally or nearly normally patent vascular segment but it is not so obvious for spasm occurring at the level of severe stenosis. In fact while in the first case an exaggerated and abnormal contraction ("spasm") limited to a definite portion of the vessel is evident, in the latter a not necessarely exaggerated (in absolute terms), even minor and possibly, "physiological" increase in tone could be responsible for the reduction or occlusion of the original patency ("dynamic stenosis"). In this instance the local abnormality would be the original narrowing rather than the segmental contraction. Between these two extremes a spectrum of combinations of functional and anatomical local abnormalities can exist. This problem is familiar to coronary angiographer who performs the ergonovine test. In some patients a very small amount of the drug is capable of provoking a focal spasm not necessarily at the site of coronary stenosis, suggesting a segmental hypersensitivity to the drug, while in others much higher doses are required to produce occlusion, generally at the level of a stenosis, when a diffuse narrowing of the entire coronary tree is evident.

A more direct evidence of varying segmental sensitivity can be appreciated comparing the low and higher dosage of ergonovine required in the same patient to produce spasm respectively in the "hot" and "cold" phase of the disease. This finding suggests that the abnormal sensitivity of some segments to the same pharmacological agent can vary spontaneously (or by drug) during time.

It is pertinent in this context to note that very little information exists as to whether during spontaneous attacks a diffuse vasoconstriction (mimicing the high-dose ergonovine hypertone) can occur. Although there is some evidence that this may be the case, this situation seems to be quite infrequent. On the other hand, the investigation of coronary vascular reactivity in patients with Prinzmetal angina using quantitative angiography and cold pressure test failed to demonstrate a greater coronary reactivity as compared to controls (5).

Thus, while coronary spasm occurring in an angiographically normal vessel, or at the level of a mild stenosis, is regarded as a pathologic phenomenon capable of producing ischemia, a lower degree of hypertone when combined with atherosclerosis may lead to the same effect. In fact the presence of the organic stenosis, per se insufficient to produce ischemia at rest, is essential to let hypertone, otherwise uneffective, to become a modulating factor of coronary flow. This of course on the condition that sufficient smooth muscle is left at the level of the plaque to contract.

The same reasoning can be applied to the hypothetical role of platelet on the genesis of coronary vasospasm as platelets seem to be activated by the presence of arterial wall damage. Even in this case, only the coexistence of two factors individually uneffective, may possibly activate a vicious cycle leading to irreversible vessel occlusion due to spasm or thrombus or both.

3.2 Remote factors

While local factors can be identified in focal hyper-contraction (hypersensitivity?) and in coronary stenosis, more complex appears the identification of remote factors. Certainly the cyclicity of spontaneous ischemia observed in many patients is the strongest argument in favor of remote influence on coronary circulation.

Spontaneous cyclic contraction has been documented in animal (6) as well as in human (7) coronary strips following various pharmacological stimulations, while cyclic decrease in flow through a severely narrowed coronary artery has been reported in the dog (8). In the latter experimental model the combination of extreme narrowing both with local platelet activation and disactivation seems to be essential to produce and maintain cyclicity, although the influence of autonomic nerve stimulation has been also suggested (9-10). In man, high frequency oscillations (with periods from less than a minute to several minutes) resembling those observed in strips and in dog, have been documented expecially in the very "unstable" phase of the disease. However the documen-tation of spontaneous oscillations with much longer periods, hardly can be explained without calling forth remote influences.

Supposing that such oscillations reflect coronary tone

response to sincronous oscillations of remote "physiological" factors, one would expect for the reasons discussed earlier, a high incidence of detectable cyclicity in those patients with more severe coronary atherosclerosis, usually revealed by episodes characterized by ST-segment depression. In contrast, hourly and daily cycling is more frequently observed in patients with Prinzmetal's angina, a population which contains patients with less atherosclerosis, as compared to those with ST-segment depression.

3.2.1 Autonomic influence

Sympathetic influence. Continous changes in vasomotor tone is a normal characteristic of arteries. Alterations in vasomotor tone are believed to be to a great extent influenced by the sympathetic nervous system. Despite the extensive literature on sympathetic control of coronary vessels, data concerning the neural control of the conduit coronary vessels as opposite to the resistence vessels is still lacking. It has been proposed that coronary spasm may be related to the neural regulation of major coronary arteries and in particular to alpha-receptor activation.

The role of alpha-adrenergic stimulation on coronary conduit artery constriction has been well established in the animal (11,12). Gerova et al. (13) investigated the neural control of the large coronaries by measuring the external diameter of the desceding artery during sympathetic stimulation. The maximal diameter decrease, provided that passive distension by blood hypertension was avoided, was 4%, a value which, if applied to man, hardly could explain myocardial ischemia in patients with normal or mildly diseased vessels. On the contrary, this mechanism could account for primary ischemia in patients with severe coronary atherosclerosis. This mechanism may also explain the increase in coronary vascular resistence during cold pressure test obtained in patients with severely damaged coronaries as opposite to controls, its reduction by phentolamine (14), and the inconsistency of provocation of ischemia by cold pressure test or epinephrine (15) or the only occasional prevention of spontaneous attacks by alpha-blockers (16). The same reasoning could be applied to the poor sensitivity of arithmetic mental stress testing applied in patients with ischemic heart disease and the consistent finding of severe atherosclerosis in those cases with positive test (3).

As far as the possibility of an abnormally high generalized sympathetic outflow in ischemic patients, the investigation by Robertson et al. (17) of autonomic function in three patients with variant angina and absent or mild stenosis does not support this view.

In a recent report by Chierchia et al. 9 patients with Prinzmetal's angina and positive ergonovine test were studied during cold pressure test and stepwise infusion of phenylephrine. Only one patient with very frequent attacks had a positive cold test, while none responded to phenylephrine. The same authors failed to document any preventing effect of phentolamine as compared to placebo in a single blind trial in 5 patients with frequent episode of Prinzmetal's angina. In addition, in patients with variant angina, an increase in heart rate and corrected Q-T interval followed or paralleled but did not precede the ST-segment changes (18).

Thus available evidence supporting the hypothesis of an adrenergically mediated coronary vasoconstriction responsible for myocardial ischemia is far from being conclusive. In addition adrenergic stimulation could likely affect not only conduit arteries synergically with organic lesions but also microcirculation, hemodynamics and myocardial metabolism.

This last consideration is important for the evaluation of the effect of beta-blocking agents in resting angina. If noradrenergic factors were a major consideration one might expect beta-blockade to increase the number of ischemic attacks enhancing alfa-adrenergic tone. This is the case in some reports (19) but not in our experience. When we compared propranolol to verapamil in a double blind crossover trial performed in 12 cases of resting or mixed angina, propranolol was not different from placebo and in one case (the one with the more severely diseased coronary arteries) was as effective as verapamil (20). This discrepancy could be explained on the basis of patient selection. It should be kept in mind that coronary flow reserve can be reduced by factors originating at the microcirculation level which is sensitive to contraction and hemodynamic changes, two parameters largely affected by alpha and beta stimulation.

In conclusion alpha-adrenergic activation can lead to ischemia only occasionally expecially in patients with severe atherosclerosis. The fact that alpha-adrenergic-blockade have been reported to occasionally reduce the attack rate does not prove that excess alpha activity is the cause of the disease, any more than for blood hypertension (17).

Parasympathetic influence. Experimental evidence exists for the functional vagal innervation of the coronary vasculature (21), however the patho-physiological significance of this innervation is far from being defined. Clinically, it is frequent to observe a higher incidence of ischemic attacks during the night which parallel the decrease in heart rate and blood pressure. This occurs expecially, but not exclusively, in patients with Prinzmetal's angina. In spite of this suggestive time-course-correlation, of the induction of coronary vasospasm by methacoline and its release by atropine reported by some Authors (23-24), in our experience, prolonged and systematic treatment with atropine did not abolish ischemic attacks.

The action of acetylcoline is supposed to be mediated through the release of norepinephrine from the post-ganglionic sympathetic nerve terminals which act on the dense population of alpha-receptors in the large coronary arteries (25).

3.2.2 Humoral factors. A part from neuro-transmitters, coronary arteries have been documented in experimental setting to be sensitive to various substances such as prostaglandins, thromboxan, hystamine, serotonin, bradikinin, leukotriene and others. On clinical grounds no definite conclusions can be drawn on a possible pathogenetic role of these substances. A focal coronary wall defect in prosta-cyclin production has been suggested as a possible cause of coronary vasospasm. However attempts to prevent episodes of variant angina by prostacyclin infusion gave inconsistent results (26). The role of serotonine seems challenged by the failure of ketanserin, a serotonin antagonist, to prevent episodes of angina (27).

3.2.3. Phsyco-emotional factors. Emotional factors have historically been recognized to interact with cardiovascular function and disease. It is possible that a so popular and entrenched conviction may have served to prevent a systematic investigation of the relationship between brain and coronary vessels.

Ernst et al (28) have demonstrated that transient coronary blood flow impairment can be operately conditioned (shock avoidance) in the dog, documenting the responsiveness of coronary vasculature to emotionally stressfull situations. In order to elucidate the mechanism of such responce, experimental studies have been performed, in which differentiation between direct and indirect (mediated by increased myocardial demand) effects of adversive conditioning on coronary vessels has been attempted by pharmacological blocking agents alone and in combination with cardiac pacing (29). The results of these studies strongly suggest that the behavioural stress results in a neurally mediated activation of the coronary alpha receptors ultimately producing an early vasoconstriction. It is interesting to note that in these experiments pacing abolished the vasoconstriction while this "protective effect" of pacing was abolished by beta-blockade.

Clinically, it has been reported that arithmetic mental test, similarly to cold-pressure test, may provoke ischemia only in a minority of cases, by decreasing coronary flow and/or increasing oxygen consumption (3). Bassan et al failed to document in coronary patient an abnormal coronary vasoconstriction in responce to mental stress (30). Once again it is possible that for a given stimulus reaching the coronary arteries, the final effect on oxygen delivery to the myocardium may be conditioned by the coronary anatomy and the magnitude of the residual reserve.

No data are available on the separate effect of psychological stress on large and small vessels neither in animals nor in humans.

3.2.4 Sleep. Differing from previous reports suggesting an association between REM phase and nocturnal ischemia, more recent and systematic studies have challenged this relationship. Nocturnal episodes occur without significant relation to sleep stages or intrasleep wake periods (31). On the other hand, the sleep pattern is markedly impaired in these patients indipendent of the presence or absence of ischemic attacks during monitoring (32).

4. CONCLUSIONS

Technology is available today to assess in man coronary resistence, the responce of large coronary arteries to various stimuli, to assess non invasively regional cardiac function and to monitor and memorize neural, cardiovascular and biochemical parameters. This procedure is complex but certainly scientifically rewarding. It is through this way that we must proceed if we are willing to understand the pathophysiology of ischemic heart disease. While the existing data in many istances are difficult to interpret or integrate, sufficient evidence does exist to warrant a concerted investigation into the involvement of psychological factors in myocardial ischemia.

ACKNOWLEDGMENTS

I am very grateful to Miss Daniela Banti for her skilled assistance in preparing the manuscript.

REFERENCES

1. Biagini A, Carpeggiani C, Mazzei MG, Michelassi C, Antonelli R, Testa R, L'Abbate A, Maseri A. 1981. Distribuzione oraria degli episodi di angina a riposo: effetti della terapia medica. G. It. Cardiol., 11: 4.
2. Heupler FA Jr, Proudfit WL, Razavi M. 1978. Ergonovine maleate provocative test for coronary arterial spasm. Am. J. Cardiol., 41: 631.
3. Specchia G, De Servi S. 1983. Provocative testing for coronary spasm. These proceedings.
4. MacAlpin RN. 1980. Relation of coronary arterial spasm to sites of organic stenosis. Am. J. Cardiol., 46: 143.
5. Raizner AE, Chahine RA, Ishimori T, Verani MS, Zacca N, Samal N, Miller RR, Luchi RJ. 1980. Provocation of coronary artery spasm by the cold pressor test. Hemodynamic, arteriographic and quantitative angiographic observation. Circulation 62: 925.

6. D'Hemecourt A, Detar R. 1978. Possible physiological basis for locally induced spasm of large coronary arteries. In Maseri A, Klassen GA, Lesch M eds. Primary and Secondary Angina Pectoris. New York, Grune & Stratton, pp. 177.

7. Ross G, Stinson E, Schroeder J, Ginsburg R. 1980. Spontaneous phasic activity of isolated human coronary arteries. Cardiovasc. Res. 14: 613.

8. Folts SD, Gallagher K, Rowe GG. 1982. Blood flow reductions in stenosed dog coronary arteries: vasospasm or platelet aggregation? Circulation, 65: 248.

9. Folts SD, Bonebrake FC. 1982. The effects of cigarette smoke and nicotine on platelet thrombus formation in stenosed dog coronary arteries: inhibition with phentolamine. Circulation 65: 465.

10. Raeder EA, Verrier RL, Lown B. 1982. Influence of the autonomic nervous system on coronary blood flow during partial stenosis. Am. Heart J., 104/ 249.

11. Feigl EO. 1967. Sympathetic control of coronary circulation. Circ. Res., 20: 262.

12. Vatner SF, Pagani M, Manders WT, Pasipoularides AD. 1980. Alpha adrenergic vasoconstriction and nitroglycerin vasodilation of large coronary arteries in conscious dogs. J. Clin. Invest., 65: 5.

13. Gerovà M, Barta E, Gero J. 1979. Sympathetic control of major coronary artery diameter in the dog. Circ. Res., 44: 459.

14. Mudge GH, Grossman W, Mills RM, Lesch M, Braunwald E. 1976. Reflex increase in coronary vascular resistance in patients with ischemic heart disease. N. Engl. J. Med., 295: 1333.

15. Yasue H, Touyama M, Shimamato M, Kato H, Tanaka S, Akiyama F. 1974. Role of autonomic nervous system in the pathogenesis of Prinzmetal's variant form of angina. Circulation, 50:

16. Ricci DR, Orlick AE, Cipriano PR, Guthaner DF, Harrison DC. 1979. Altered adrenergic activity in coronary arterial spasm: insight into mechanism based on study of coronary hemodynamics and electrocardiogram. Am. J. Card., 43: 1073.

17. Robertson D, Robertson RM, Nies AS, Oates JA, Friesinger GC. 1979. Variant angina pectoris: investigation of indexes of sympathetic nervous system function. Am. J. Card., 43: 1080.

18. Chierchia S., Crea F., Davies GJ, Berkenboom G, Crean P, Maseri A. 1982. Coronary spasm: any role for alpha receptors? Circulation, 66 (Suppl. II) 247 (abstract).

19. Robertson D, Alastair JJW, Vaughn WK, Robertson RM.

1982. Exacerbation of vasospastic angina pectoris by propranolol. Circulation, 65: 281.

20. Parodi O, Simonetti I, L'Abbate A, Maseri A. 1982. Verapamil versus propranolol for angina at rest. Am. J. Card., 50: 923.

21. Feigl EO. 1969. Parasympathetic control of coronary blood flow in dogs. Circ. Res., 23: 509.

22. Yasue H, Touyama M, Kato H, Tanaka S, Akiyama F. 1976. Prinzmetal's variant form of angina as a manifestation of alpha-adrenergic receptor-mediated coronary artery spasm: documentation by coronary arteriography. Am. Heart J., 91: 148.

23. Endo M, Hirosawa K, Kaneko N, Hase K, Inoue Y, Konno S. 1976. Prinzmetal's variant angina. Coronary arteriogram and left ventriculogram during angina attack induced by methacoline. N. Engl. J. Med., 294: 252.

24. Stang JM, Kolibash AJ, Schorling JB, Bush CA. 1982. Methacoline provocation of Prinzmetal's variant angina pectoris: a revised perspective. Clin. Cardiol., 5: 393.

25. Levy MN. 1971. Sympathetic - parasympathetic interactions in the heart. Circ. Res., 29: 437.

26. Chierchia S, Patrono C, Crea F, Ciabattoni G, De Caterina R, Cinatti GA, Distante A, Maseri A. 1982. Effects of intravenous prostacyclin in variant angina. Circulation, 65: 470.

27. De Caterina R, Carpeggiani C, L'Abbate A. 1983. A double-blind, placebo-controlled study of ketanserin, a serotonin arterial and platelet receptor blocker, in vasospastic angina. International Congress on Thrombosis. Stockholm (abstract).

28. Ernst FA, Kordenat RK, Sandman CA. 1979. Learned control of coronary blood flow. Psychomat. Med., 41: 79.

29. Billman GE, Randall DC. 1980. Mechanism mediating the coronary vascular response to behavioural stress in the dog. Circ. Res., 48: 214.

30. Bassan MM, Marcus MS, Ganz W. 1980. The effect of mild-to-moderate mental stress on coronary hemodynamics in patients with coronary artery disease. Circulation, 62: 933.

31. Maggini C, Guazzelli M, Mauri M, Chierchia S, Cassano GB, Maseri A. 1978. In Primary and secondary angina pectoris. Maseri A, Klassen GA and Lesch M eds. Grune & Stratton Inc. New York, p. 157.

32. Maggini C, Guazzelli M, L'Abbate A, Biagini A, Chierchia A, Pieri M, Rocca R, Carpeggiani C, Emdin M. 1983. Fourth International Congress of APSS, Bologna (abstract).

PROVOCATIVE TESTING FOR CORONARY SPASM.

G.SPECCHIA, S. de SERVI
Division of Cardiology, Policlinic "S. Matteo", University of
Pavia, I - 27100 Pavia, Italy.

Two pathogenetic mechanisms play a role in triggering acute myocardial
ischemia. The first occurs when a fixed coronary stenosis limits the increase
of coronary blood flow in front of augmented myocardial metabolic requirements.
The second mechanism takes place when the ischemic event is produced by a
primary acute decline in coronary flow, generally because of coronary
vasospasm(1).

The Knowledge of pathogenetic mechanism could be of definite importance
in the evaluation of an ischemic syndrome. The demonstration of a vasospastic
mechanism could mainly affect the management of the patients, influencing
therapeutic decision in favour of medical or surgical approach.

Provocative test. The difficulties in documenting spontaneous coronary
spasm during angiographic studies had led to the induction of spasm by
physical or pharmacological means and to the use of these diagnostic tests
in suspected vasospastic angina.Provocative tests of coronary spasm have
a diagnostic value and utility if they are able to induce spasm only in
patients with"primary" vasospastic angina and not in normals. High
specificity and sensitivity of these tests is required and the induced ischemi
episodes should be similar to the spontaneous ones.Finally they must have few
and non dangerous side-effects,the greatest potential risk being associated
with the ischemic episodes they may induce.

Ergonovine test. Ergonovine Maleate,an ergot alkaloid,with a constrictive a
tion on the vascular smooth muscles,has been used since 1949 because of
its ability to provoke chest pain associated with ST changes in patients
with angina pectoris. Recent angiographic studies have demonstrated that
Ergonovine can induce myocardial ischemia in patients with angina at rest

through its capacity to trigger a coronary vasospasm (2) (3).

The mechanism of Ergonovine action is not yet completely clarified.

The effect of the drug on the arterial wall may be direct or mediated by alpha-adrenergic,serotonergic or dopaminergic receptors.

A positive intravenous Ergonovine test shows significant ischemic E.C.G. changes with or without typical chest pain,and coronary arteriography performed at that time reveals coronary vasoconstriction,which may result in severe or total occlusion of the coronary vessels due to the spasm.

In negative tests,clinical signs and E.C.G. changes do not occur and coronary vessels are only mildly narrowed as normal physiological response to the drug's pharmacological action.

In Ergonovine-positive patients,clinical,electrocardiographic,haemodynamic and angiographic features of the induced ischemic episodes have been shown to be similar to those observed during spontaneous attacks. (4)

A coronary vasospasm induced by Ergonovine may completely occlude a major coronary vessel either superimposed or not to an organic fixed stenosis, namely an atherosclerotic lesion. In these cases the E.C.G. shows transient ST-segment elevation which represents transmural myocardial ischemia. This pattern is the hallmark of the so called Prinzmetal's variant angina.

In fewer cases,Ergonovine-induced coronary spasm may occlude a distal portion of the vessel,a smaller coronary branch,or determine a severe,diffuse narrowing of the coronary tree:in these cases the E.C.G. shows ST-segment depression,representig subendocardial ischemia.(5)

Thus Ergonovine may induce myocardial ischemia in at least two different ways:
1-Causing a severe coronary spasm and,frequently, a complete transient vasospastic occlusion of a coronary vessel. The spasm may be superimposed to a fixed coronary stenosis or affect a completely normal coronary vessel.
2-Narrowing a coronary vessel which presents a non significant organic stenosis.

These mechanisms of determining myocardial ischemia may be also effective in in the spontaneous attacks in patients with angina at rest.

The Ergonovine test is highly sensitive and specific for the detection of vasospastic angina.In patients with documented episodes of angina at rest,the sensitivity of the test is higher than 80% and is about 100% in patients with Prinzmetal's angina.

Because of the well known variation in the natural course of vasospastic angina,the sensitivity of the test should be lesser in the phases of spontaneous remission of the disease.(6)

Other vasoactive substances,like Histamine, Methacoline and Epinephrine have been shown to be able to induce coronary spasm and many separate membrane receptors have been postulated to mediate coronary vasoconstriction.

The mechanism of spontaneous coronary vasospasm may be beyond the normal level of interaction between each agonist and its membrane receptor.

Alternatively, the membrane receptors, which mediate vasomotory response, may be altered resulting in vasospasm by activation of any of them.

Alkalosis test. In normal subjects alkalosis decreases coronary blood flow and increases coronary arteriolar resistance whereas it could induce coronary vasospasm in patients with abnormal sensitivity to such coronary vasomotory response.

Alkalosis has been induced by infusion of alkaline solution or by sustained hyperventilation or hyperventilation plus buffer infusion.(7)

The effect of alkalosis has been attributed to the reduction in H-ions concentration facilitating the inflow of Calcium-ions through the cellular membrane and the smooth muscle contraction.

The fact that coronary spasm induced only by hyperventilation is often delayed suggests that different mechanisms could play a role in coronary vasoconstriction determined by buffer infusion or by hyperventilation.(8)

In our experience the sensitivity of alkalosis test is very low and remarkable side-effects are common (dizziness, lightheadness,nausea,paresthesia skin rush).

Cold pressor test.Cold pressor test is performed by submersing the patient's free hand in ice water for one minute.The cold stimulus elicits a simpathetic reflex, mediated by alpha adrenergic receptors,which is able to increase coronary resistance and reduce coronary blood flow in patients with coronary artery disease.The effect of cold pressor test has been ascribed to inappropriate vasoconstriction unopposed by metabolic vasodilatation. (9)

Cold-induced vasoconstriction may provoke myocardial ischemia when an increase of myocardial oxygen consumption is not compensated by a corresponding increase in coronary flow.

Coronary spasm during cold pressor test has been also described in patients with variant angina. (10) The sensitivity of this test in angina at rest is low (8%) but it is higher in patients with Prinzmetal's angina.

Mental stress. Emotional stress,which elicits a strong sympathoadrenal discharge,may produce myocardial ischemia in patients with coronary artery disease. Although mental arithmetic stress test has been proposed as a simple method of inducing emotional reactions on cardiac patients, there is a substantial lack of information about the ability of the test to provoke myocardial ischemia in patients with coronary artery disease. Moreover it is not clear whether an inappropriate vasoconstriction may play a role in the production of an ischemic response to emotional stress.

In a recent study (11),13 patients with variant angina underwent a mental arithmetic stress test. The patients were asked by one physician,whom they had never seen before,to substract a two-figure number,say 17,serially from a four-figure number,say 1013.

The physician,having a chronometer, required the patient to give the correct answer every two seconds, while another physician exhorted the patient to concentrate about the calculations and reproached them for their lack of effort if the answer was not corrected. The procedure lasted one and a half minute and was stopped before its completion if angina, ST-segment elevation 2 mm or ST-segment depression 3mm occurred. A twelve-lead electrocardiogram was recorded at the end of the test and for five minutes thereafter.

Three patients showed ST-segment elevation during the test, four patients
ST-segment depression while six patients had a negative test and did not
show ST-segment abnormalities. One patient, who showed ST-segment elevation
during mental stress, developed a ventricular tachycardia that required
electrical defibrillation. To investigate the pathogenetic mechanism of
mental-stress induced myocardial ischemia, great cardiac vein flow was
measured using the thermodilution technique in 2 patients with isolated
left anterior descending artery disease who showed ST-segment depression
in response to mental stress. In both patients coronary resistance increa=
sed suggesting coronary vasoconstriction.These data demonstrate that coronary
constrictive stimuli may be elicited by mental stress in susceptible patients.

Exercise testing. Exertional angina is considered to be the result of
increased myocardial metabolic requirements in the presence of organic
obstructions of major coronary vessels. Thus, exercise testing may induce
myocardial ischemia by increasing heart rate and blood pressure, two im=
portant determinants of myocardial oxygen consumption. However, recent
data (12,13) have shown that exercise may trigger coronary spasm in
susceptible patients. In our experience, about one third of patients with
Prinzmetal's angina developed during or just after exercise ST-segment
elevation in the same leads where it occurred during spontaneous attacks.
In all patients studied, a completely occlusive coronary spasm was demo=
nstrated by coronary angiography performed at the time of this ECG pheno=
menon. However the response to exercise of patients with variant angina
is frequently variable,changing from one day to another. This variability
of angina threshold is a clinical aspect which is frequently observed
in patients with vasospastic angina: in recent studies (14,15) we provided
the objective demonstration of the variable occurrence of chest pain and
ST-segment abnormalities in such patients, by repeating serial bycicle
exercise tests on different days. We believe that the variable threshold
of angina during exertion should be secondary to the difference in tone of
the coronary arteries at the start of exercise. When coronary arterial tone
is low, these patients can tolerate a greater whorkload because myocardial

oxygen supply is limited only by organic obstructions of the coronary arteries, but when coronary tone is high, oxygen supply is further reduced and even moderate physical activities can precipitate chest pain and myocardial ischemia. A substantial support to this hypothesis has been provided by the observation that nifedipine, a calcium antagonist drug , may enhance angina threshold in these patients by increasing coronary flow and reducing coronary resistance during exercise. However, further studies are needed to clarify if coronary tone changes may play a role in the production of myocardial ischemia in patients with stable exertional angina, who do not seem to be susceptible to vasospastic influences.

Conclusions . Our understanding of the pathophysiology of the spontaneous attacks of vasospastic angina has been enhanced by the widespread use of provocative tests and by the observation of clinical, ecg and angiographic patterns during Ergonovine-induced myocardial ischemia.

Provocative testing for coronary artery spasm is also of high clinical value in the diagnosis of vasospastic angina . The indication for provocative tests is well established when vasospastic angina is suspected but all diagnostic means (especially Holter monitoring) are negative. In such cases the risk of misleading an important ischemic disease would be greater than the potential risk associated with the provocative tests.

References

1. Maseri A, Pesola A, Marzilli M et al. Coronary vasospasm in angina pectoris. Lancet 1977; 1: 713
2. Specchia G, Angoli L, De Servi S et al. Spasmo coronarico indotto dalla somministrazione di ergonovina maleato in soggetti affetti da angina spontanea. G Ital Cardiol 1976; 6: 1777
3. Heupler FA, Proudfit WL, Razawi M et al. Ergonovine maleate provocative test for coronary-artery spasm. Am J Cardiol 1978; 41: 631.
4. Specchia G, Bramucci F, Angoli L et al. Spontaneous and provoked coronary artery spasm: are they the same? Eur J Cardiol 1978; 8: 581.
5. L'Abbate A, Ballestra A, Maseri A et al. Morphology of coronary spasm , its relation to coronary atherosclerosis, clinical and electrocardiographic findings. In: Coronary Arterial Spasm. Bertrand MF editor , Lille 1979, p 109.

6. Waters DD, Szlachcic J, Théroux P et al. Ergonovine testing to detect spontaneous remission of variant angina during long-term treatment with calcium antagonist drugs. Am J Cardiol 1981; 47: 179.

7. Yasue H, Nagao M, Omote S, et al. Coronary arterial spasm and Prinzmetal's variant form of angina induced by Hyperventilation and Tris-Buffer infusion. Circulation 1978; 58: 56.

8. Girotti LA, Crosatto JR, Messuti H et al. The hyperventilation test as a method for developing successful therapy in Prinzmetal's angina. Am J Cardiol 1982; 49: 834.

9. Mudge GH, Grossman WW, Mills RM et al. Reflex increase in coronary vascular resistance in patients with ischemic heart disease. N ENgl J Med 1976; 295: 1333.

10. Raizner AF, Chahine RA, Ishimori T et al. Provocation of coronary artery spasm by cold pressor test. Circulation 1980; 62: 925.

11. Specchia G, De Servi S, Falcone C et al. Mental arithmetic stress testing in patients with coronary artery disease. In press.

12. Specchia G, De Servi S, Falcone C et al. Coronary arterial spasm as a cause of exercise-induced ST-segment elevation in patients with variant angina. Circulation 1979; 59: 391.

13. Specchia G, De Servi S, Falcone C et al. Significance of exercise-induced ST-segment elevation in patients without myocardial infarction. Circulation 1981; 63: 46.

14. De Servi S, Specchia G, Curti MT et al. Variable threshold of angina during exercise: a clinical manifestation of some patients with vasospasti angina. Am J Cardiol 1981; 48: 188.

15. De Servi S, Specchia G, Falcone C et al. Variable threshold exertional angina in patients with vasospastic myocardial ischemia. Am J Cardiol 1983; 51: 397.

HEMODYNAMIC CHARACTERIZATION OF DIFFERENT MENTAL STRESS TESTS.

L.TAVAZZI, G.MAZZUERO, A.GIORDANO, A.M.ZOTTI, G.BERTOLOTTI
Centro di Riabilitazione di Veruno - Fondazione Clinica del Lavoro - VERUNO (NO) - Italy

INTRODUCTION

The main purpose of this study was to assess whether different types of mental stress would induce cardiac activation in patients with recent myocardial infarction and, in the affirmative, to analyze the response principally about the occurrence of arrhythmias, myocardial ischemia or ventricular disfunction in order to identify 1) subjects at potential risk for mental stress 2) psychological and cardiovascular markers predictive of such response.

PATIENT POPULATION

Forty eight men aged 26-73 (mean 54.5) were studied 40 days (range 20 to 126 days) after acute myocardial infarction. The site of infarction was anterior in 18 patients, inferior in 25 and antero-inferior in 5. All patients were functionally classified as NYHA I or II.

Specific cardiovascular treatment was discontinued before the tests (nitrates, calcium-antagonists 24-48 hours, beta-blockers 72 hours). No patient were treated with digitalis glucosides or amiodarone.

METHOD

Psychological profile. The psychological profile of each patient was established in the day before the administration of the stressors, by means of an interview and a battery of tests including Wechsler-Bellevue F1, to evaluate the Intelligence Quotient (IO); Coronary Prone Behavior Questionnaire (1), Rathus Assertiveness Schedule (2,3), IPAT Anxiety Scale, IPAT Depression Scale, Middlesex Hospital Questionnaire and Life Experience Survey (4,5) to identify the psychological profile; State-Trait Anxiety Inventory X1 (STAI X1) (6,7) to evaluate anxiety state before and after the stress tests.

Stressors. Mental stress was induced by administration of 4 different stressors: mental arithmetic, Sack's test (8,9), Raven progressive matrices PM 47 and a white noise. Twenty two patients underwent the arithmetic test. The 26 subsequent patients were tested with 4 stressors, according to a randomized pattern. The mental arithmetic was graded into three different levels of difficulty in relation with the patient IQ (1013-17...etc. for IQ $\gg$ 110, 251-7 for 110$\rangle$=IQ$\rangle$=90, 101-3 for IQ$\langle$90). Each stress lasted 3' with a time interval of 5' between one stress and the following one.

Before and after the stressors administration patients filled in the STAI X1.

ECG and hemodynamic monitoring. A 7F Swan-Ganz thermodilution catheter was percutaneously introduced and located in a branch of the right pulmonary artery under fluoroscopic control. A 12 lead ECG was recorded before the stressors administration and at 1' intervals during the whole duration of the study. Two ECG leads were continuously visualized on a monitor. Blood pressure was measured by cuff at 1' intervals. An ECG lead, the right atrial pressure and pulmonary arterial pressure curves were visualized on an oscilloscope and continuously recorded on a magnetic tape. Mean pulmonary wedge pressure and mean right atrial pressure were measured at 1' interval. Cardiac output was determined by thermodilution before and at the end of each stress test. The administration of stressors started after 5' rest values recording.

All patients underwent 24 hours Holter monitoring after mental stress testing.

Statistical analysis. Analysis of variance with repeated measures and post hoc Neuman-Keuls test were applied for the intra and interstressor analysis. The correlation between hemodynamic and psychological parameters was tested by stepwise multiple regression analysis.

RESULTS

Hemodynamic activation during mental arithmetic. Table 1 shows basal and peak hemodynamic mean values observed during mental arithmetic in

48 patients.

Table I: Hemodynamic changes induced by mental arithmetic in 48 pts 40±10 days after acute myocardial infarction (mean ±SD)

Abbreviations: HR: heart rate, BP: systemic blood pressure, RAP: mean right atrial pressure, PWP: mean pulmonary wedge pressure, CO: cardiac output.

	CONTROL		STRESS
HR	75.0±13.5	*	90.8±15.5
SystBP	137.4±20.7	*	161.8±27.0
DiastBP	91.4±11.2	*	104.0±14.2
RAP	6.7±2.3	*	9.7±2.5
PWP	15.2±5.8	*	22.0±7.2
CO	4.85±.87	°	5.35±1.63

* p .001   ° p .05

Heart rate, systemic blood pressure and filling pressures of both ventricles significantly increased. On the average the left ventricular filling pressure reached values indicative of left ventricular disfunction (22mmHg); the increments were > 5mmHg in 73% of the patients and > 10mmHg in 19%. The mean increase in cardiac output at the end of the stress test was small but statistically significant. Heart rate increased by > 10beats/min in 75% of patients and by > 20beats/min in 27%. The increment significantly correlates with the increase of systolic (r=0.44) and diastolic (r=0.46) blood pressure and of left ventricular filling pressure (r=0.63). Systolic blood pressure increased > 20mmHg in 73% of patients and > 40mmHg in 17%.

Hemodynamic activation during 4 stressors. Table II shows basal and peak hemodynamic values observed in 26 patients who underwent 4 stress tests according to a randomized succession pattern.

Table II: Hemodynamic changes due to mental arithmetic, Sacks test, Raven PM and noxious noise administered to 26 pts 44±12 days after acute myocardial infarction (mean ±SD)   * p < .001
Abbreviations as in table I. C: control, S: stress   ° p < .05

	arithmetic		Sacks test		Raven PM47		Noise	
	C	S	C	S	C	S	C	S
HR	75±14 *	90±16	73±13 *	83±13	75±12 *	84±12	74±13 °	76±13
SystBP	132±21*	152±28	131±22*	151±27	131±21*	147±26	132±22*	141±26
DiastBP	89±12*	101±16	88±11*	100±14	89±13*	98±14	88±12*	93±12
RAP	7±2 *	10±2	7±2 *	9±2	7±2 *	10±3	8±2	8±3
PWP	15±5 *	22±8	15±6 *	20±7	15±6 *	19±8	16±6 *	17±7
CO	4.7±.9 °	5.3±1.7	4.9±.9	4.7±1.2	4.8±.9	4.9±1.0	4.8±1.1°	5.1±.9

All stressors induced hemodynamic activation, as showed by the significant differences between basal and stress-induced values for most parameters considered.

Furthermore, analysis of the variance showed highly significant differences in activation among the 4 stressors for most the parameters. Mental arithmetic induced the highest activation, noise the minimal, while incomplete sentences (Sack's test) and the problem solving (Raven PM 47) yielded intermediate results (table III)

Table III: Interstress statistical significance.
Abbreviations: as in table I; St: stressor, A: mental arithmetic, S: Sacks test, R: Raven PM47, N: noise.

| HR | | SystBP | | DiastBP | | $\overline{RAP}$ | | $\overline{PWP}$ | |
St	p	St	p	St	p	St	p	St	p
A-R	<.01	A-R	<.01	A-R	<.05	A-S	<.01	A-S	<.01
A-N	<.01	A-N	<.01	A-N	<.01	A-N	<.01	A-R	<.01
S-N	<.01	S-N	<.01	S-R	<.05	S-N	<.01	A-N	<.01
				S-N	<.01	R-N	<.01	S-N	<.01
								R-N	<.01

The order in the stressors sequence did not influence the hemodynamic response: no significant differences were found both among basal values and among the hemodynamic changes induced by first, second, third or fourth stress test.

The age of the patients, the site of infarction and the basal hemodynamic values did not appear to be correlated with the hemodynamic changes induced by mental stress.

Even though the hemodynamic response was in general homogenous in intensity, a few exception could be noted. Two patients experienced acute pulmonary oedema within the first ten minutes after stressor administration. One of the two showed during arithmetic calculation an increase by 30% of heart rate (from 66 to 86 beats/min), by 10% in systolic blood pressure (from 145 to 160mmHg), by 83% in pulmonary wedge pressure (from 23 to 42mmHg). The other patient had only moderate increases, which induced an acute left ventricular failure probably related to the elevated basal values of pulmonary wedge pressure. Both patients did not show electrocardiographic changes. This was the sole episode of pulmonary oedema that both patients experienced.

By contrast, other patients with marked hemodynamic changes did not have subjective symptoms. For example, the oldest patient (73 years old) showed during arithmetic calculation an increase in heart rate by 69% (from 77 to 130beats/min), in systolic blood pressure by 25% (from 160 to 200mmHg), in pulmonary wedge pressure by 40% (from 18 to 30mmHg) while rimaining symptomless.

Eletrocardiographic aspects. a) Arrhythmias. Table IV reports the arrhythmias observed during Holter monitoring according to the Lown classification.

Table IV: Arrhythmias observed during 24 hours ECG recording.

Pts n.	Lown classes
7	0
11	1
2	2
8	3
11	4 A
8	4 B
1	5

During stress testing only 2 patients had rhythm disturbances, both of them during mental arithmetic. One patient who had ventricular premature beats at baseline, showed several coupled ventricular beats during stress testing; another had polymorphous atrial and ventricular premature beats.

b) Ventricular repolarization. Only two patients showed ST-segment elevation in leads with abnormal Q waves during mental arithmetic.

None of the patients showed ST-segment depression or anginal pain during mental stress.

Correlation between psychological profile and hemodynamic response to mental stress. The STAI X1 scores were not significantly different before and after the study completion. The changes in hemodynamic parameters occurring during mental stress were poorly related to the psychological profiles. With regard to mental arithmetic (48 pts), we found the best correlations between left ventricular filling pressure and both Middlesex's Hysteria factor (r=0.39, F=8.2) or IPAT's external Anxiety Factor (r=0.33, F=5.3); heart rate and Middlesex's Hysteria Factor (r=0.33, F=5.3); systolic blood pressure and Middlesex's free floating anxiety factor (r=0.30, F=5.3). The last correlation improved if IPAT total scores and Coronary Prone Behavior were added (r=0.57); when considering all the psychological variables r increased to 0.72

DISCUSSION

Many studies are available about mental stress in normotensive or hypertensive individuals, but few reports have been published on this kind of experiences in patients with coronary artery disease (10-16).

Our investigation has shown that mental stress in patients with recent myocardial infarction without clinically evident complications represents a challenge for the heart pumping function. This is evident when considering the mean levels of left ventricular filling pressure and the two patients with stress-induced pulmonary oedema.

It is interesting to note the marked difference of hemodynamic activation induced by the stressors tested: invariably mental arithmetic appeared to be the most active, while noise was the less effective. This is in accord with the sensory intake-rejection theory (17,18). The noise is a typical sensory-intake, mental arithmetic a sensory-rejection, Raven PM47 and Sacks test mixed sensory intake-rejection. The great difference among stressors showed therefore that only some types of mental stress could be a possible risk factor for the ischemic patients.

The cardiovascular changes provoked by an acute stressor could be different from those induced by prolonged stress. For this purpose, it is to be noted on one hand, that a serial combination of stressors did not seem to cause changes in the response, the latter being highly conditioned by the single stressor and not by its temporal position during serial stimulations; on the other hand the two patients who had acute pulmonary oedema underwent 4 stress tests in a session lasted over 30 minutes (including the recovery intervals): this might be one of the factors causing the heart failure.

By contrast, in this study mental stressors did not appear as risk factors for arrhythmias or myocardial ischemia. This could not be true for patients with different clinical patterns of ischemic heart disease (15) or with very low ischemic threshold (19).

The attempt on identifying patients at risk, i.e. patients with a high hemodynamic response during mental stress, by means of the study of the individual psychological profile failed. A poor correlation between the results of the psychological investigation and the data of the hemodynamic study was noted. Only some traits of anxiety seemed to indicate a trand of a greater hemodynamic response. As mental stress has shown to be effective in compromising the hemodynamic equilibrium it must be concluded that in patients with recent myocardial infarction either the hemodynamic response is indipendent from the psychological pattern or the psychological tests adopted have poor accuracy in the identification of at-risk patients. Further investigations are needed on this topic.

REFERENCES
1. Young L, Barboriak JJ, Anderson AA; Hoffman RG: Attitudinal and behavioral correlates of coronary heart disease. J.Psychosom.Res. 24:311,1980
2. Rathus SA: A 30 item schedule for assessing assertive behavior. Behavior Ther. 4: 398, 1973
3. Campanelli M, Tamburello A: L'inventario di Rathus. G.Ital.Analisi e Modificazioni del Comportamento. I: 60, 1979
4. Sarason IG, Johnson JH, Siegel JM: Assessing the impact of life change: development of the life experience survey. J.Consulting and Clin.Psychol. XLVI: 932, 1978
5. Pancheri P: Stress emozioni malattia. Ed. A.Mondadori, Milano, 1980
6. Spielberger CD, Gorsuch RL, Lushene RE: Manual for the state-trait anxiety inventory. Ed. Consulting Psychology Press, Palo Alto, 1970
7. Lazzari R, Pancheri P: S.T.A.I.(State-Trait Anxiety Inventory). Questionario di autovalutazione per l'ansia di stato e di tratto. Manuale di istruzioni. Ed. Organizzazioni Speciali, Firenze, 1981
8. Sacks JM, Levy S: The sentence completion test. In: ABT e BELLAK: Projective Psychology, Knopf, New York, 1950
9. Riva A: Il reattivo delle frasi da completare di Sacks e Coll. nell'uso della psicodiagnostica di tipo "clinico". Ed. Organizzazioni Speciali, Firenze, 1969
10. Buckwalsky R: Hemodynamics before and after physical endurance training in patients with myocardial infarction under various physical

and psychomotor test. Clin.Cardiol. 5: 332, 1982

11. DeBusk R, Barr Taylor C, Stewart Agras W: Comparison of treadmill exercise testing and psychologic stress testing soon after myocardial infarction. Am.J.Cardiol. 54: 907, 1979

12. Krenauer P, Toth L, Konig W: Erhohter diastolischer Pulmonalarteriendruck Herzfrequenz und Blutdruck bei Koronarkranken unter psychischer Belastung. Verh.Dtsch.Ges.Kreisl.Forsch. 45: 6, 1979

13. Taylor CB, Davidson DM, Houston N, Agras WS, DeBusk RF: The effect of a standardized psychological stressor on the cardiovascular response to physical effort soon after uncomplicated myocardial infarction. J.Psychosom.Res. 26: 263, 1982

14. Robinson BF: Relation of heart rate and systolic blood pressure to the onset of pain in angina pectoris. Circulation XXXV: 1073, 1967

15. Schiffer F, Howard Hartley L, Schulman CL, Abelman WH: Evidence for emotionally-induced coronary arterial spasm in patients with angina pectoris. Br.Heart J. 44: 62, 1980

16. Bassan MM, Marcus HS, Ganz W: The effect of Mild-to-moderate mental stress on coronary hemodynamics in patients with coronary artery disease. Circulation 62: 933, 1980

17. Lacey JI, Lacey BC: Verification and extension of the principles of autonomic response-stereotypy. Am.J.Psychol. 71: 50, 1958

18. Green J: A review of the Lacey's physiological hypothesis of heart rate change. Biological Psychol. 11: 63, 1980

19. Specchia G, De Servi S, Falcone G, Salerno J, Gavazzi A, Mussini A, Angoli L, Bramucci E: I test provocativi dello spasmo coronarico: quando e come eseguirli. In: "La cardiopatia ischemica silente". Ed. P.L.Prati, 1982, p.176

THORACIC AUTONOMIC NERVES REGULATING THE CANINE HEART

J. A. ARMOUR
Dalhousie University, Department of Physiology and Biophysics
Halifax, Nuova Scotia B3H 4H7, Canada

1. INTRODUCTION

As the topic of neural regulation of the heart has been reviewed recently (16,30,31,40), the present discussion will be confined to the thoracic elements involved in such regulation. Firstly, some of the functions of cardiac afferents will be described. This will be followed by a discussion of some of the functions of efferent cardiac nerves. Then some of the interactions between these components - the reflex regulation of the heart - will be discussed.

2. ACTIVITY OF AFFERENT FIBERS IN CARDIOPULMONARY NERVES ORIGINATING FROM THORACIC CARDIOVASCULAR RECEPTORS

There are a variety of afferent receptors which transduce cardiac, thoracic vascular and pulmonary events into patterns of afferent nerve activity. This activity varies according to the location of the receptor (1,31,38,39,45). It has been postulated that mechanosensitive and chemosensitive receptors are located in the heart and great thoracic vessels (1,31,38). Chemoreceptors- One type of receptor has afferent traffic which is relatively constant and unrelated to the cardiac cycle (Fig. 1) (1,41). It increases its traffic following coronary artery ligation and injections of certain chemicals, but not following mechanical distortion of the heart (41). This type has been tentatively identified as being non-mechanoreceptive and may be chemoreceptive (1).

The constantly firing receptor is activated to the greatest degree following intravenous infusion of a noxious chemical, like cyanide (Fig. 1, panel 5). It is important to note that noxious chemicals also activate mechanoreceptors which were not previously active in a physiological preparation (Fig. 1, panel 5). These later receptors do respond to mechanical distortion and should be classified as mechanosensitive and not chemosensitive. Thus, the identification of non-mechanosensitive (tentatively

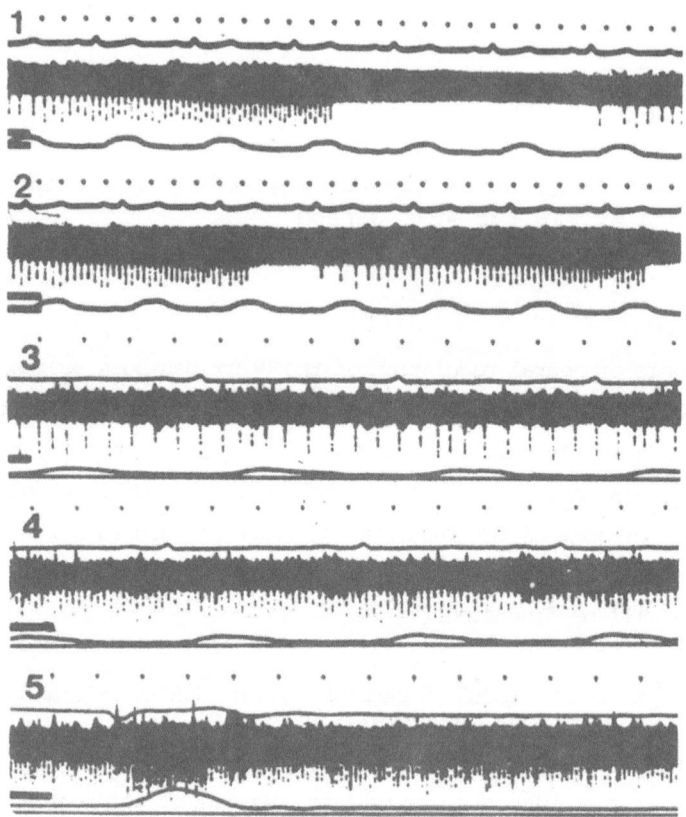

FIGURE 1. Two different types of afferent nerve traffic recorded from two slips of the recurrent cardiac nerve, one being pulmonary inflation (1 and 2) and the other a constantly firing receptor (3,4,5). When the respiratory rate changed so did the period of firing of the pulmonary inflation receptor (1,2). In the other afferent fiber, traffic was relatively constant during control states (30 Hz). Following 30 seconds of respiratory cessation firing increased to 70 Hz (4). The respiration was restored and cyanide injected (0.025 mg/kg I.V.); the traffic increased to 120 Hz (5). Note that a ventricular mechanoreceptor, which was previously inactive, became active during ventricular systole after cyanide (5). Each panel has an EKG, nerve recording and right ventricular pressure. The black bars on the left of 1 and 2 are 0 and 25 mmHg pressure calibrations. In the lower trace, the black bar represents 25 mmHg and 0 mmHg is represented by the bottom line across each panel. Timing dots are]00 msec apart.

chemosensitive) receptors may have been confused in the past, as silent mechanosensitive receptors can be activated by noxious chemicals.

Mechanoreceptors - These form the other major category of thoracic cardio-vascular receptors. They are located in the atria, ventricular endocardium, myocardium and epicardium as well as the thoracic vessels. They act as sensitive transducers of local dynamics (1,31,38,41). Atrial length receptors have been characterized as being active during one portion of the cardiac cycle - that is during atrial systole or diastole (38). However, as they are sometimes activated during systole and at other times during diastole, such categorization is unsatisfactory. Traffic from atrial receptors appears to reflect the dynamics of the region in which they are located, and this can change from cycle to cycle (1). The specificity of atrial receptors is further demonstrated by recording their afferent traffic during atrial fibrillation. In such a state many atrial receptors display periods of regular phasic activity, even though no cyclic pressure is being generated in the atrial chamber (Fig. 2)(41). These data imply that

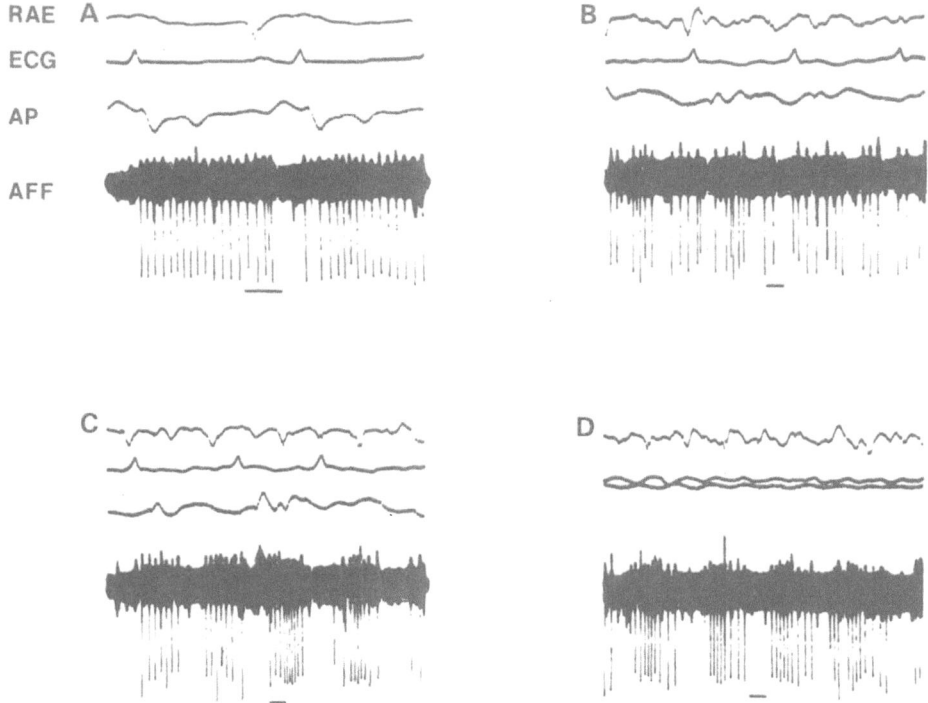

FIGURE 2. Each panel represents, from top to bottom, a right atrial electrogram (RAE), lead II electrocardiogram (ECG), right atrial pressure (AP) and afferent traffic from a twig of a canine recurrent cardiac nerve (AFF). Control conditions (A) are compared with periods of atrial fibrillation (B & C), as well as atrial and ventricular fibrillation (D). Rhythmic traffic occurred during atrial (C) as well as atrial and ventricular (D) fibrillation. Timing bars = 100 msec.

a local region of the atrium, which contains a receptor field, may contract regularly during atrial fibrillation. Thus, it appears that atrial mechano-receptors respond to local dynamics and are not easily placed into specific categories.

This is equally true of ventricular mechanoreceptors which can be divided physiologically into epicardial, myocardial and endocardial types (1,38,41,45). Canine endocardial ventricular receptors have traffic which is specifically related to the cardiac cycle (1). Afferent fibers from these receptors sometimes arise from mechanoreceptors located in two different ventricular regions (1). Mechanoreceptors are also located in the mid-wall of the left ventricle. These may or may not display activity which is readily correlated to the cardiac cycle. The third type, ventricular epicardial receptors, usually has afferent traffic which changes from being related to the cardiac cycle to periods when it is not. An afferent fiber can arise from mechanoreceptors located in two different regions of the ventricular epicardium. Sometimes such a receptor field may include the epicardium of an atrium, further complicating our under-standing of such a receptor. The majority of canine ventricular mechano-receptors have afferent fibers of the C fiber category (1). Atrial and aortic receptors have afferent fibers of the A fiber category (1). Different cardiopulmonary nerves contain differing populations of afferent fibers (1). The variety of receptors, the specificity of the nerves con-taining their afferents, and the fact that an afferent can arise from mechanoreceptors in two separate regions of the heart (including one on an atrium and the other on a ventricle) demonstrates some of the complexities of these structures. A lot more research needs to be done in order to understand the complexity of these receptors and to determine which type of receptor has afferent fibers in the vagi and which type in the sympathetic nerves. Sympathetic rami have afferents from a few aortic receptors, many epicardial and myocardial ventricular mechanoreceptors (31,39) and a few of the constantly firing receptors. No ventricular endocardial receptors afferent have been detected in the sympathetic rami or ansae. A large population of aortic mechanoreceptor and some atrial and ventricular afferents exist in the vagi (38).

3. REGULATION OF THE HEART BY EFFERENT CARDIOPULMONARY NERVES

This subject has been reviewed extensively (32,41), therefore only a brief overview will be given here. Efferent preganglionic parasympathetic (9,23,41) and postganglionic sympathetic (6,41) fibers in specific cardiopulmonary nerves (8) influence dromotropism (28) and inotropism (40) of specific cardiac regions. The majority of canine sympathetic efferent postganglionic neurons which regulate the heart and lungs are located in the middle cervical ganglia (6,34). Specific cardiac nerves can modify chronotropism (41) and initiate tachydysrhythmias (21,22). The sympathetic nerves augment intramyocardial (41) and intraventricular pressures as well as coronary artery (26) and venous (27) pressures. Stellate ganglion stimulation increases coronary artery (12,26) and coronary venous resistances (10). Efferent sympathetic neurons are known to regulate the levels of catecholamines in the coronary circulation (29,46). Therefore, it is obvious that efferent nerves modify the coronary vasculature, heart rate and cardiac inotropism.

4. REFLEX REGULATION OF THE HEART

4.1 Cardiovascular Responses Following Electrical Stimulation of Afferent Fibers in Cardiopulmonary Nerves

Stimulation of the afferent components of certain cardiopulmonary nerves in an anesthetised dog can initiate bradycardia, atrial force supression, augmentation of ventricular force and hypertension (8). Stimulation of the central end of a sectioned cardiopulmonary nerve in a conscious dog results in hypotension or hypertension, depending on stimulus parameters and the nerve stimulated (7). Afferents in vagal components of the cervical vagosympathetic complex can initiate reflex bradycardia and hypertension (Fig. 3,A). Stimulation of the afferents in the sympathetic component of this nerve complex initiates bradycardia and hypotension (Fig. 3,B). Following atropine administration, stimulation of the afferent vagal component initiates tachycardia, positive atrial and ventricular inotropism and hypertension (Fig. 3,C). Stimulation of the sympathetic afferent component following atropinization causes negative inotropism in the atria and ventricles and hypotension (Fig. 3,D). These data suggest that efferent parasympathetic cardiovascular neurons can be activated or inhibited reflexly depending on the type of stimulation. Stimulation of afferents in a cardiopulmonary nerve can modify coronary

936

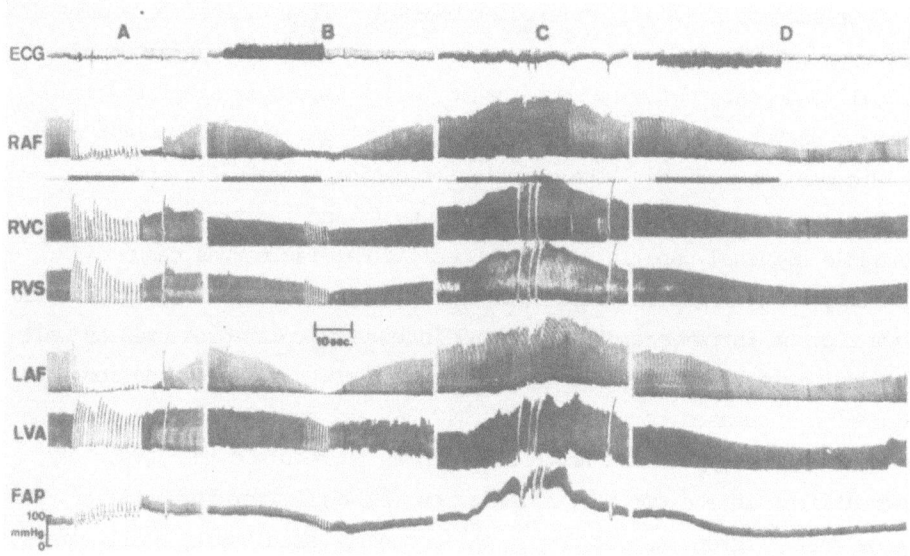

FIGURE 3. Afferent stimulation of the vagal component of the right vagosympathetic complex initiated bradycardia, right (RAF) and left (LAF) atrial force supression and augmentation of right ventricular conus (RVC) and sinus (RVS) forces, left ventricular force (LVA) and blood pressure (FAP) (panel A). Stimulation of afferents in the sympathetic component of the vagosympathetic complex initiated bradycardia, cardiac negative inotropism, and hypotension (panel B). Following atropinization, stimulation of the vagal afferents initiated positive cardiac inotropism and hypertension (panel C). Following atropinization the sympathetic afferents initiated negative inotropism and hypotension, presumably by withdrawal of sympathetic activity.

artery flow via reflexes. Even though afferent stimulation of a cardiopulmonary nerve augments left ventricular force, intramyocardial pressure, left ventricular pressure and systemic arterial pressure, coronary artery flow (phasic and mean) can remain relatively constant until stimulation ceases (Fig. 4). Upon cessation of such a stimulation coronary artery flow increases significantly. These data imply that, despite increased intramyocardial, aortic and presumably central coronary artery pressure, coronary artery resistance was modified such that coronary artery flow remained relatively unchanged. Coronary artery resistance can be modified reflexly by carotid sinus mechanisms (19). Thus, not only cardiac chronotropism and inotropism are modulated via autonomic reflexes, but also coronary vascular flow.

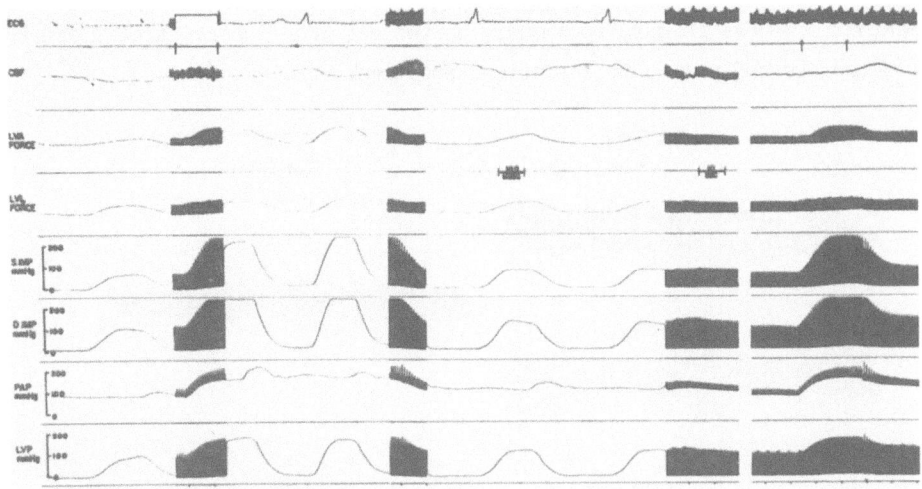

FIGURE 4. Stimulation of afferent components in a sectioned recurrent cardiac nerve (between arrows) augmented anterior (LVA) and lateral (LVL) left ventricular epicardial forces, superficial (S. IMP) and deep (D. IMP) anterior left ventricular intramyocardial pressures, femoral artery pressure (PAP) and left ventricular cavity pressure (LVP) in an anesthetised dog. There was little change in the left ventral descending coronary artery phasic (left panel) or mean (right panel) flow (CBF) during the stimulation. After stimulation ceased phasic (left) and mean (right) artery flow was augmented.

4.2. Activation of Efferent Fibers in Cardiopulmonary Nerves Following Stimulation of Afferent Fibers in Other Cardiopulmonary Nerves

4.2.1. Small Nerve Bundle Preparations. Parasympathetic preganglionic efferent axons in cardiopulmonary nerves sometimes have traffic which fires in a pattern related to the cardiac cycle (2,24,41), particularly if the animal preparation is in a good physiological state (2). The traffic in other of these axons can be unrelated to the cardiac cycle (Fig. 5). Efferent parasympathetic traffic in the vagi has been divided into at least 7 categories one of which is correlated to the cardiac cycle (24). Since this traffic differs throughout a physiological experiment (2) it may be very difficult to precisely categorize it. Most of the efferent parasympathetic fibers in canine cardiopulmonary nerves were found to be activated reflexly by mechanical distortion to the carotid artery (Fig. 5,C and D), none were found which were activated by mechanical distortion of the heart or great thoracic vessels (2). Since activity in these fibers persisted after severing the ansa subclavia, but was abolished by sectioning

938

the ipsilateral vagi, it is presumed that they were efferent parasympathetic fibers. It has been suggested that the majority of canine efferent parasympathetic fibers in cardiopulmonary nerves are modified by carotid artery baroreceptors and pulmonary inflation receptors.

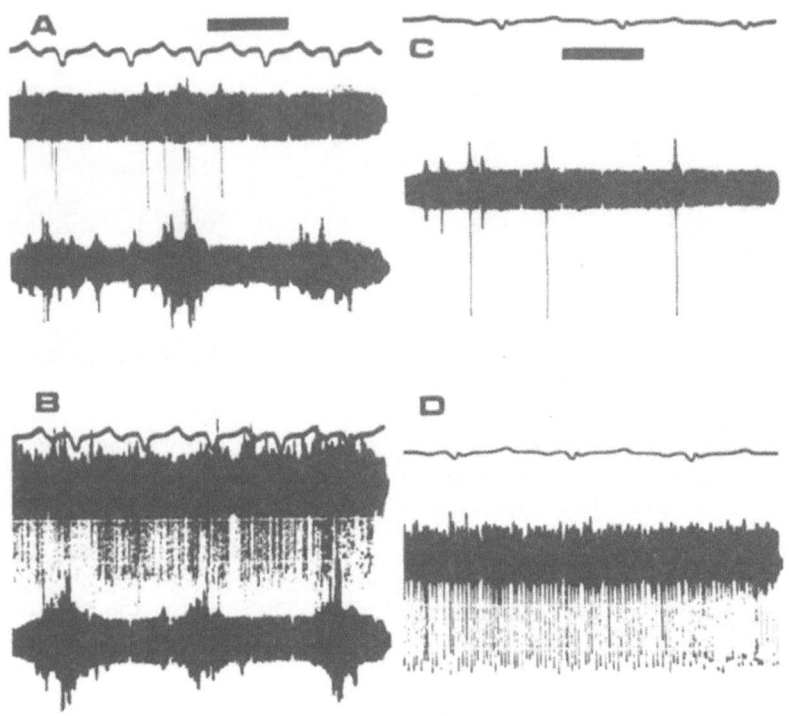

FIGURE 5. Recordings of an EKG (upper trace, panel A), an efferent parasympathetic nerve bundle (middle trace, panel A) and efferent sympathetic nerve bundle (lower trace, panel A), both from the recurrent cardiac nerve, demonstrated that at two different times within one minute (panels A and B) the sympathetic traffic was relatively constant but the parasympathetic traffic was not. Time bar in panel A is 400 msec and is the same for panel B. Efferent parasympathetic traffic recorded from the innominate nerve of another animal was random (panel C), and augmented by touching the carotid bulb (panel D). Time bar in panel C is 200 msec and is the same for panel D.

It is commonly thought that the parasympathetic and sympathetic efferent nerves function in a reciprocal fashion. That is, as the activity increases in one it decreases in the other. If one records efferent fibers in a cardiac nerve which simultaneously displays traffic from preganglionic parasympathetic and postganglionic sympathetic efferent neurons, the traffic in one can increase while that of the other remains essentially unchanged or even increases (Fig. 5,A and B). Thus, the parasympathetic and sympathetic efferent neurons do not always regulate the heart in a reciprocal fashion. Activity of efferent sympathetic neurons can also be grouped into that which is correlated to the cardiac cycle and that which is not (2,44). It is certainly incorrect to think that efferent post-ganglionic sympathetic fibers regulating the heart must have traffic which is cyclically correlated to the cardiac cycle. Another generally held view is that the autonomic nervous system is a functionally diffuse system, having no highly localized regulatory capacity. However, traffic in efferent postganglionic sympathetic fibers is, in many instances, activated by distorting one highly localized region of the heart, but not the rest of the heart (Fig. 6,C and I). Such traffic, which can be very specific in its pattern, in some instances is very similar to that of cardiac afferent fibers (2). These data suggest the presence of discrete reflex mechanisms regulating the heart. Efferent sympathetic neurons can be activated by electrical stimulation of afferent cardiopulmonary nerves, each afferent nerve activates efferent nerves after differing latencies (Fig. 6,D,E and F). Complex reflex mechanisms regulate the heart.

4.2.2. Whole Nerve Preparations. Stimulation of an afferent cardio-pulmonary nerve can result in the generation of a compound action potential (CAP) in an ipsilateral efferent cardiopulmonary nerve (Fig. 7,A)(4). These CAPs can be altered by changing the frequency of stimulation and by various pharmacological agents (4), indicating that they are the result of synaptic mechanisms. Some of these C.A.P.s persist following acute decentralization of the thoracic autonomic ganglia, demonstrating that reflexes exist in thoracic autonomic ganglia independent of the central nervous system (2,4,41). In the dog the majority of synapses involved in these reflexes exist within the MCG (4), where the majority of efferent sympathetic postganglionic neurons regulating the heart and lungs exist (6). Occasionally, the CAPs are augmented following decentralization of the thoracic ganglia (Fig. 7,B) and supressed following stimulation of the sympathetic fibers which are in the rami (Fig. 7,D). These data indicate that efferent preganglionic sympathetic neurons may regulate these local reflexes. This regulation can occur at very low stimulation frequencies, indicating that a little change of activity of spinal cord neurons may

940

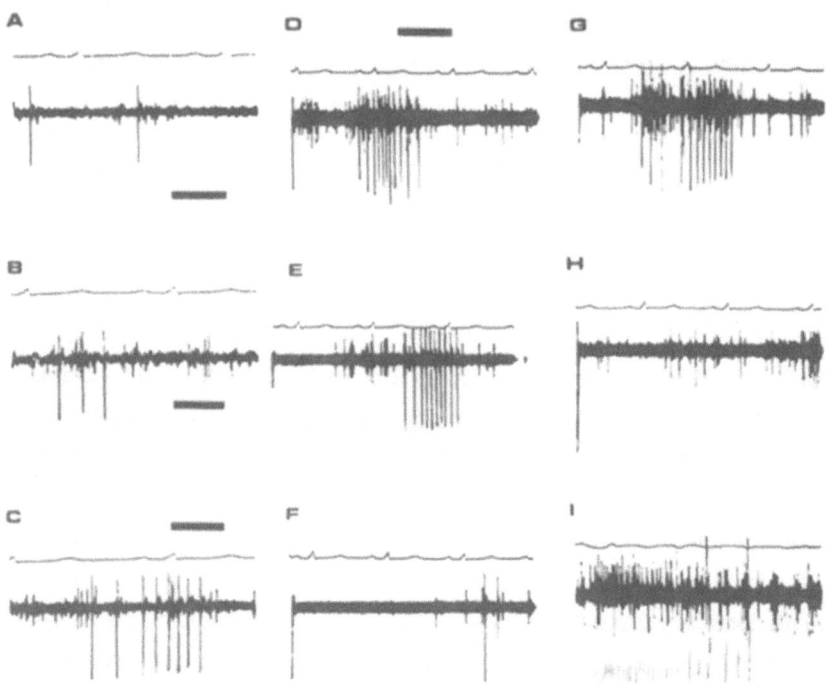

FIGURE 6. In control conditions (A) an EKG and efferent sympathetic traffic from a slip of the right ventral ansa were photographed (timing bar of 100 msec is the same for A, B & C). When the aorta was partially occluded traffic increased (B), as it did during mechanical distortion of the interventricular septum (C). Afferent stimulation of the recurrent cardiac nerve before (D) and after (G) sectioning of the cervical vagi activated the efferent neuron (time bar = 200 msec for all subsequent panels). Stimulation of the afferent innominate nerve activated the efferent sympathetic neuron before (E), but not after (H), bilateral cervical vagotomy. Afferent stimulation of the dorsal nerve generated activity before the vagi was cut (F). Note the different latencies of activation in D, E and F. When the heart was fibrillated and the ventricles opened, mechanical distortion of the interventricular septum activated the efferent sympathetic neuron (I). Distortion of the rest of the heart did not activate this neuron. These data indicate that efferent sympathetic neurons can be activated reflexly by cardiac receptors in highly localized regions of the heart. Afferents involved in this activation travelled in the vagi and the ansae.

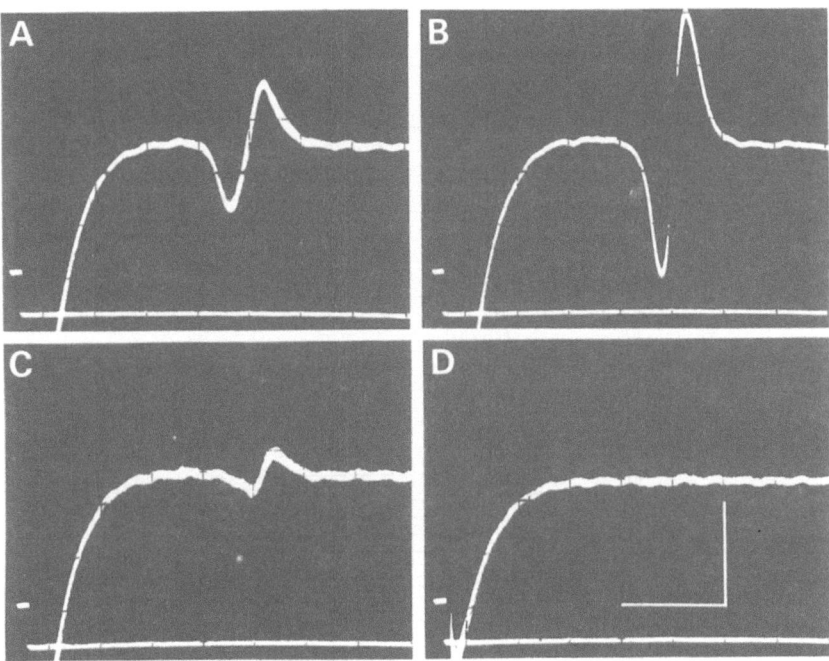

FIGURE 7. Ten stimuli (9V, 5 msec, lower trace) delivered to the left cranial medial nerve at 1 Hz produced smaller CAPs in the caudal pole nerve before (A) than after (B) decentralization of the left stellate ganglion. Increasing the stimulus frequency in the decentralized MCG preparation to 10 Hz reduced the CAPs by over 75% (C). Ten 1-Hz stimuli delivered immediately after a 30 sec stimulation of the T1 and T2 rami elicited no CAPs (D), indicating that preganglionic neurons can supress the reflexly generated synapses in thoracic autonomic ganglia. Vertical calibration bar for CAPs: 2 mV; horizontal time bar = 20 msec.

profoundly affect local thoracic cardiovascular neural regulation. The synapses involved in local reflexes can be modified by a number of pharmacological agents (4,5). Hexamethonium (18,30) and atropine (20,25) block synaptic transmission in the autonomic nervous system. Such transmission can be modified by adrenergic blocking agents as well (17). Some of the reflexly generated C.A.P.s, in the canine MCG preparation, can be abolished by hexamethonium (Fig. 8), atropine (Fig. 9), phentolamine (Fig. 10) or propranalol (Fig. 11)(5). Some of these reflexly generated CAPs persist after the administration of these cholinergic and adrenergic pharmacological blocking agents (4,5). These CAPs can be transiently abolished by a local injection of chymotrypsin into the MCG (Fig. 12,A) or abolished for

942

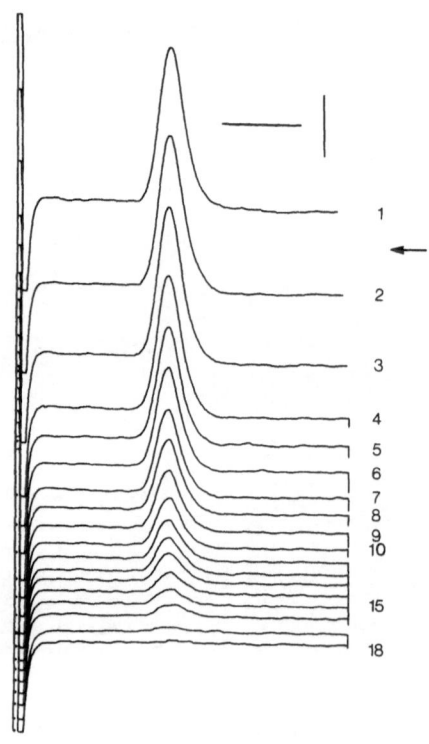

FIGURE 8. Stimulation (1 Hz) of the right thoracic vagus, in a decentralized MCG preparation, generated a CAP in the right recurrent cardiac nerve ~ 75 msec after the stimulus (line 1). Hexamethonium (1 mg/kg) (arrow) abolished the CAP within 17 s (line 18). Calibrations: 2 mV and 50 ms.

long periods of time by a local injection of manganese into the MCG (Fig. 12,C)(4). Injections of an equal volume of normal saline into the ganglion does not grossly modify the CAP (Fig. 12,B)(4). As manganese is known to block synaptic transmission, but not transmission along axons (11,35), it is presumed that synaptic activity persists in acutely decentralized thoracic autonomic ganglia after cholinergic and adrenergic pharmacological blockade. Histological (42) and physiological (36) evidence has implicated substance P as a neurotransmitter in autonomic ganglia. As chymotrypsin transiently modifies the generated CAPs in thoracic autonomic ganglia, it is inferred that peptinergic synaptic transmission may exist in thoracic autonomic ganglia (4). It appears that synaptic activity generated in mediastinal ganglion following activation of cardiopulmonary afferents may be regulated in part by efferent neurons in the cord. Such synaptic activity may be very complex, involving a number of neurotransmitters.

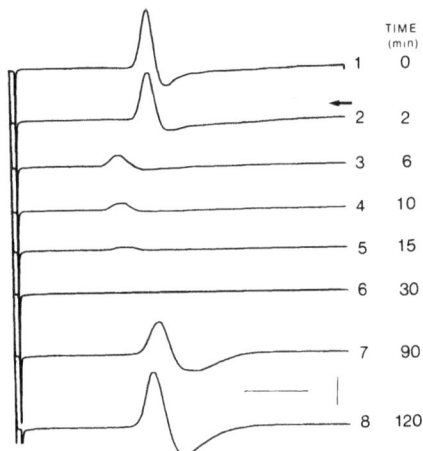

FIGURE 9. Stimulation of the left vagus (0.5 Hz) in a decentralized MCG preparation resulted in a CAP in the caudal medial nerve (line 1). After atropine (0.1 mg/kg) (arrow) the latency between stimulus time and CAP onset and peak shifted. The CAP was gradually reduced (lines 2-5) and then at 30 min was absent (line 6), before reappearing (lines 7 and 8). The right hand column gives the time after injection. Calibrations: 0.5 mV and 20 ms.

944

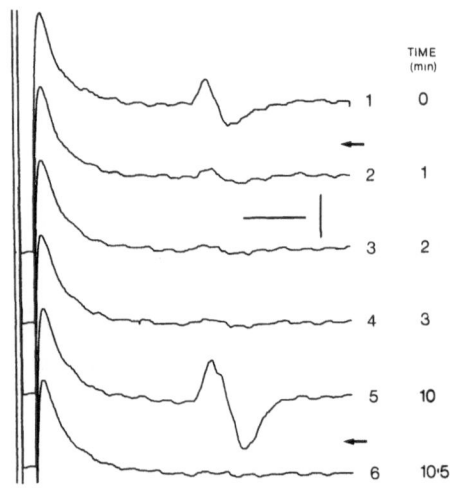

TIME
(min)

FIGURE 10. Stimulation (0.5 Hz) of the left thoracic vagus in a decentralized MCG preparation resulted in a CAP (latency, 43 ms) in the caudal pole nerve (trace 1). Phentolamine (0.1 mg/kg) administered (first arrow) resulted in a great reduction of the CAP within 3 min (line 4). By 10 min the CAP had returned to a larger size than control (line 5). A 1 mg/kg injection of phentolamine (second arrow) abolished the CAP (line 6). The time after the start of the experiment appears in the right column. Calibrations: 0.5 mV and 15 ms.

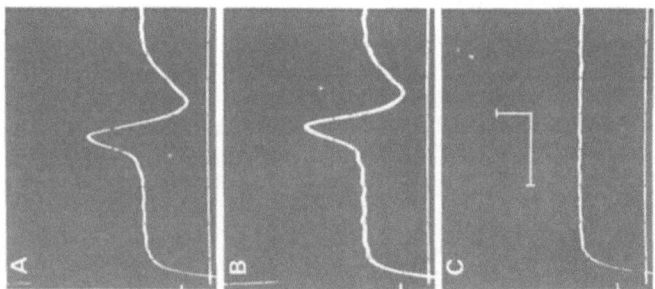

FIGURE 11. Ten 1 Hz stimuli delivered to the left thoracic vagus of a decentralized MCG preparation generated 10 CAPs (latency 35 ms) in the intermediate medial cardiac nerve (A, upper trace). The CAPs were unchanged by hexamethonium, atropine and phentolamine (B), but were abolished within 30 s of propranalol (1 mg/kg) administration (C). Each panel shows recordings following 10 stimuli. Calibrations: 2 mV and 20 ms.

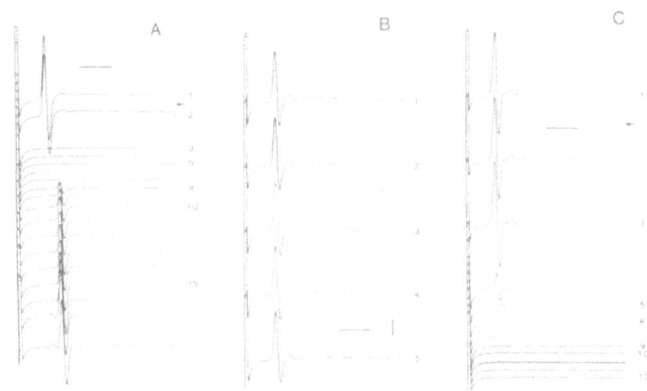

FIGURE 12. Three sequences of stimulations were recorded when chymotrypsin (A), saline (B) and manganese (C) were injected into the MCG. (A) Eighteen stimuli (10 V, 2 msec, 1 Hz) were delivered to the caudal medial nerve and a CAP recorded from the caudal pole nerve. After the first stimulus 0.5 μg of chymotrypsin was injected into the MCG (arrow). Transient abolition of the CAP occurred within 3 sec. (B) The injection thereafter of 0.5 ml saline into the same region of the MCG did not affect the CAP. Each line represents four superimposed CAPs resulting from four sequential stimuli, the total representing twenty sequential stimuli. The saline was injected after the first stimulus. (C) In the same preparation, manganese (0.1 mg) abolished the CAP 5 sec after it was injected into the MCG site (arrow). Calibrations: 20 msec and 1 mV.

5. SUMMARY

Efferent sympathetic and parasympathetic neurons regulate chronotropism, dromotropism, inotropism and diastolic tension of the myocardium, as well as the coronary vasculature. Thoracic afferents arising from vascular and myocardial receptors have cell bodies in dorsal root ganglia and the nodose ganglia which modify medullary (23,33) and spinal cord neurons (15,41). Spinal cord neurons receive inputs from many levels of the cord (13). The central nervous system neurons regulate the preganglionic sympathetic and parasympathetic neurons. Spinal cord neurons regulate some of the sympathetic neurons in thoracic autonomic ganglia. Others are regulated by afferents arising from thoracic contents - local autonomic reflexes (2,4,41). Some of these local reflexes are modified by efferent preganglionic sympathetic neurons, others function relatively independently of the central nervous system (2,4,14). Recent research

has dispelled the notion that the organization of the peripheral autonomic nervous system is similar to that of the central nervous system. As
well, the numerous cardiac and thoracic vascular receptors are complex
and difficult to lump into rigid categories. These receptors can activate
both negative and positive feedback mechanisms, regulating the heart via
neurons located either in the central nervous system and in the thoracic
autonomic ganglia or in the thoracic autonomic nervous system alone. At
present the hierarchy of control between thoracic autonomic neurons and
central nervous system neurons is unknown. However, it may be that local
reflexes regulate the heart from beat to beat, whereas the central nervous
system neurons may act to modulate overall integration of the cardiovascular
system. The complex anatomy and function of neural elements regulating
the heart must be fully appreciated before significant progress is made
in understanding this system in humans. Information accumulated through
the efforts of many investigators is re-establishing the importance of
the autonomic nervous system in the genesis of cardiac pathology (3,37).
However, we are still a long way from understanding the complex anatomy
and physiology of the neurons which regulate the heart.

REFERENCES

1. Armour, J.A. Physiological behaviour of thoracic cardiovascular
    receptors. Am. J. Physiol. 225:177-185, 1973.
2. Armour, J.A. Instant-to-instant reflex cardiac regulation. Cardiology,
    61:309-328, 1976.
3. Armour, J.A. Implication of neural regulation of the heart in health
    and disease. Can. Med. Ass. Journal, 123:91-93, 1980.
4. Armour, J.A. Studies on synaptic transmission in the decentralized
    middle cervical ganglion of the dog. Brain Res. Bull., 10:103-109,
    1983.
5. Armour, J.A. Synaptic transmission in thoracic autonomic ganglia
    of the dog. Accepted for publication, Can. J. Physiol. Pharmacol.,
    1983.
6. Armour, J.A. and D.A. Hopkins. Localization of sympathetic post-
    ganglionic neurons of physiologically identified cardiac nerves in
    the dog. J. Comp. Neurol., 202:169-184, 1981.
7. Armour, J.A. and J.B. Pace. Cardiovascular effects of thoracic afferent
    nerve stimulation in conscious dogs. Can. J. Physiol. Pharmacol.,
    60:1193-1199, 1982.
8. Armour, J.A. and W.C. Randall. Functional anatomy of canine cardiac
    nerves. Acta. Anat. 91:510-528, 1975.
9. Armour, J.A., W.C. Randall and S. Sinha. Localized myocardial responses
    to stimulation of small cardiac branches of the vagus. Am. J. Physiol.
    228:141-148, 1975.

10. Armour, J.A. and G.A. Klassen. Pressure and flow in epicardial coronary veins of the dog heart. Accepted, Can. J. Physiol. Pharmacol. 1983.

11. Bagust, J. and G.A. Kerkut. The use of the transition elements manganese, cobalt and nickel as synaptic blocking agents on isolated hemisected, mouse spinal cord. Brain Res. 182:474-477, 1980.

12. Berne, R.M., H. DeGeest and M.N. Levy. Influence of the cardiac nerves on coronary resistance. Am. J. Physiol., 208:763-769, 1965.

13. Blair, R.W., R.N. Weber and R.D. Foreman. Characteristics of primate spinothalmic tract neurons receiving viscerosomatic convergent inputs in T3-T5 segments. J. Neurophysiol., 46:797-811, 1981.

14. Bosnjak, Z.L., J.L. Seagard and J.P. Kampine. Peripheral neural input to neurons of the stellate ganglion of the dog. Amer. J. Physiol., 242:R237-R243, 1982.

15. Brooks, C. McC., K. Koizume and A. Sato. Integrative functions of the autonomic nervous system. University of Tokyo Press, Japan, 1979.

16. Brown, A.M. Cardiac reflexes in: The Heart, volume 1 of The Cardio-vascular System. Ed - R.M. Berne, N. Sperelakis and S.R. Geiger. American Physiol. Society, Maryland, 1979.

17. Chen, S.S. Transmission in superior cervical ganglion of the dog after cholinergic supression. Amer. J. Physiol., 221:209-213, 1971.

18. Eccles, R.M. and B. Libet. Origin and blockade of the synaptic responses of curarized sympathetic ganglia. J. Physiol., 484-503, 1961.

19. Feigl, E.O. Carotid sinus reflex control of coronary blood flow. Circ. Res. 23:223-237, 1968.

20. Flacke, W. and R.A. Gillis. Impulse transmission via nicotinic and muscarinic pathways in the stellate ganglion of the dog. J. Pharmacol. Exp. Ther., 163:266-276, 1968.

21. Hageman, G.R., J.M. Goldberg, J.A. Armour and W.C. Randall. Cardiac dysrhythmias induced by autonomic nerve stimulation. Am. J. Cardiol. 32:823-830, 1973.

22. Hageman, G.R., W.C. Randall and J.A. Armour. Direct and reflex cardiac bradydysrhythmias from small vagal nerve stimulations. Am. Heart J. 89:338-348, 1975.

23. Hopkins, D.A. and J.A. Armour. Medullary cells of origin of physio-logically identified cardiac nerves in the dog. Brain Res. Bull. 8:359-365, 1982.

24. Jewett, D.L. Activity of single efferent fibers in the cervical vagus nerve of the dog, with special reference to possible cardio-inhibitory fibres. J. Physiol. 175:321-357, 1964.

25. Jones, A. Ganglionic actions of muscarinic substances. J. Pharmacol. Exp. Ther. 141:195-205, 1963.

26. Kelley, K.O. and E.O. Feigl. Segmental α-receptor-mediated vaso-constriction in the canine coronary circulation. Circ. Res. 43:908-917, 1978.

27. Klassen, G.A. and J.A. Armour. Epicardial coronary venous pressures: autonomic responses. Can. J. Physiol. Pharmacol. 60:698-706, 1982.

28. Kralios, F.A., L. Martin, M.J. Burgess and K. Millar. Local ventricular repolarization changes due to sympathetic nerve-branch stimulation. Am. J. Physiol. 228:1621-1629, 1975.

29. Lavalle, M., C. Laurencin, J. deChamplain and R.A. Nadeau. Liberation cyclic AMP and catecholamine from the heart during left stellate stimulation in the anesthetized dog. Can. J. Physiol. Pharmacol. 59:533-540, 1981.

30. Libet, B. and T. Tosaka. Slow inhibitory and exactatory postsynaptic responses in single cells of mammalian sympathetic ganglia. J. Neurophysiol. 32:43-50, 1969.

31. Malliani, A. Cardiovascular sympathetic afferent fibers. Rev. Physiol. Biochem. Pharmacol. 94:11-74, 1982.

32. Manger, W.M. Catecholamine in normal and abnormal cardiac function. Karger, Basel, 1982.

33. McAllen, R.M. and K.M. Spyer. The location of cardiac vagal preganglionic motorneurons in the medulla of the cat. J. Physiol., London 258: 187-204, 1976.

34. McGill, M., D.A. Hopkins and J.A. Armour. Physiological studies of canine sympathetic ganglia and cardiac nerves. J. Autonom. Nerv. Syst. 6:157-171, 1982.

35. Meiri, U. and R. Rahamimoff. Neuromuscular transmission: inhibition by manganese ions. Science 176:308-309, 1972.

36. Morita, K., R.A. North and Y. Katayama. Evidence that substance P is a neurotransmitter in the myenteric plexus. Nature, London. 287:151-152, 1980.

37. Osler, W. The Lumleian lectures on angina pectoris. Lancet. 1:839, 1910.

38. Paintal, A.S. Vagal sensory receptors and their reflex effects. Physiol. Nerv. 53:159-227, 1973.

39. Peters, S.R., D.R. Kostreva, J.A. Armour, E.J. Zuperku, F.O. Igler, R.L. Coon and J.P. Kampine. Cardiac, aortic, pericardial, and pulmonary vascular receptors in the dog. Cardiology: 15:85-100, 1980.

40. Randall, W.C., J.A. Armour, W.P. Geis and D.B. Lippincott. Regional cardiac distribution of the sympathetic nerves. Fed. Proc. 31:1199-1208, 1972.

41. Randall, W.C. (ed). Neural regulation of the heart. Oxford University Press, New York, 1977.

42. Reinecke, M., E. Weike and W.G. Forssmann. Substance P - Immunoreactive nerve fibers in the heart. Neuroscience Letters 20:265-269, 1980.

43. Schwartz, P.J., A.M. Brown, A. Malliani and A. Zanchetti, editors. Neural mechanisms in cardiac arrhythmias. Raven Press, N.Y., 1978.

44. Seller, H. The discharge pattern of single units in thoracic and lumbar white rami in relation to cardiovascular events. Pflugers Arch: 343:317-330, 1973.

45. Thoren, P. Role of cardiac vagal c-fibers in cardiovascular control. Rev. Physiol. Biochem. Pharmacol. 86:1-94, 1979.

46. Yamaguchi, N., J. deChamplain and R. Nadeau. Correlation between the responses of the heart to sympathetic stimulation and the release of endogenous catechelamines into the coronary sinus of the dog. Circ. Res. 36:662-668, 1975.

ACKNOWLEDGEMENT

The author acknowledges the typing assistance of Joan MacNeil.

NERVOUS CORONARY CONSTRICTION VIA α-ADRENOCEPTORS: COUNTERACTED BY METABOLIC REGULATION, BY CORONARY β-ADRENOCEPTOR STIMULATION OR BY FLOW DEPENDENT, ENDOTHELIUM-MEDIATED DILATION? *

E. BASSENGE, J. HOLTZ, R. BUSSE and M. GIESLER
Albert Ludwigs Universität, Herman Herder Str. 7, D-7800 Freiburg 1 - FRG.

1. INTRODUCTION

The effects of psychologic stress on the coronary system are transmitted by nervous and humoral signals. The coronary arteries of the human heart receive a dual innervation of the autonomic nervous system: both adrenergic and acetylcholine-esterase positive nerve fibers have been demonstrated perivascularly (for references see recent reviews: 2, 6).

The direct effects of the autonomic coronary innervation on myocardial arterioles (where coronary flow normally is regulated) are generally considered as of minor importance, since these arterioles are under a very strong metabolic control. Any changes in myocardial activity, induced by psychologic stress via nervous and humoral signals, will cause changes in myocardial metabolic demands. Breakdown products of myocardial metabolism, such as adenosine, hydrogen ions, potassium ions, carbon dioxide and others, dilate the coronary arterioles and are released by the surrounding myocardial cells. These metabolic influences constitute an effective feedback system, linking coronary flow tightly to myocardial metabolism (2, 6).

However, in the large epicardial coronary arteries, which do not contribute significantly to the regulation of coronary flow under physiologic conditions, large increments in coronary smooth muscle tone and dramatic declines in coronary diameter can occur in some patients with coronary heart disease (13). These coronary spasms cause temporal deficits in myocardial perfusion, deterioration of myocardial contraction, and, sometimes, myocardial infarction (13). Thus, the question of the role of psychologic

---

*Supported by Fritz Thyssen-Stiftung

stress on the coronary system should be focussed to the effects of the autonomic nervous system on the large, epicardial coronary arteries.

In patients, coronary spasms have been provoked pharmacologically by adrenergic and cholinergic stimulants (4, 9, 18, 22, 23) Recently, however, it was shown that the coronary spasms provoked by parasympathomimetic drugs occur via the customary reflex adrenergic response to drug-induced hypotension (18), pointing to the role of sympathetic innervation of the coronary arteries. The functional existance and innervation of constrictive α-adrenoceptors in epicardial arteries has been demonstrated in animal models (1, 5, 8, 10, 20). In patients, coronary spasms could be elicited by pharmacologic stimulation of α-adrenoceptors (22, 23) and by reflex activations of sympathetic vasoconstriction by the cold pressor test (14, 15, 21).

In most animal studies, the constriction of epicardial arterie by α-stimulation, induced by sympathetic nerve stimulation or norepinephrine injection, was demonstrated after pharmacological β-blockade (1, 5, 10). Similarly, coronary spasms in patients wer elicited by adrenaline injection only after pretreatment with propranolol (22, 23). In some patients, it was shown that chronic β-blockade by propranolol caused an exacerbation of vasotonic angina pectoris (16). This may indicate that the net effect of the activation of sympathetic coronary innervation is the result of a balance between α-adrenoceptor mediated constriction and β-adrenoceptor mediated dilation. In intact, conscious dogs it was claimed that large epicardial arteries are regulated by β-adrenergic mechanisms (19). Furthermore, it was shown that increments in myocardial metabolic demands are followed by dilations o large epicardial arteries by an unknown mechanism (12). These dat support the concept that stimulation of coronary vascular β-adrenoceptors and, additionally, of myocardial β-adrenoceptors, form a protective mechanism against the deleterious effects of coronar α-stimulation during psychological stress.

In this study, we propose an additional mechanism contributing to this protection by its dilator action. This action arises from the endothelial cells, which release a vasoactive factor in res-

ponse to altered shear stress at their luminal surface. We describe experiments designed to demonstrate that an increase in flow
through a large epicardial artery, caused by a dilation of the
small intramyocardial arterioles (by whatsoever mechanism) induces
a progressively developing dilation of this large artery by a
mechanism which involves the endothelial cells.

2. METHODS

2.1. Experiments in conscious dogs

Ten healthy mongrel dogs (24 - 36 kg) of either sex were in
this study. The dogs were trained to lie quietly on an experimental table for several hours. They were chronically instrumented to
continuously register coronary flow in the left circumflex coronary artery (electromagnetic flowmeter) and the external vascular
diameter of the left circumflex and the left anterior descending
coronary artery (ultrasonic transit time technique, using perivascularly implanted piezoelectric microcrystals, for details see
20). A pneumatic cuff implanted distally to the flowmeter was used
to reversibly occlude the artery or to limit increases in coronary
flow at any desired level by creating a graded stenosis by partial
inflation of the pneumatic cuff. The preparation is shown schema-
tically in figure 1.

2.2. Experiments in isolated coronary arteries in vitro

Segments (6-8 mm length, 1-2 mm outer diameter, n=12) of branches of the left circumflex or the left anterior descending coronary artery were excised from canine hearts, which were removed
after the dogs had been sacrificed in pentobarbital anesthesia.
The segments were cannulated with 2 hypodermic needles, stretched
to their in-situ length and fixed within a perfusion chamber.
Intraluminal perfusion and extraluminal superfusion of the vessels
could be regulated separately, as shown in figure 2. The diameter
of the vessels was registered continuously by a photoelectrical
system, while flow and transmural pressure were controlled independently. Technical details have been published previously (3).

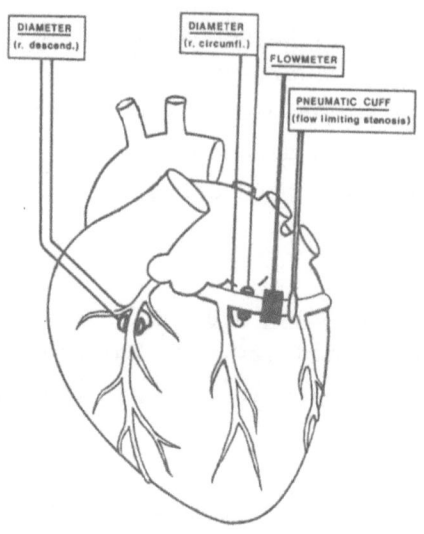

FIGURE 1. Experimental preparation of the dogs for the in vivo studies. Partial or complete inflation of the pneumatic cuff resulted in a temporal stenosis or occlusion of the circumflex branch of the left coronary artery.

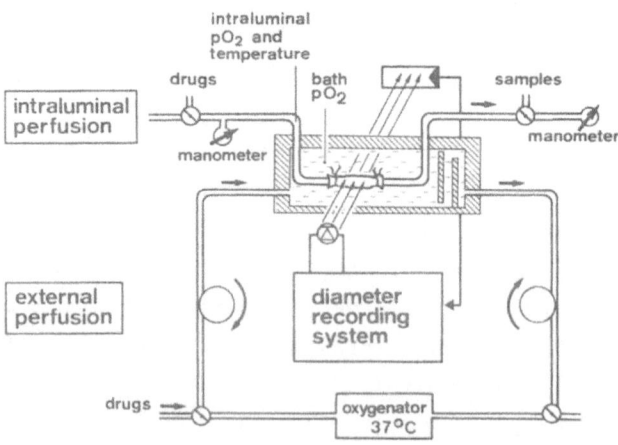

FIGURE 2. Experimental set up for the analysis of vasomotion of isolated coronary arteries in vitro.

3. RESULTS

3.1. Experiments in conscious dogs

Augmentations of coronary flow by more than 300 % of control flow were elicited in the conscious resting dogs by either intravenous injections of adenosine (1-3 µM/kg) or by temporal occlusion of the circumflex coronary artery for 15 - 40 seconds, resulting in a postischemic reactive hyperemia. Typical examples of those two experimental procedures are shown in figures 3 and 4. Adenosine injections caused a slowly developing dilation in both coronary branches, reaching a maximum after 90 seconds. The increase in external diameter was 102 ± 23 µm in the circumflex artery and 112 ± 36 µm in the descending artery, while mean arterial pressure was lowered by 11 ± 4 mmHg. When the adenosine-induced increase in coronary flow in the circumflex artery was prevented by partial inflation of the cuff (creating a flow limiting stenosis), the dilation of the epicardial circumflex artery prior to this stenosis was completely abolished, while the dilation in the unaffected descending artery (control vessel) was well preserved (figure 3).

Parallel observations were made during postischemic reactive hyperemia (figure 4). Occlusion of the circumflex artery for 30 - 40 seconds caused a decline in the diameter of the vessel prior to the occlusion site by 32 ± 12 µm. Upon release of the occlusion, coronary flow increased by more than 400 %, and 90 seconds after release of the occlusion, the diameter of the vessel was increased by 90 ± 19 µm above the pre-occlusion value. Both, during the occlusion and during the post-occlusion reactive hyperemia no significant changes in the diameter of the descending artery (control vessel) were observed (figure 4).

In an additional experimental series in the same dogs, dilations of the epicardial arteries were induced by nitroglycerin injections (0.8 - 80 µg/kg i.v.). It was tested, whether the nitroglycerin-induced dilations were influenced by creating a flow-limiting stenosis similarly as in the experiments with adenosine or reactive hyperemia. Nitroglycerin caused significant dilations of both epicardial branches at a dosage (0.8 µg/kg i.v.) which did not cause any detectable increase in coronary flow

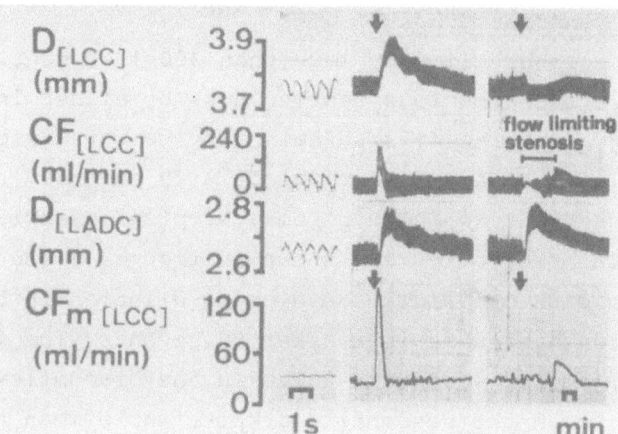

FIGURE 3. Dilation of epicardial arteries following adenosin injections. Tracings of a typical experiment. Abbrevations: $D_{[LCC]}$: diameter of left circumflex coronary artery; $CF_{[LCC]}$: coronary flow of left circumflex coronary artery; $D_{[LADC]}$: diameter of left anterior descending coronary artery; $CF_{m[LCC]}$: mean coronary flow of left circumflex coronary artery.

Arrows indicate the injection of adenosine (1.5 µM/kg i.v.). Note that the dilation of the circumflex coronary artery is completely abolished, when the increase in flow is limited by a stenosis (established by the pneumatic cuff, see figure 1).

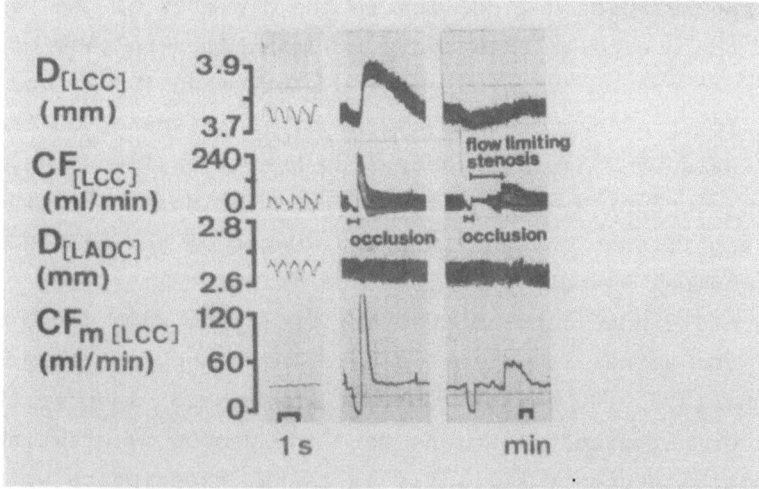

FIGURE 4. Dilation of an epicardial artery following postischemic reactive hyperemia. Abbreviations see figure 3. When the increase in flow upon release of the occlusion is prevented by an appropriate stenosis, the dilation of the circumflex artery is abolished.

(figure 5). Dilations induced by higher dosages of nitroglycerin in both branches of the coronary artery were not affected, whether the increase in flow in the circumflex branch was limited by a stenosis or not (figure 5).

3.2. Experiments in isolated coronary arteries

The isolated segments were perfused at a rate of 0.5 - 1.0 and activated by serotonin (10^{-7} M) in the superfusate (thus acting on the outer surface of the vessel wall). Augmentation of pulsatile flow (5-fold increase of mean flow) caused an increase in outer diameter by 74 ± 26 μm (=4.8 ± 1.1 % of outer diameter, n = 12) within 3 - 7 min after onset of the flow increase. This dilation was reversible, when flow was restored back to the control value, the diameter returned to the control value (figure 6).

After endothelial removal (15 min perfusion with collagenase, 2 U/ml, 0.5 ml/min), the diameter of the arteries was slightly increased by 35 ± 21 μm (15 min after washout of the collagenase). Under this condition, the identical increase in luminal perfusion rate did no longer induce an increase in outer diameter. Functional integrity of the arteries after collagenase treatment was documented by the well preserved responsiveness to constrictive (serotonine) or dilatory (nitroglycerin) stimuli applied from the outside of the vessels. Removal of the endothelium was verified histologically (serial transversal sections of the vessels, endothelial nuclei below 2 % of control value) and functionally (no dilatory response to luminal hypoxia with P_{O_2} below 40 mmHg and to luminal application of 10^{-6} M acetylcholine).

4. DISCUSSION

Our in vitro results demonstrate that a dilatory signal from the endothelium is transmitted to the smooth muscle cells of coronary arteries, when the flow through these arteries is increased. Changes in transmural pressure as the cause of this dilation could be excluded by appropriate control of all variables in the experimental set up. Recently, it was shown that signals from the endothelium are involved in the dilation induced by luminal hypoxia (3) and acetylcholine (7). We used these two

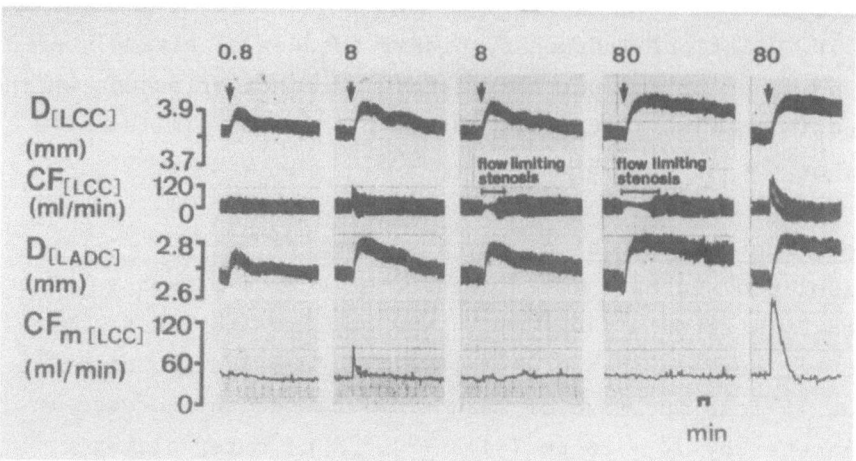

FIGURE 5. Nitroglycerin-induced dilations of epicardial arteries. Abbreviations see figure 3. The numbers on top of the arrows indicate the nitroglycerin dosage (in µg/kg i.v.). At the lowest dosage, dilations of the arteries without any increase in flow were elicited. With higher dosages, the dilations of the epicar= dial arteries were not affected, when the increase in flow was prevented by the stenosis.

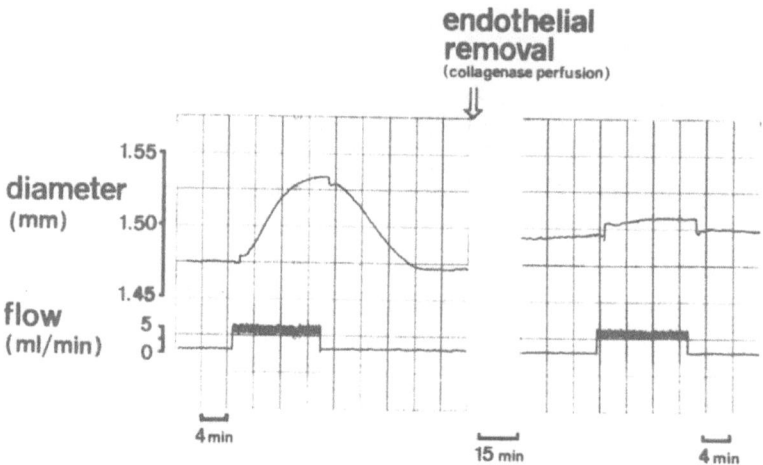

FIGURE 6. Flow-dependent dilation of an isolated coronary artery. The removal of the endothelium by collagenase perfusion demonstra- tes that this dilation is mediated by the endothelium.

reactions to demonstrate the loss of a functionally active endo-thelium after collagenase treatment (figure 6).

Presently, the nature of this vasoactive endothelial signal is not identified. Though cell-to-cell conduction via junctions between endothelial and muscular cells cannot be ruled out the formation of a chemical messenger is a more likely explanation. Altered shear stress on the luminal surface of the endothelial cells (caused by changes in the perfusion rate) may cause libera-tion of arachidonic acid by stimulation of phospholipase A_2. The unidentified messenger could be a cyclooxygenase - or a lipoxyge-nase - metabolit of arachidonic acid.

Our results in the conscious dogs demonstrate that this mecha-nism is operative in vivo. During the temporal stop of coronary flow, a decline in diameter proximal to the occlusion was observed while in the parallel undisturbed vessel this decline did not occur (figure 4). This may indicate that in normal canine coronary arteries without atherosclerosis, a dilatory stimulus is trans-mitted to the smooth muscles even at basal flow rates in the resting animals. This stimulus is increased, when the flow is aug-mented (figures 3 and 4). The nitroglycerin-experiments demon-strate that this antianginal drug dilates the epicardial arteries without this flow-dependent, endothelium-mediated mechanism. In addition, these experiments demonstrate that the dilatory response of an epicardial artery is not affected by a flow limiting steno-sis, provided that the dilation is not caused by this flow-dependent mechanism (figure 6).

Presently, we do not know whether such a flow-dependent mecha-nism is operative in man. Furthermore, the effect of endothelial damage during atherosclerosis and the role of altered shear stresses at vessel sites stenosed by lesions on this dilator mechanism must remain speculative. However, a pathophysiological contribution of this mechanism in context with several clinical observations is an attractive hypothesis. Thus, an explanation may be given why in patients with Prinzmetal's angina, the coro-nary spasms occur at night, when myocardial demands and hence coronary flow are minimal: The flow-dependent dilation as a pro-tection against exaggerated constrictive responsiveness of the

coronaries should also be minimal at night. Similarly, this flow
dependent dilation might be involved in the resolution of coronar
spasms during the so-called "walk-through-phenomen" (spasms at th
onset of exercise are resolved with continuation of exercise).
In speculating about the possible clinical role of the endotheliu
mediated dilation it is an important question, whether this dilation is strong enough to counteract constrictions of coronary
arteries. In preliminary experiments (HOLTZ et al, unpublished)
we observed a complete abolition of serotonin-induced constrictions of epicardial arteries by this mechanism in vivo. Serotonin
is a very effective constrictor of coronary arteries in vitro.
Coronary constrictions by sympathetic nerve stimulation in dogs
could only be demonstrated, when the concomitant increase in
coronary flow was suppressed by β-blockade (see introduction).
Only in the experiments by GEROVA et al (8), an α-adrenergic
sympathetic coronary constriction was obtained without β-blockade
Interestingly, in this model no increase in coronary flow occurre
due to the characteristics of the model (8). Thus, it can be
deduced that normally the increase in myocardial metabolism and
in coronary flow during sympathetic activation counteract the
constrictive effect of the sympathetic nerves on the epicardial
arteries by the endothelial mechanism.

This may have implications for the response of coronary arteries to sympathetic activation during psychologic stress. During
the evolution of man, sympathetic activation was frequently
connected with situations of "fight and flight". In the course o
both reactions large increments in myocardial performance
(several fold increase in cardiac output!) and in coronary flow
occur, resulting in the described protective dilation of the
epicardial arteries through the flow dependent endothelial mechanism (which is also the physiological explanation for the unexplained ascending dilation described in other vascular beds; 11,17
When reactions such as fight or flight do not occur in most
situations of sympathetic activation, the conditions for un-
favorable sympathetic coronary constrictions might be given.

5. REFERENCES

1. Bassenge E, Holtz J, Müller C, Kinadeter H, Kolin A. 1980. Experimental evaluation of coronary artery vasomotion: possible significance for myocardial ischemia in coronary heart disease. Adv.Clin.Cardiol.1, 300.
2. Berne RM, Rubio R. 1979. Coronary circulation. In: The cardiovascular system. Handbook of Physiology, Section 2, Vol.I. American Physiol.Soc. Bethesda, Maryland, p.873.
3. Busse R, Pohl U, Kellner C, Klemm u. 1983. Endothelial cells are involved in the vasodilatory response to hypoxia. Pflügers Arch 397, 78.
4. Endo M, Hirosawa K. Kaneko N, Hase K, Inoue Y, Konno S. 1976. Prinzmetal's variant angina. Coronary arteriogram and left ventriculogram during angina attack induced by methacholine. N.Engl.J.Med. 294, 252.
5. Ertl G, Fuchs M. 1980. Alpha-adrenergic vasoconstriction in arterial and arteriolar sections of the canine coronary circulation. Basic Res.Cardiol 75, 600.
6. Feigl E. 1983. Coronary Circulation. Physiol.Rev. 63, 1.
7. Furchgott RF, Zawadzki JV. 1980. The obligatory role of endothelial cells in the relaxation of arterial smooth muscle by acetylcholine. Nature 288, 373.
8. Gerova M, Barta E, Gero J. 1979. Sympathetic control of major coronary artery diameter in the dog. Circulat.Res. 44, 459.
9. Gerson MC, Noble RJ, Wann LS, Faris JV, Morris SN. 1979. Noninvasive documentation of Prinzmetal's angina. Am.J. Cardiol. 43, 329.
10. Kelley KO, Feigl EO. 1978. Segmental α-receptor-mediated vasoconstriction in the canine coronary circulation. Circulat.Res. 43, 908.
11. Lie M, Sejersted OM, Kiil F. 1970. Local regulation of vascular cross section during changes in femoral arterial blood flow in dogs. Circulat.Res. 27, 727.
12. Marcho P, Hintze TH, Vatner SF. 1981. Regulation of large coronary arteries by increases in myocardial metabolic demands in conscious dogs. Circulat.Res. 49, 594.
13. Maseri A, Chierchia S, L'Abbate A. 1980. Pathogenetic mechanisms underlying the clinical events associated with atherosclerotic heart disease. Circulation 62, V3.
14. Mudge Jr. GH, Grossman W, Mills Jr. RM, Lesch M, Braunwald E. 1976. Reflex increase in coronary vascular resistance in patients with ischemic heart disease. New Engl. J.Med. 295, 1333.
15. Raizner AE, Chahine RA, Ishimori T, Verani MS, Zacca N, Jamal N, Miller RR, Luchi, RJ. 1980. Provocation of coronary artery spasm by the cold pressor test. Circulation 62, 925.
16. Robertson RM, Wood AJJ, Vaughn WK, Robertson D. 1982. Exacerbation of vasotonic angina pectoris by propranolol. Circulation 65, 281.
17. Schretzenmayr A. 1933. Über kreislaufregulatorische Vorgänge an den großen Arterien bei der Muskelarbeit. Pflügers Arch.Ges.Physiol. 232, 743.

18. Stang JM, Kolibash AJ, Schorling JB, Bush CA. 1982. Methacholine provocation of Prinzmetal's variant angina pectoris: a revised perspective. Clin.Cardiol. <u>5</u>, 393.
19. Vatner SF, Hintze TH, Macho P. 1982. Regulation of large coronary arteries by β-adrenergic mechanisms in the conscious dog. Circulat.Res. <u>51</u>, 56.
20. Vatner SF, Pagani M, Manders WT, Pasipularides AD. 1980. Alpha-adrenergic vasoconstriction and nitroglycerin vasodilation of large coronary arteries in the conscious dog. J.Clin.Invest. <u>65</u>, 5.
21. Waters DD, Szlachcic J, Bonan R, Miller, DD, Dauwe F, Theroux P. 1983. Comparative sensitivity of exercise, cold pressor and ergonovine testing in provoking attacks of variant angina in patients with active disease. Circulation <u>67</u>, 310.
22. Yasue H, Touyama M, Shimamoto M, Kato H, Tanaka S, Akiyama F. 1974. Role of autonomic nervous system in the pathogenesis of Prinzmetal's variant form of angina. Circulation <u>50</u>, 534.
23. Yasue H, Touyama M, Kato H, Tanaka S, Akiyama F. 1976. Prinzmetal's variant form of angina as a manifestation of alpha-adrenergic receptor-mediated coronary artery spasm: documentation by coronary arteriography. Am.Heart J. <u>91</u>, 148.

CLINICAL CLUES TO PSYCHOLOGICAL AND NEURO-HUMORAL MECHANISMS OF ARRHYTHMOGENESIS

D.C. RUSSELL

Cardiovascular Research Unit, University of Edinburgh, UK.

INTRODUCTION

Clinical evidence relating acute psychological stress to arrhythmogenesis is in large part circumstantial, anecdotal and difficult to evaluate.   Nevertheless a clear association would appear to exist between the two.

This was recognised even in antiquity. Pliny for example describes several instances of sudden death following such situations as intense emotional stress, receipt of news of a great victory, public speaking in the senate or impassionate acting. Interestingly he himself later sustained an acute, presumably arrhythmic death whilst escaping from Pompeii and the eruption of Vesuvius in A.D. 79. Similarly in 1776 Fothergill reported the case of an apprehensive patient who died "in a sudden and violent transport of anger". The post-mortem was performed by the celebrated anatomist John Hunter who subsequently died suddenly presumably of ventricular fibrillation during the course of a heated argument in the board room of St. George's Hospital in London having previously declared "my life is in the hands of any rascal who chooses to annoy and tease me".

There is nothing new in being frightened to death and every modern clinician has a fund of experiences linking situations of acute emotional stress to arrhythmogenesis. An example of this occurred in Edinburgh when following initiation of resuscitation procedures for cardiac arrest in the coronary care unit, a patient in an adjacent bed and alarmed at the situation, himself fibrillated and required resuscitation. This led to the somewhat flippant suggestion that ventricular fibrillation is an "infectious" condition! Again quite recently one of our hospital

porters collapsed with a ventricular tachycardia on being called
to rush a cardiac arrest trolley to one of the wards he himself
required cardioversion. Indeed, accounts are legion of induction (
arrhythmias such as paroxysmal tachycardia, symptomatic extra-
systoles or ventricular fibrillation in association with acute
emotions of for example, joy or grief, exuberation or anger,
frustration and disappointments. The possibility arises therefore
as suggested by Lown that certain psychological or emotional
factors may "trigger" malignant arrhythmias particularly if the
heart is chronically predisposed to electrical instability by
ischaemic myocardial disease.

Undoubtedly a variety of pathophysiological mechanisms may be
involved with effects on both sympathetic and parasympathetic
autonomic nervous systems and modulation of humoral mechanisms.
Although most knowledge in this area derives from experimental
studies clinical evidence may provide some important clues to
mechanisms of arrhythmogenesis in man. This may best be approache(
by consideration of induction of ventricular fibrillation and of
ventricular premature beats or tachycardia separately as differin
electrophysiological mechanisms may be involved.

PSYCHOLOGICAL STRESS AND SUDDEN CARDIAC DEATH

In general sudden cardiac death is thought to result from
ventricular tachyarrhythmias and ventricular fibrillation althoug
a minority of documented cases of profound bradycardia or asystol
are reported. Furthermore quite different mechanisms may be
operative during early and later periods of myocardial ischaemia
and on coronary reperfusion. Clinical studies relating psychological stress to induction of ventricular fibrillation must be
viewed therefore against this background of a probable spectrum
of pathophysiological processes. Unfortunately, however, most
clinical evidence is anecdotal or derived from uncontrolled case
series and not combined with either electrophysiological or
metabolic measurements. Nevertheless a wide variety of stressful
situations have been linked to sudden cardiac death and some are
outlined in Table 1.

Table 1. Psycho-social factors associated with sudden cardiac death.

Acute emotions anger, fear, joy
Fatigue
Fights or arguments
Bereavement
Low socio-economic class
Low educational level
Type 'A' behaviour
Depression

Premonitory symptoms and stress factors preceding sudden death have been evaluated retrospectively in a number of studies, either from interviewing of next of kin of victims or from patients resuscitated successfully from out-of-hospital ventricular fibrillation. Rissanon et al (1978) found an incidence of "acute" emotional stress immediately preceding death in 19% of cases half of whom died suddenly. More "chronic" stress preceded the event in 25% of patients and was associated with a higher incidence of later arrhythmias and myocardial infarction. The commonest antecedent factor in 32% of cases was of unusual fatigue.

The immediate antecedent events of sudden death have been assessed in detail by Hinkle (1981) in a 5 year prospective study of 1839 deaths from coronary heart disease in a group of 260,000 employed men. Stressful events were subdivided into those associated with probable vagal mediated effects, such as urination, defaecation, lifting or straining, diving into a swimming pool, shaving or drinking iced tea, and into those associated with probable sympathetic mediated effects, such as physical exercise, emotional arousal, or driving. It is of interest that "vagal" events preceded 17.1% of deaths whereas "sympathetic" events 34.1% of deaths. No obvious premonitary events occurred in 34.1% of deaths. Regards the state of arousal of the victims it may be relevant that 88% were awake and 42% active whereas 8.5% were asleep at the time of death. The phase of sleep indeed may be of additional importance as evidence of arrhythmias is as common in REM sleep as in the awake state.

A more detailed assessment of possible psychological "trigger" factors has been obtained by Reich et al (1981) in a study of

117 patients 53% of whom were resuscitated from ventricular fibrillation. "Trigger factors" were identified in 21% of cases. These occurred usually within less than an hour of the event and included feelings of intense anger, fear or excitement, job-related or marital stress or bereavement. A lower incidence of demonstrable coronary arterial disease was found in the group with "trigger factors" than in the total population. Conversely in those patients with no demonstrable coronary disease the incidence of "trigger factors" was doubled. Findings of this nature have led Lown and co-workers to postulate that ventricular fibrillation may result from an interplay netween the neuro-humoral stimulation of an acute stressful event and underlying myocardial electrical instability. The more normal the myocardium therefore, the greater the psychological stress necessary to induce arrhythmias.

Acute events may be modulated however by a background of more chronic psychological effects, often associated with some major life-event. Bereavement of either a spouse or close friend or relative within the previous 6 months significantly increases the risk of sudden death (Jacobs et al 1977). In a careful case control study by Cottington et al (1980) this was the only significant factor affecting mortality, which was increased six-fold with respect to controls. Similar effects were not observed for work-related problems such as marital, financial or legal problems. Engel (1978) furthermore found 20% of 275 sudden death cases to have developed within 3 weeks of onset of a grief reaction.

Danger or a threat to life may be a potent risk factor. In a well-controlled survey (Trichopaulos et al 1983) of the incidence of sudden deaths in the first 5 days following the 1981 Athens earthquake cardiac mortality was found to be significantly increased on the third day after the major earthquake with a 50% increase in the short term probability of occurrence of a fatal cardiac event. Other factors include divorce (Rahe et al 1978), job loss, disappointments with family members and poor educationa level (Kitagawa et al 1973, Vinblatt et al 1978).

Psychometric testing suggests that a combination of ongoing psychological and social stress may be more relevant than any one single factor. Greene et al (1972) found this in half of the 26 sudden deaths studied, the greatest risk apparently deriving from an association between depression and a state of arousal. Certainly both depression (Bruhn et al) and extrovert or "Type A" behaviour may correlate with sudden death. On the basis of retrospective studies Bruhn et al suggest that a triad of high depression score, a sense of "joyless striving" at work and of "Type A" behaviour is predictive and represents a chronic state of "drained" emotions.

Engel (1978) suggests a possible pathophysiological correlate of these seemingly contradictory groups of observations. From animal studies it is known that sudden stress may induce one of two responses - either the well-known "flight-flight response" or alternatively what has been termed a "conservation-withdrawal response" mediated by a hypothalamic reflex during which the animal "plays dead" and develops a profound bradycardia and hypotension. In man it is possible that either of these two responses may occur, the flight-flight response being manifest in acute excitement or anger and the conservation-withdrawal response by "emotional shock", vaso-vagal responses or possibly by chronic depression or what Engel terms "seething anger". He suggests that sudden death may result when both these responses are triggered simultaneously as for example by an acute emotional stress superimposed on a background state of psychological uncertainty or fear of death.

The initiation of a sympathetic mediated flight-flight response certainly may be associated with sudden death. Myers et al (1977) found significant stress factors in 91 of 100 men who died suddenly, occurring within 24 hours of the event in 40 men and within 30 minutes in 23. Situations included involvement in a traffic accident, attack by dogs, fights, notification of divorce and so forth. In a study of 497 homicides in Ohio post-mortem revealed no evidence of traumatic death in 15 cases in 10 of whom severe arguments or fights preceded death (Cebelin et al 1980). Subendocardial myofibrillar degeneration was found consistent with exposure to high levels of catecholamines.

By contrast clinical evidence for vagally mediated responses is less clear. The bizarre phenomenon of voodoo death may fall into this category. Death can result within 24 hours of the witch doctor pronouncing the "crime" and is literally from fright. It has been suggested this may be mediated by activation of a primitive "diving reflex" normally elicited by exposure of the face to cold with production of bradycardias and asystole. No documented clinical evidence however supports this contention.

PSYCHOLOGICAL STRESS AND VENTRICULAR ARRHYTHMIAS.

Acute psychological stress may precipitate benign as well as malignant ventricular arrhythmias. A selection of stress factors known to be associated with induction of ventricular premature beats is outlined in Table 2.

Table 2.   Stress factors associated with benign ventricular
            arrhythmias.

            Emotional trauma
            Public speaking
            Motor car racing
            Driving
            Personal danger
            Arousal from sleep
            Psychological stress testing

It appears possible therefore to evoke ventricular premature beats by a wide range of forms of stress, those best documented being such situations as public speaking, driving in traffic, motor car racing and acute anxiety, excitement or fear or anger. Many reports are however anecdotal and uncontrolled. Wellens describes the case of a young girl who developed recurrent ventricular fibrillation on arousal from sleep by an alarm clock, the initial episode being precipitated when woken by a clap of thunder. Lown et al describe a patient previously resuscitated from ventricular fibrillation who developed ventricular tachycardias on being reminded of the event. Many similar instances could be cited.

Clinical evidence suggests an important role of sympatho-
adrenal activation and catecholamines. Taggart et al (1969, 1970)
have shown increases in catecholamine levels in association with
arrhythmias induced by public speaking or motor racing and further
that these effects are abolished by pre-treatment with a beta-
blocking agent, oxprenolol. Similarly, Coumel (1980) describes
6 cases of life-threatening ventricular tachyarrhythmias in young
or middle-aged men precipitated by extra or psychological stress.
These appeared related to some form of autonomic imbalance and
some were catecholamine induced.

More direct evidence for a link between stress and arrhythmogenesis might derive from psychological stress testing. Although
not entirely reproducible Lown et al (1978) were able to double
the incidence of premature beats in 19 patients with complex
arrhythmias or previously resuscitated from ventricular
fibrillation by this means and were able to induce ventricular
tachycardia in one subject.

Finally the special case of ventricular arrhythmias associated
with the long QT syndrome deserves mention. As described elsewhere
in this volume, its clinical features are explicable on the basis
of a chronic imbalance between the right and left sided sympathetic
innervation to the heart. Such arrhythmias may be triggered by
emotional stimuli and successfully abolished by pre-treatment with
a beta-blocking agent or surgical ablation of the left stellate
ganglion.

CLINICAL CLUES TO MECHANISMS OF ARRHYTHMOGENESIS

Clinical evidence would be in support therefore of the concept
of a strong relationship between stress and genesis of both benign
and malignant ventricular arrhythmias. This may be attributed
perhaps simplistically to direct electrophysiological effects of
stimulation of either the sympathetic or para-sympathetic neural
input to the heart or imbalance between the two. Certainly
similar phenomena may be induced in a variety of experimental
models and clinical parallels found.

Undoubtedly direct neural effects are operative under certain
circumstances. Effects of stress are more complex however and may

include a variety of important metabolic changes which directly or indirectly could modulate the vulnerability of the myocardium (in particular the ischaemic myocardium) to development of arrhythmias. In recent years interest in Edinburgh has focussed on this role of metabolic factors in the genesis of lethal ventricular arrhythmias with particular reference to changes in free plasma free fatty acids, glucose and catecholamines. It would seem pertinent therefore at this point of review some of the clinical evidence in support of this view and to suggest some possible pathophysiological mechanisms which might on this basis link stress, metabolism and arrhythmias.

METABOLIC EFFECTS OF STRESS.

A major systemic response may result from an acute stressful event which may be psychogenic or associated with the pain of acute myocardial ischaemia or infarction. Circulating levels of catecholamines, in particular of noradrenaline are elevated as are cortisol, glucagon and growth hormone. Insulin secretion may be depressed partly as a result of direct inhibitory action of adrenaline or by beta stimulation. Together with the glycogeno-lytic effect of catecholamines on the liver, this may promote hyperglycaemia. Adipose tissue and also intramyocardial lipolysis may be stimulated largely by the elevation of plasma catecholamines with release of free fatty acids (FFA) and glycerol into the circulation. In addition, significant but as yet ill-defined effects of other substances including endorphins enkephalins polypeptides or prostaglandins may result. The severity of these responses will relate to the stimulus and may be inappropriate to the metabolic needs of the electrophysiologically vulnerable myocardium.

The relative importance of these multiple metabolic effects in terms of arrhythmogenesis is unclear. In 1970 Kurien and Oliver suggested in their "fatty acid hypothesis" that marked elevations in plasma free fatty acids under such conditions might be an important independent arrhythmogenic factor particularly in the presence of enhanced adrenergic activity. The background to this hypothesis was the clinical observation of a significant correlation between high levels of plasma FFA and arrhythmias and deaths in a series of 200 patients with acute myocardial infarction. 33% of patients with plasma FFA > 1200 m Eq/l died compared with 5% of patients with plasma levels < 800 m Eq/l (Oliver et al 1968). It was not established however whether the changes were causally related to plasma FFA or whether both FFA elevation and arrhythmogenesis themselves followed an increase in plasma levels of catecholamines or enhancement of sympathetic activity.

Several early clinical studies confirmed the association between elevated plasma FFA and arrhythmias during the myocardial ischaemia of acute infarction (Gupta et al 1969, Reiman and Schwandt 1971; Prakash et al 1972) but other studies failed to do so (Nelson 1970; Revans 1972). Unfortunately in many of these studies plasma FFA were grossly overestimated as a result of undetected in vitro lipolysis (Riemersma et al 1982). Equally controversial have been a series of experimental studies and the question remains unresolved. No clinical studies are available linking psychological stress to plasma FFA and arrhythmogenesis directly.

Indirect clinical evidence in favour of an independent role of FFA in arrhythmogenesis derives from observations on the use of antilipolytic therapy within the first six hours of acute myocardial infarction. Using a nicotinic acid analogue (5-fluoro-nictonic acid) Rowe et al showed a significant reduction in incidence of ventricular tachycardia in a subgroup of patients with the greatest degree of reduction of plasma FFA by the drug. (See Table 3).

Such arrhythmogenic effects of FFA, if substantiated might be mediated by a variety of mechanisms including toxic effects of accumulation of fatty acids or fatty acyl derivatives within the cell either on the cell membrane or by inhibition of enzyme systems. Conversely they might adversely modulate energy metabolism by increasing myocardial oxygen consumption or stimulating wasteful energy cycling. These effects are likely to be enhanced by increased adrenergic stimulation. An alternative suggestion (Opie et al 1977) is that a vicious cycle may be

Table 3.   Incidence of V.T. during acute myocardial infarction
(Rowe, Neilson and Oliver 1975)

Delay from onset Free Fatty Acid Levels	Placebo (71) 3-12 hrs.	Nicotinic Acid (32) Within 5 hrs	Analogue (36) 5-12 hrs.
>50% fall in 4 hrs <800 uEq/l for 20 hrs	42%	0%**	70%
<50% fall in 4 hrs >800 uEq/l	70%	29%*	77%

*P<0.01   **P<0.003

initiated, at least during myocardial ischaemia, whereby
enhanced sympathetic drive might increase plasma FFA levels,
thereby reducing glucose utilisation by the myocardial cell,
further increasing ischaemic injury and stimulating a further
increase in sympathetic activity and peripheral lipolysis.
Psychogenic stress might be expected to exacerbate such a
mechanism.

The role of glucose in arrhythmogenesis is equally controversial. Sodi-Pallares first suggested some 30 years ago that
administration of hypertonic glucose-insulin-potassium regimes
might be anti-arrhythmic during myocardial ischaemia following
both clinical and experimental studies. Attempts to reproduce
these findings however were not uniformly successful. Mittra
(1965, 1967) performed two clinical trials with positive result,
but this was not confirmed in a large multicentre MRC trial in
1968. Smaller doses of glucose were given, orally and not intravenously and often later than the periods now recognised as those
of ventricular arrhythmias. More recently the findings of Rogers
and co-workers are of particular interest. Hospital mortality rate
within the first few hours of onset of symptoms of acute
myocardial infarction was approximately halved in 70 patients by
administering glucose-insulin-potassium. The greatest reduction
in mortality was achieved in patients with heart complications
and single vessel coronary disease. Such infusions are also antilipolytic and cause a dramatic fall in plasma FFA levels. It is
difficult therefore to dissociate possible effects of increased

glucose availability from reduction in plasma FFA. The possibility remains however that modulation of plasma glucose levels by stress may have some protective action against arrhythmias and counteract concomitant opposite arrhythmogenic effects of catecholamines, autonomic imbalance, or elevation in plasma FFA. Current experimental studies from our laboratory suggest indeed that patterns of glycolytic activity in the ischaemic zone may be critical in determining patterns of re-entrant excitation and arrhythmogenesis at least for early ventricular arrhythmias or fibrillation.

Detailed mapping studies at the time of onset of early ventricular arrhythmias suggest that rapid impulse propagation may occur largely in areas of residual aerobic glycolytic metabolic activity. Furthermore administration of glucose may rapidly ameliorate abnormalities in cardiac refractoriness or repolarisation. Antilipolytic therapy has similar although less striking beneficial effects in the presence of high background catecholamine levels. Clearly these effects may be modulated by alterations in substrate availability as well as circulating catecholamines or direct sympathetic activation.

Full discussion of the role of elevation in circulating adrenaline or noradrenaline which may accompany acute psychological stress is beyond the scope of this review. Deserving mention however is the possibility that elevated levels of circulating catecholamines and in particular adrenaline may act presynaptically on the sympathetic nerve terminal to further enhance local noradrenaline release and hence metabolic and electrophysiological effects of sympathetic stimulation.

CAN THE EVIDENCE BE IGNORED?

We should like to suggest therefore that myocardial electrophysiological activity may be critically modulated during stressful situations by acute alterations in cardiac metabolism, particularly against a background of myocardial ischaemia and infarction and that this could tip the balance in certain situations in favour of a lethal versus non-lethal arrhythmia. This could operate at three levels, not only within the myocardial cell itself as a result of derangement of normal homeostatic biochemical mechanisms, but also indirectly due to metabolic effects of neurogenic stimulation, or from alterations in substrate utilisation or exposure to circulating humoral agents. These effects would run parallel with direct neural mediated effects and could significantly modify the electro-physiological response of the myocardium. Clinical evidence in favour of these concepts is as yet fragmentary but should not be ignored.

REFERENCES

1. Bruhn JG, Paredes A, Adsett CA, Wolf S. Psychological predictors of sudden death in myocardial infarction. J. Psychosom Res. 1978; 18, 187-191.

2. Cebelin MS, Hirsch RL. Human stress cardiomyopathy. Hum. Pathology 1980; 11, 123-132.

3. Cottington EM, Matthews KA, Talbott E, Kuller LH. Environmental events preceding sudden death in women. Psychosom Med. 1980; 42, 567-574.

4. Coumel P, Fidelle J, Lucet V. Catecholamine induced severe ventricular arrhythmias with Adams-Stokes syndrome in children. Br. Heart J., 1978; 40, 28-37.

5. Engel GL. Psychological stress, vasodepressor (vaso-vagal) syncope and sudden death. Ann. Int. Med. 1978; 89, 403-412.

6. Jacobs S, Ostfield A. An epidemiologic review of mortality of bereavement. Psychosom Med. 1977; 39, 344-357.

7. Greene WA, Moss AJ, Goldstein A. Delay, denial and death in coronary heart disease. In: Stress and the Heart. Elliot RL, ed. Mt. Kisco, New Y. Futura 1964; 143-162.

8. Greene WA, Goldstein S. Moss AJ. Psychosocial aspects of sudden death. Arch. Int. Med. 1972; 129, 725.

9. Gupta DK, Young R, Jewitt DE. Increased plasma free fatty acids and their significance in patients with acute myocardial infarction. Lancet 1969; 2, 1209-1213.

10. Hinkle LE. The immediate antecedents of sudden death. Acta Med. Scand. 1981; 51, 210, 207-217.

11. Kurien VA, Oliver MF. A metabolic cause of arrhythmias during acute hypoxia. Lancet, 1970; 1, 813-816.

12. Lown B, deSilva RA. Rules of psychological stress and autonomic nervous system changes in provocation of

ventricular premature complexes. Am. J. Cardiol. 1978; 41, 979-985.

13. Myers A, Dewar HA. Circumstances attending 100 sudden deaths from coronary artery disease with coroners necropsis. Br. Heart J., 1975; 37, 1133.

14. Mittra B. Potassium glucose and insulin in treatment of myocardial infarction. Lancet, 1965; 2, 607-609.

15. Nelson PG. Free fatty acids and cardiac arrhythmias. Lancet, 1970; 1, 733.

16. Oliver MF, Kurien VA, Greenwood TW. Relation between serum free fatty acids and death after acute myocardial infarction. Lancet, 1968; 1, 710-715.

17. Opie, L.H. Metabolism of free fatty acids glucose and catecholamines in acute myocardial infarction. Am. J. Cardiol., 1975; 36, 938-953.

18. Prakash R, Parmley W, Horvat M. Serum cortisol, plasma free fatty acids and urinary catecholamines as indicators of complications in acute myocardial infarction. Circulation 1972; 45, 736-745.

19. Rahe RH, Romo M, Bennet L, Siltanen P. Recent life changes, myocardial infarction and abrupt coronary death. Arch. Int. Med. 1974; 133, 221.

20. Reich P, deSilva RA, Lown B, Murawski BJ. Acute psychological disturbances preceding life threatening ventricular arrhythmias. J. Amer. Med. Ass., 1981; 246, 233-235.

21. Reimann R, Schwandt P. Frischer wezinfarct und freie fertsauren. D. Med. Wochenschr. 1971; 96, 93-96.

22. Rissanen V, Romo M, Siltanen P. Premonitoring symptoms and stress factors preceding sudden death from ischaemic heart disease. Acta Med. Scand. 1978; 204, 389.

23. Rogers WJ, Stanley AW, Breining JB, et al. Reduction of hospital mortality rate of acute myocardial infarction with glucose-insulin-potassium regime. Am. Heart J. 1976; 92, 441.

24. Rowe MJ, Neilson JMM, Oliver MF. Control of ventricular arrhythmias during myocardial infarction by antilipolytic therapy using a nicotinic acid analogue. Lancet, 1975; 2, 295-302.

25. Russell DC, Lawrie JS, Riemersma RA, Oliver MF. Metabolic aspects of rhythm disturbances. Acta Med. Scand. 1982; 51, 210, 71-80.

26.   Sodi-Pallares D, Ponee de les J, Bisteni A, Medrano G.
      Potassium, glucose and insulin in myocardial infarction
      Lancet, 1969; 1, 1315-1316.

27.   Taggart P, Gibbons D, Somerville W.   Some effects of
      motor car driving on the normal and abnormal heart.
      Brit. Med. J., 1969; 4, 130-134.

28.   Taggart P, Carruthers M, Somerville W.   Electrocardio-
      gram, plasma catecholamines and lipids and their
      modification by oxprenolol when speaking before and
      audience.   Lancet, 1970; 2, 341-346.

29.   Weinblatt E, Ruberman W, Goldberg JD.   Relation of
      education to sudden death after myocardial infarction.
      New Engl. J. Med. 1978; 299, 66.

30.   Wellens HJJ, Vermeulen A, Durrer D.   Ventricular
      fibrillation, occurring on arousal from sleep by
      auditory stimuli.   Circulation, 1972; 46, 661-665.

CLINICAL CLUES AND EXPERIMENTAL EVIDENCES OF THE NEURO-HUMORAL INTERPRETATION OF CARDIAC ARRHYTHMIAS.

P.J.Schwartz, E.Vanoli, A.Zaza
Istituto di Clinica Medica IV,Università di Milano e Centro di Fisiologia Clinica e Ipertensione, Ospedale Maggiore, Via F.Sforza 35 - 20122 Milano

The critical role of the autonomic nervous system in the genesis of cardiac arrhythmias has been repeatedly demonstrated at clinical and at experimental level. This is particularly true for life-threatening arrhythmias.During the last five years issue has been reviewed in detail (1-4).

While in most cases the neural effects are mediated by direct changes in cardiac electrophysiology, in the arrhythmias induced by myocardial ischemia effects on coronary circulation mediated by the sympathetic nerves play an important contributory role (5).

A new concept, which has prompted to steps forward in the understanding of the complex relationship between the autonomic nervous system and cardiac arrhythmias, is that left sided cardiac sympathetic nerves carry a particularly high arrhythmogenic potential (4). As a consequence, it became apparent that any type of imbalance in cardiac sympathetic innervation leading to dominance of left sided nerves could impair cardiac electrical stability and favour arrhythmias (6).

In this chapter, the evidence linking cardiac arrhythmias and the autonomic nervous system will be presented by briefly analyzing the effects of increasing or decreasing cardiac sympathetic activity in a variety of conditions. Therefore, we will discuss the effects of stimulation of the left stellate ganglion and of removal of either the right or left stellate ganglia, which represent the most important relay stations for the cardiac sympathetic nerve fibres.

Effects of sympathethic stimulation

Activation of localized nerve pathways or unilateral sympathetic discharges may be more arrhythmogenic than massive but homogenous activation of the sympathetic nervous system.

Armour (7) et al.in 1972 and Hageman et al.(8) in 1973 have indicated that while right sided stimulation induces only sinus tachycardia, stimulation of nerves originating from the left stellate ganglion results in a variety of rhythm disturbances.Most of these arrhythmias are supraventricular.However, stimulation of the left stellate ganglion may induce major arrhythmias as ventricular tachycardia even in an intact heart (see Fig.1).

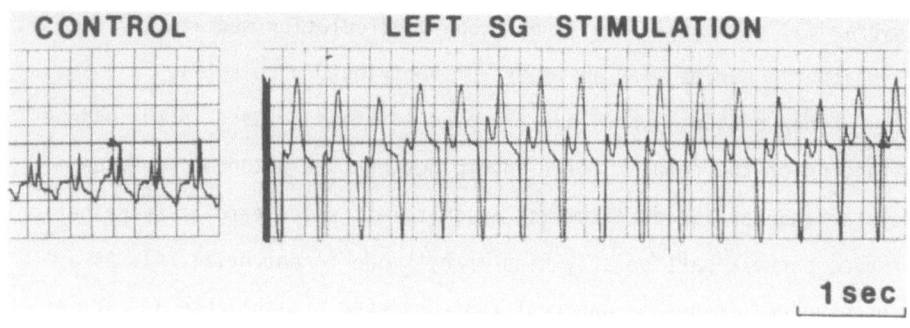

FIGURE 1. Anesthetized cat.Ventricular tachycardia induced by left stellate ganglion stimulation.

Not surprisingly this arrhythmogenic potential of left sympathetic stimulation becomes much more evident in the presence of a myocardial infarction or of acute myocardial ischemia.As an example Fig.2 illustrates the effect of mechanical stimulation of the left stellate ganglion in an anesthetized man with a prior myocardial infarction.The patient was in sinus rhythm until the left stellate ganglion was mechan-

ically stimulated with a blunt instrument producing ventricular premature beats and couplets.

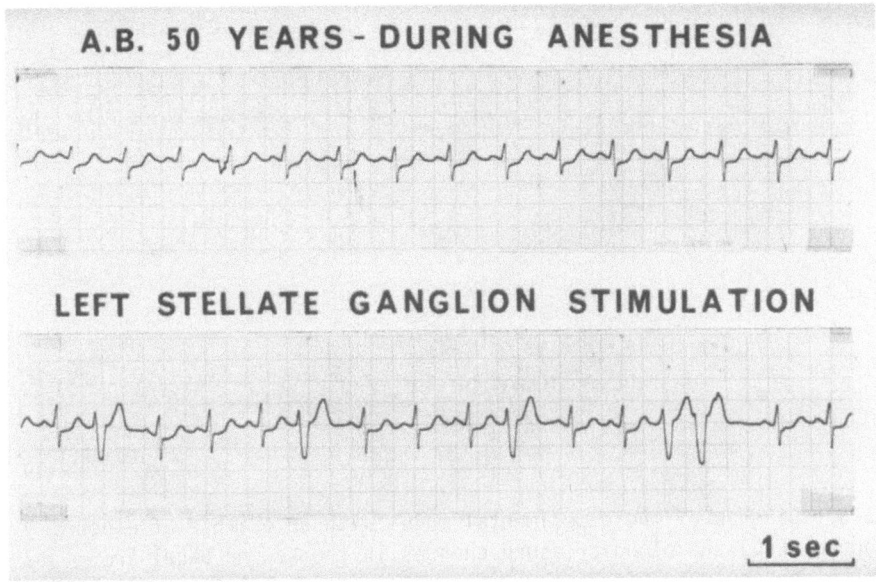

FIGURE 2. Electrocardiographic tracing in an anesthetized man 50 days
after occurrence of an anterior myocardial infarction complicated by
ventricular fibrillation.The patient was in sinus rhythm until the left
stellate ganglion was mechanically stimulated, leading to frequent
premature ventricular beats and couplets. This patient is partecipating
in a multicenter trial in which survivors of an anterior myocardial
infarction complicated by ventricular fibrillation are randomly allocat-
ed to treatment with either placebo,beta-adrenergic blockade with
oxprenolol, or left stellectomy (high thoracic left sympathectomy).
(From Schwartz and Stone, ref.4).

Cardiac arrhythmias depending upon interaction between acute
myocardial ischemia and sympathetic hyperactivity may be repeatedly reproduced in an experimental model already described (9).In these experiments in anesthetized cats a brief stimulation of the left stellate
ganglion is superimposed on a short lasting coronary artery occlusion.
While the two stimuli per se do not induce any arrhythmias, their interaction provokes in six-eight consecutive trials ventricular arrhythmias,
particularly ventricular tachycardia and fibrillation, in almost 70% of
animals studied. (see Fig.3 and Fig. 4).

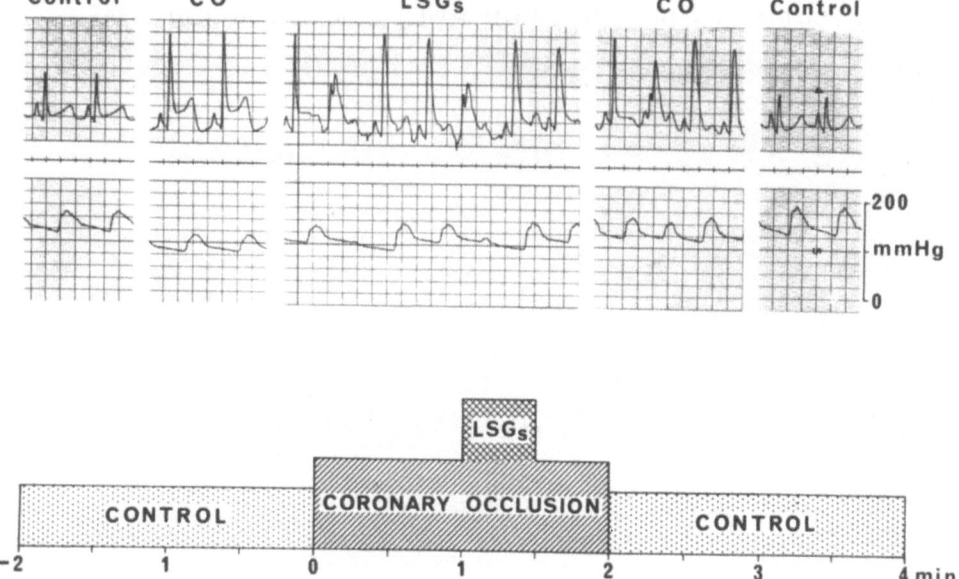

FIGURE 3. Diagram of the experimental model with an example of electro-cardiogram (ECG) and blood pressure changes.In the second panel are shown the changes occurring after 1 min of coronary occlusion (CO) just prior to sympathetic stimulation: blood pressure is lower, and classic ischemic changes are obvious.In the third panel, electrical stimulation of the left stellate ganglion (LSG$_s$) induced the occurrence of premature ventricular beats and a slight increase in blood pressure ;stimulation artifacts are visible in the ECG tracing.In the fourth panel, after cessation of sympathetic stimulation while coronary occlusion is still maintained, ventricular premature beats are still present.In the fifth panel, a few seconds after release of occlusion both ECG and pressure have returned to normal. (From Schwartz P.J.and Vanoli, E. ref.9).

Thus, activation of left -sided cardiac sympathetic nerves facilitates arrhythmias in a variety of conditions, and particularly in ischemic hearts.

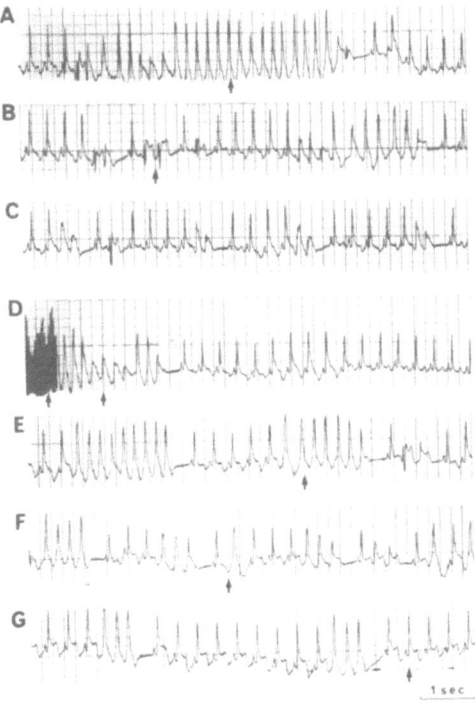

FIGURE 4. Seven consecutive trials in a cat in which no arrhythmias would be elicited by separate coronary artery occlusion and left stellate ganglion stimulation. The arrows in A, B,E,F, and G indicate the moment of release of a 2 min coronary occlusion, 30s after the end of electrical stimulation of the left stellate ganglion.C is 10s after release of occlusion.In D, ventricular tachycardia (VT) was already present at the end of sympathetic stimulation and release of coronary occlusion was anticipated.Runs of VT and/or frequent multifocal premature ventricular contractions (PVCs) are regularly elicited.(From Schwartz P.J. and Vanoli E. ref.9)

Effects of sympathetic denervation

Ablation of either the right or left stellate ganglion produc

es a state of sympathetic imbalance with left or right dominance,

respectively. In these conditions, the modifications on the propensity

to cardiac arrhythmias provide critical information on the contribution,

or lack of it, of sympathetic nerves to arrhythmogenesis.

Exercise is a condition in which conscious animals may be studied repeatedly and which high level of sympathetic activity may be obtained. Schwartz and Stone (10) have tested the arrhythmogenic effect of right stellectomy in dogs with normal heart subjected to sub-maximal exercise. In these experiments they observed an higher incidence of ventricular arrhythmias in dogs with right stellectomy (86%) compared to control (8%) and to dogs with left stellectomy (11%). The same type of inter-vention may also facilitate the development of life-threatening arrhythmias in conscious cats exposed to psychologic stress(11).

The importance of these data is in showing that susceptibility to ventricular arrhythmias can be enormously increased by simply removing the right stellate ganglion in otherwise normal heart.

Furthermore the arrhythmogenic effect of right stellectomy has been confirmed in man (12).Seventy one patients affected by Raynaud's Syndrome, without cardiac involvement, and treated with unilateral and bilateral stellectomy have been studied with a sub-maximal exercise stress test and compared to a group of healthy controls. Exercise induc ed ventricular arrhythmias in 32% of patients with right stellectomy,in contrast with patients with left stellectomy (10%)and with controls(3%).

The role of imbalance of sympathetic activity in increasing the propensity of a normal heart to ventricular tachiarrhythmias has also been confirmed by Schwartz et al.(13) using the ventricular fibrillation threshold (VFT) as index of electrical instability of the heart .In these experiments they found that left stellectomy or blockade by cold produced a major increase in VFT (+72%), i.e. decreased the vulnerabil-ity to ventricular fibrillation, while right stellectomy or blockade significantly lowered it (-48% VFT).

The role of sympathetic activity in the genesis of cardiac tachy-arrhythmias, becomes critical in situations of derangement of the elec-trophysiological status of the myocardium as acute myocardial ischemia.

Short lasting coronary artery occlusion in conscious animals represents a model in which ischemia induced arrhythmias are greatly enhanced by high sympathetic activity; indeed it is well known that

acute myoardial ischemia, in the first few minutes, is commonly accompanied by an increase in sympathetic activity: acute myocardial is -
chemia excites sympathetic afferent fibers (14) and elicits a powerful
cardio-cardiac sympathetic reflex(15) which plays a critical role in the
genesis of these arrhythmias(16).Alternate blockade of either right or
left stellate ganglia during a coronary artery occlusion allows to investigate the effect of the unilateral sympathetic discharge in condition of high electrical instability of the heart. Experiments by
Schwartz and Stone (17) have shown that blockade of the right stellate
ganglion greatly enhances the development of ventricular arrhythmias
during acute myocardial ischemia while blockade of the left stellate
ganglion has a major protective effect. Fig. 5 shows an example of
ventricular tachiarrhythmias induced by myocardial ischemia only during
blockade of the right stellate ganglion but not when both ganglia are
normally functioning.

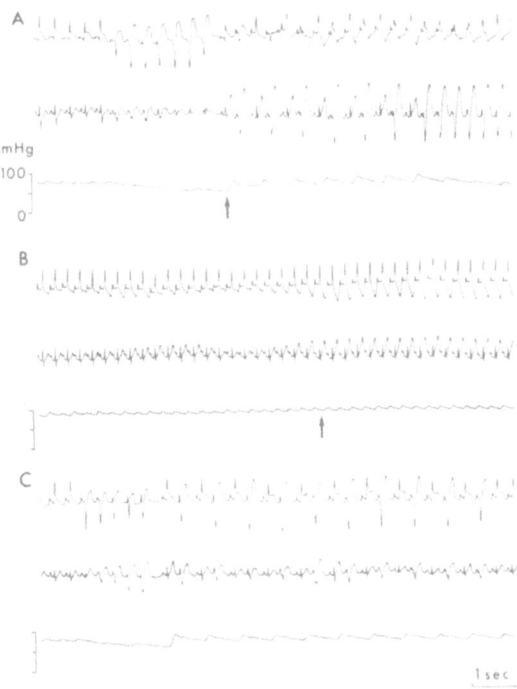

FIGURE 5. Tracings, from top to bottom, are:ECG($D_1$and D_2),aortic BP. Paper speed 25 mm./sec.Effect of a 20 sec. occlusion of both circumflex and descending coronary artery during the right stellate ganglion block-ade (SGB) and in control. The arrows indicate the release of the oc-clusion; in C the occlusion was released 2 sec. after the end of the strip.A: CAO during right SGB.An episode of VT followed by ventricular bigeminy is evident. B: CAO in control condition.No arrhythmias.C:CAO during right SGB.Again VT and ventricular arrhythmias. (from Schwartz P.J., Stone H.L. and Brown A.M. ref. 16).

Thus, despite the fact that the heart was exposed to less sympathetic activity, the incidence of life-threatening arrhythmias was higher. Similarly, the incidence of ventricular fibrillation was higher during coronary occlusion performed with simultaneous right stellate blockade than during control occlusion.

Thus in a variety of experimental conditions and animal species it has been shown that removal of right cardiac sympathetic nerves facilitates the occurrence of cardiac arrhythmias.

This paradoxical effect of right stellectomy depends on the pres-ence of an intact left stellate ganglion (18) and it has been analyzed in detail elsewhere (4).These findings and the growing evidence of the arrhythmogenic role of left sided nerves have prompted the evaluation of the potential protective effect of left stellectomy. Already some of the experiments discussed above had pointed in this direction.

Two additional studies provide almost conclusive, and clinically relevant, evidence of the antifibrillatory effect of left stellectomy. A recent study (19) has been performed in conscious dogs with a prior anterior myocardial infarction undergoing an occlusion of the circumflex coronary artery lasting ten minutes. In these experiments ventricular fibrillation occurred in 11 of 17 (65%) control dogs and in 5 of 15(33%) experimental dogs, which underwent left stellectomized 3 weeks before the trial.This study indicates that left stellectomy exerts a major protective effect in reducing the incidence of ventricular fibrillation when conscious dogs with a previous anterior myocardial infarction are exposed to an episode of acute myocardial ischemia.

Short lasting coronary occlusion (30-90 seconds) may not produce cardiac arrhythmias at rest, however when it is coupled with exercise it results in a high incidence of life-threatening arrhythmias. A model in which dogs with a prior anterior myocardial infarction undergo brief occlusion of circumflex coronary artery during exercise represents a further step in the analysis of the relationship between myocardial ischemia , sympathetic activity and ventricular arrhythmias. In experiments by Schwartz,Billman and Stone (20) dogs are subjected to submaximal exercise on a motor-driven treadmill for 18 minutes. At the 17th minute a balloon occluder previously positioned around the circum- flex coronary artery is inflated producing an acute myocardial ischemia for 2 minutes. This protocol begins 3 weeks after initial surgery and is repeated after production of an anterior myocardial infarction. Grossly, this model resembles what may happen to a patient with a prior myocardial infarction who engages in a physical activity and has a brief reduction in coronary flow (spasm?) leading to acute myocardial ische- mia, cardiac pain, and arrest of exercise. Using this protocol, ventricular tachyarrhytmias occurred in 8 of 15 control dogs (53%), culminating in ventricular fibrillation in 6 (40%).Thirty-five dogs have been studied after production of an anterior myocardial infarction and in this group the incidence of ventricular arrhythmias and of ventri- cular fibrillation was higher (74% and 66% respectively). 14 of these dogs have been studied, after left stellectomy; ventricular arrhythmias oc- curred in only two dogs (14%) and ventricular fibrillation in none. These data confirm the major antifibrillatory effect of left stellectomy during acute myocardial ischemia associated with high sympathetic activity.

Thus there is ample evidence that removal of left cardiac sympathe- tic nerves can confer a significant protection from malignant ventri- cular arrhythmias. Indeed, a clinical trial involving the comparison between a placebo, a beta-blocker and high thoracic left sympathectomy in patients with a myocardial infarction at high risk for sudden death, is presently ongoing in Northern Italy.

REFERENCES

1. Schwartz PJ,Brown AM,Malliani A, Zanchetti A. 1978. Neural mecha-
    nisms in cardiac arrhythmias. New York, Raven Press.
2. Malliani A, Schwartz PJ, Zanchetti A. 1980, Neural mechanisms in
    life-threatening arrhythmias. Am.Heart J. 100:705-715
3. Lown B. 1979. Sudden cardiac death: the major challenge confront-
    ing contemporary cardiology. Am.J.Cardiol. 43:313-328.
4. Schwartz PJ, Stone HL. 1982. The role of autonomic nervous system
    in sudden coronary death. Ann.N.Y.Acad.Sci. 382:162.
5. Schwartz PJ, Billman E, Stone HL. 1983.Analysis of the mechanisms
    by which left stellectomy might be useful in secondary prevention
    after myocardial infarction. In First Year post Myocardial
    Infarction, Kulbertus HE and Wellens HJJ eds.,New York Futura(in press)
6. Schwartz PJ. ,1983. Sympathetic imbalance and cardiac arrhythmias
    In:Nervous control of cardiovascular function. W.C.Randall ed.
    Oxford Univ.Press (in press).
7. Armour JA, Hageman GR, Randall WC. 1972. Arrhythmias induced by
    local cardiac nerve stimulation. Am J Physiol 223:1068
8. Hageman GR, Goldberg JM,Armour JA, Randall WC.1973. Cardiac dys-
    rhythmias induced by autonomic nerve stimulation. Am J Cardiol 32:
    823
9. Schwartz PJ,Vanoli E. 1981. Cardiac arrhythmias elicited by inter-
    action between acute myocardial ischemia and sympathetic hyperacti-
    vity: a new experimental model for the study of antiarrhythmic
    drugs. J.Cardiovasc Pharmacol.3:1251-1259
10. Schwartz PJ, Stone HL. 1979. Effects of unilateral stellectomy upon
    cardiac performance during exercise in dogs. Circ Res 44:637-645
11. Schwartz PJ. 1978. Experimental reproduction of long QT Syndrome.
    Am J Cardiol 41:374.
12. Austoni P, Rosati R, Gregorini L, Bianchi E, Bortolani E,Schwartz
    PJ. 1979. Stellectomy and exercise in man. Am J Cardiol 43:399
13. Schwartz PJ, Snebold NG,Brown AM. 1976. Effects of unilateral
    cardiac sympathetic denervation on the ventricular fibrillation
    threshold. Am J Cardiol 37:1034-1040
14. Malliani A, Recordati G, Schwartz PJ. 1973. Nervous activity of
    afferent cardiac sympathetic fibres with atrial and ventricular
    endings. J Physiol (Lond) 229:457
15. Malliani A, Schwartz PJ, Zanchetti A./A sympathetic reflex elicited
    by experimental coronary occlusion.Am J Physiol 217:703-709, 1969
16. Schwartz PJ, Foreman RD, Stone HL,Brown AM.1976.Effect of dorsal root
    section on the arrhythmias associated with coronary occlusion.
    Am J Physiol 231: 923-28
17. Schwartz PJ,Stone HL,Brown AM.1976.Effect of unilateral stellate
    ganglion blockade on the arrhythmias associated with coronary oc-

clusion.Am Heart J.92:589-99

18. Schwartz PJ,Verrier RL, Lown B. 1977. Effect of stellectomy and vagotomy on ventricular refractoriness in dogs. Circ Res 40:536-540

19. Schwartz PJ, Stone HL. 1980.Left Stellectomy in the prevention of ventricular fibrillation caused by acute myocardial ischemia in conscious dogs with anterior myocardial infarction.Circulation 62: 1256-1265.

20. Schwartz PJ,Billman GE,Stone HL. Autonomic mechanisms in ventricular fibrillation induced by myocardial ischemia during exercise in dogs with a healed myocardial infarction.An experimental model for sudden cardiac death. Submitted for publication.

BLOOD PRESSURE CONTROL DURING MENTAL STRESS.

JAMES CONWAY, NICHOLAS BOON, JOHN VANN JONES AND PETER SLEIGHT.

CARDIAC DEPARTMENT, JOHN RADCLIFFE HOSPITAL, HEADINGTON, OXFORD, U.K.

There is a widely held belief that neurogenically mediated fluctuations in blood pressure may lead ultimately to sustained elevation in blood pressure (Brown et al 1974). Thus psychological stress may be an important aetiological factor in hypertension. This hypothesis has been tested directly in spontaneously hypertensive rats. These animals have an exaggerated pressor response to psychological stress (Hälback and Folkow 1974) and if they are shielded from environmental stimuli the rise in blood pressure is delayed and ultimately the evolution of hypertension is modified. (Hälback 1975). This then is the clearest evidence that stress plays a role in hypertension.

No comparable studies are possible in humans. The evidence supporting a role for stress in essential hypertension depends upon several lines of evidence, none, however, have yet been absolutely convincing and further evidence is required.

To examine the possibility that the autonomic nervous system might be involved in hypertension two broadlines of approach have generally been adopted.

(1) Evidence for increased sympathetic tone
(2) Evidence for increased variability of blood pressure

The evidence for an increased sympathetic tone has been reviewed extensively elsewhere (Abboud 1982) and further support is to be found in the reports of increased plasma catechols in hypertension particularly amongst the younger subjects. The data on this has recently been extensively reviewed (Goldstein 1983). While a greater number of studies have reported elevations of noradrenaline, adrenaline has also been found to be elevated in some groups of hypertensives. This will be discussed further below.

If increased blood pressure variability were shown to occur in man it could indicate man may share with the spontaneously hypertensive animal an increase responsiveness to psychological stimuli.

Although labile blood pressure has commonly been believed to be a stage in the development of hypertension. there is very little evidence to support this concept. It seems likely that the reports of increased lability have stemmed from the fact that in mild hypertension subjects sometimes appear to be hypertensive and at other times not. That is to say, that blood pressure fluctuates not widely but around the upper limit of normal, i.e: 140/90 (Conway 1970). In fact, most studies which have used direct recordings of blood pressure, have shown that subjects with mild or borderline hypertension do not show greater variability than those with established hypertension. Rather is the reverse true; variability in blood pressure (in absolute terms) increases with the level of pressure (Watson et al 1980, Floras and Sleight 1982).

Although spontaneous variability in blood pressure may not occur in mild hypertension there are reports of an increased sensitivity to mental stress in borderline hypertension, since these subjects show a greater and more prolonged pressor response to mental arithmetic. Furthermore this appears to be in some degree genetically determined and those who have these characteristics tend to develop permanent hypertension (Falkner et al 1979 and 1981).

Blood pressure variability therefore presents an opportunity to study the involvement of the autonomic

ervous system in hypertension since psychological stress
and physical exercise are the factors which determine day
to day variability in blood pressure.

To do this, studies have been made of the blood
pressure changes which occur with sleep and mental arousal
in 13 subjects with borderline blood pressure elevation.
From sleep, the 3 stages of mental arousal have been
studied, a drowsy state, reading a simple newspaper and
finally mental arithmetic.

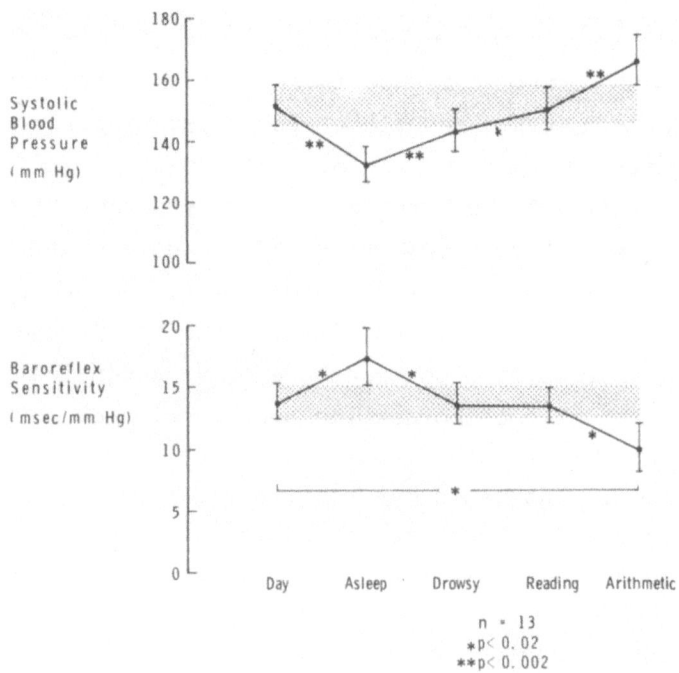

FIGURE 1. The changes in blood pressure and baroreflex
sensitivity with sleep and mental arousal. Baroreflex
sensitivity was determined by measuring the stage of the
regression line between pulse interval (R-R interval) and
the rise in systolic pressure following the intravenous
injection of phenylephrine. (Smyth et al 1969).

If the subject remains in a drowsy state, blood pressure rises a little after awakening but it does not return to the previous daytime value. Thereafter it increases further with mental arousal (Fig. 1). With these changes in blood pressure plasma noradrenaline does not increase significantly though plasma adrenaline does. In fact systolic pressure may rise from a mean of 134 mm Hg during sleep to 166 mmHg with mental arithmetic without significant change in plasma noradrenaline. At the same time however adrenaline increased significantly from 0.21 ± 0.006 to 0.37 ± 0.052 μmol/L. This indicated that a search for the involvement of the autonomic nervous system in hypertension should use measurements of adrenaline and not noradrenaline. There has indeed been some evidence for modestly elevated plasma adrenaline levels in a proportion of hypertensive subjects. (Franco-Morcelli et al 1976, Buhler et al 1980, de Champlain et al 1981), but the numbers of subjects studied is small and further work in this area is required.

Another change has been seen as subjects pass through these stages of mental arousal. The baroreceptor reflex has been shown to operate at its most sensitive level during sleep (Smyth et al 1969) and then progressively becomes less sensitive with mental arousal (Fig. 1). It can also be shown that the variability in blood pressure is at its lowest during sleep and increases with mental arousal. Blood pressure variability has also been shown to be greater when the sensitivity of the baroreflex is reduced (Watson et al 1980, Floras and Sleight 1982).

Although there are many factors which affect baroreflex sensitivity, these findings indicate that throughout the day mental stress induces varying degrees of inhibition of the reflex and this parallels the rise in blood pressure. The autonomic pathways responsible for this appear to favour stimulation of the adrenal medulla.

REFERENCES

Brown JJ, Lever AF, Robertson JIS and Schalekamp MA.
Pathogenesis of essential hypertension.
Lancet 1, 1217, 1976.
Hallbäck M and Folkow B.
Cardiovascular response to acute mental stress in
spontaneously hypertensive rats.
Acta Physiol Scand. 90: 684, 1974
Hallbäck M.
Consequence of social isolation on blood pressure and
cardiovascular design in spontaneously hypertensive rats.
Acta Physiol Scand. 93: 455, 1975.
Abboud F.
The sympathetic system in hypertension
Hypertension 4 : II, 208, 1982.
Goldstein DS.
Plasma catecholamines and essential hypertension.
Hypertension 5: 86, 1983.
Conway J.
Labile Hypertension: The problem.
Circulation Research 24 - 27 I 43, 1970.
Watson RDS, Stallard TJ, Flinn RM and Littler WA.
Factors determining direct arterial pressure and its
variability in hypertensive man.
Hypertension 2: 333, 1980.
Floras JS and Sleight P.
The lability of blood pressure
IN: Hypertensive Cardiovascular Disease
Ed: A. Amery, R. Fagard, P.Lijnen and J. Straessen,
M. Nijhoff pg 104, 1982.
Falkner B, Onesti G, Angelakos ET, Fernandes M and
Langman C.
Cardiovascular response to mental stress in adolescents.
Hypertension I, 23, 1979.
Falkner B, Onesti G, Hamstra B.
Stress response characteristics of adolescents with
high genetic risk for essential hypertension. A five
year follow-up.
Clin and Exper Hypertension 3: 583, 1981.
Franco-Morselli R, Elghozi JL, Joly E, DiGuilio S and
Meyer P.
Increased plasma adrenaline concentrations in benign
essential hypertension. Brit Med J, 2: 1251, 1977.
De Champlain J, Cousineau D and Lapointe L.
Evidences supporting an increased sympathetic tone and
reactivity in a group of patients with essential
hypertension.
Clin & Exper Hypertension 2, 359, 1980
Buhler FR, Kioski W, van Brummelen P, Amann FW, Bertel O,
Landmann R, Lutold BE and Bolli P.
Plasma catecholamines and cardiac, renal and peripheral
vascular adrenoceptor-mediated responses in different
age groups of normal and hypertensive subjects.
Clin & Exper Hypertension, 2: 409, 1980.

Smyth HS, Sleight P, Pickering GW,
Reflex regulation of arterial pressure during sleep in
man: a quantitative method of assessing baroreflex
sensitivity.
Circulation Research, 24: 109, 1969.

SOMATIC RESPONSES TO ACUTE STRESS AND THE   RELEVANCE FOR THE STUDY
OF THEIR MECHANISMS

A.W. VON EIFF, H. NEUS, W. SCHULTE
Medizinische Universitätsklinik, Venusberg, D-5300 BONN I,
FRG.

The somatic responses to emotional stress are varied and related to
reactions of the autonomic nervous system as well as to reactions of
the endocrine system (1,2). Our team has been working on the influence
of stress on circadian rhythm, basal metabolism, respiration, muscular
tone and circulatory regulation (3-5). The following is confined to four
aspects of the circulatory regulation: 1. The blood pressure (BP)
reactivity and essential hypertension. 2. The adaptation and the therapeutic influences on BP reactivity. 3. The stimulus specifity in cardiovascular reactions to noise. 4. The sex specifity in BP regulation.

Blood pressure reactivity and essential hypertension

In a series of experiments the cardiovascular responsiveness to emotional stress has been studied. Emotional stress testing using mental arith-
metics as stimulus has been reported by several investigators, but the
performance differed in details (6-1o). In our experiments the stress
situation has been elicited by adding one- und two-digit numbers during
an exposition to noise at 90 dB (A).

BP under basal conditions and emotional stress was studied in 48
subjects with elevated casual BP but without hypertensive cardiovascular
complications and in 48 age-matched normotensive controls (11). Under
basal conditions the BP values of the hypertensive group were within the
normal range, but markedly higher than the values of the normotensive
group. During the stress exposition the difference in BP between both
groups increased because the hypertensive group had a stronger rise of
blood pressure. After dividing both groups into two subgroups according
to age, the older subjects demonstrated a higher increase in BP than the
younger ones. This was true for the hypertensive as well as for the
normotensive groups.

One result of this study was the stronger reactivity of BP to emotional stress observed in hypertension. Similarly, in other mental stress tests such as sorting of steel ball bearings by their size (12) or solving visual puzzles (13) this hyperreactivity of BP in hypertensives has also been described. The important factor shared by these tests may be their attention demanding or challenging effect. Thus, the inconsistent results on hyperreactivity of hypertensives to cold exposition may be understandable (14,15).

Furthermore our investigation demonstrated that the increased reactivity of BP exists in hypertensives of both sexes and is not age-dependent. That the BP response to emotional stress increases with age was already described (16). But the effect of age can be designated as additional factor, independent from hypertension.

In a second investigation the BP reactivity to emotional stress was studied in borderline hypertension and two groups of hypertensives, namely of stage I and II (17). It could be shown that even in borderline hypertensives the reactivity of BP is stronger when compared to normotensives. A further increase of the reactivity could be demonstrated for the hypertensives. Besides, the hypertensives of stage II had a stronger reaction than those of stage I. This concerned both systolic and diastolic pressure (fig. 1).

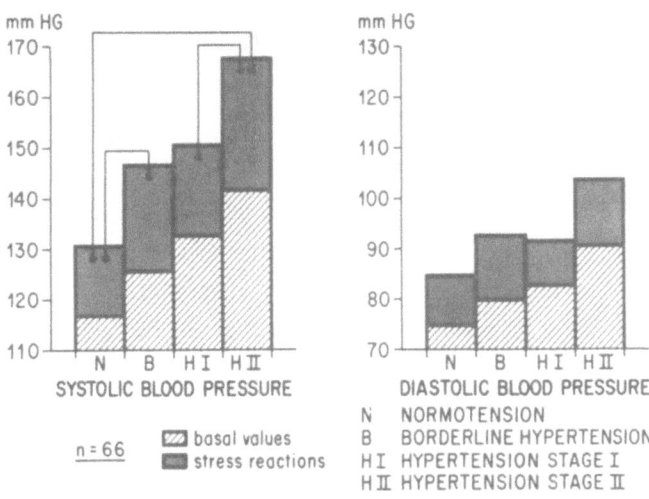

Fig. 1. Basal values and BP reactivity to mental stress in normotensives, borderline hypertensives and hypertensives of stage I and II(17).

From this investigation it could be concluded that the phenomenon of BP hyperreactivity exists throughout the course of hypertension and may be due to secondary blood vessel changes, especially in higher grade hypertension. But a striking result was that even borderline hypertensives have such an increased reactivity of BP which was later confirmed by Jern (1o). Thus, the hyperreactivity of BP exists before the manifestation of hypertension, too. In this case it probably was not due to hypertensive complications, but should be termed as functional.

This component of BP hyperresponsiveness was especially demonstrated in studies in which normotensives with genetic risk of hypertension were studied In such an investigation (18) we divided a group of healthy male subjects according to the occurence of hypertension in their families. The genetic risk group did not differ from that with normotensive parents in casual as well as in basal BP values. However, the response of BP to mental stress was stronger in the group with genetic risk of hypertension.

This result, which is in accordance with a similar investigation in adolescents (19),shows that the hyperreactivity of BP exists before the development of hypertension and may depend on a genetic factor. Apparently organic changes of the blood vessels could be excluded, but the functional component of hyperreactivity was obvious. Light and Obrist (20) observed that the incidence of parental hypertension was greater among young males with high reactions of heart rate (HR) to stress than among those with a low HR reactivity, emphasizing the importance of HR in predicting future hypertension. Similarly, we had found higher reactions of HR to emotional stress in borderline hypertensives with stronger reactivity of BP (21).

Hemodynamics under mental arithmetic has been studied by Brod (6), but BP hyperreactivity was not described. Thus, to get further insight into the regulation of BP hyperreactivity, the hemodynamic response to emotional stress in the developmental stages of hypertention was studied (22). In borderline and stage I hypertensives the stronger responses of BP in comparison to normotensives were elicited by increased responses of HR and cardiac output (CO). Mental stress did not affect the stroke volume (SV) and only a slight decrease of the total peripheral resistance (TPR) was noted which was approximately the same in both groups.

Thus, it may be concluded that during the development of hypertension, the

increased BP reactivity to emotional stress is due to higher reactions of
HR and CO provoked by a stronger stimulation of the heart. Very similar results were obtained in animal studies (23,24), in which spontaneously hypertensive rats exposed to "mental stress" showed higher increases of BP and
CO than normotensive Wistar rats. An accentuated neurogenic excitation of
the SHR heart during increased alertness was assumed.

According to the results of these studies the BP hyperreactivity can be
designated as a genetically determined precursor of hypertension which is
elicited in challenging situations. Stressful daily life activities of
persons with genetic risk of hypertension may provoke a stronger autonomic
especially sympathoadrenergic stimulation of the circulation.

Adaptation and therapeutic influences in blood pressure reactivity

Under clinical aspects the question arises as to how the hyperreactivity
of BP to stress can be effectively treated, especially in established hypertension. This may be important in lowering the BP throughout the day. The
effects of beta-blockers are of particular interest, as hyperreactivity is
mediated by an increase of CO.

Because of adaptation phenomena the appropriate clinical studies could
not be performed in an intraindividual comparison. In a study on hypertensives and normotensives the emotional stress test was carried out twice, with
1 to 7 day intervals between tests (25). This study demonstrated that the
hypertensive group in the first experiment exhibited enhanced reactions of
systolic and diastolic BP as expected. However, in contrast to the normotensive group in which no adaptation phenomena were observed, there was a strong
adaptation in the hypertensive group regarding the stress reaction of systolic BP and HR. Thus, in regard to cardiovascular reactivity there were no
longer any differences between both groups in the second experiment (fig.2).
An independent study of hypertensives on placebo therapy had similar results
(26).

One conclusion of these investigations is that a therapeutic trial in
which emotional stress reactions are an important factor can only be carried
out in an interindividual and not in an intraindividual comparison. As effects
of structural vessel alterations (27) are the same in repeated experiments,
these results also underline the importance of the functional component in

ADAPTATION OF STRESS-REACTIONS
IN NORMOTENSIVES AND HYPERTENSIVES

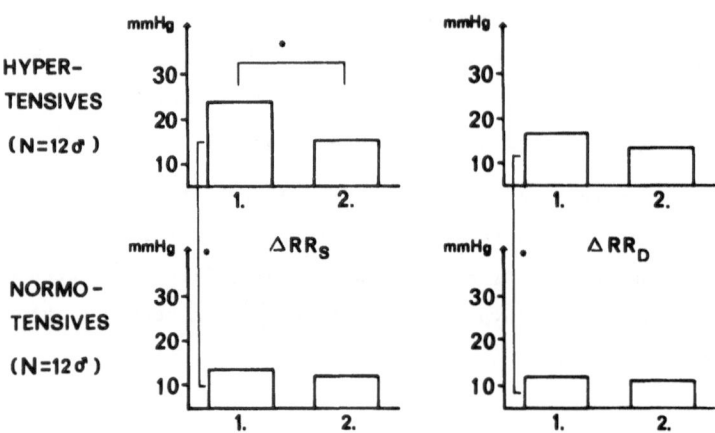

Fig. 2 Adaptation of BP reactivity to mental stress in normotensives and hypertensives (25).

in the hyperreactivity of hypertensives. This hyperreactivity is only apparent as long as the stimulus has a challenging effect. From this it may be concluded that the diagnosis of BP hyperreactivity is only possible when using a sufficiently strong experimental stressor. The hyperreactivity cannot be a pathogenetically important factor in the development of hypertension when coping mechanisms induce adaptation to stress situations in every-day life.

For this reason we investigated the effect of an antihypertensive treatment on BP reactivity in an interindividual comparison of beta-blockers and diuretics (28). As a result, no differences between the two treatments were found regarding stress reactivity of BP, although the reaction of HR tended to be attenuated by beta-blockers. Thus, it was concluded that an effective antihypertensive treatment does not necessarily influence BP reactivity to emotional stress. Also the acute administration of a beta-blocker did not influence BP reactions to stress (9,29-32).

The specificity of emotional stress tests is clearly apparent when comparing these results with analogue studies using an ergometric exercise test. Here, beta-blockers effectively lower the increase of BP as well as of HR (33,34).

Although, as outlined above, under physiological conditions the BP hyperreactivity to stress in essential hypertension is mediated by a stronger increase in CO, pharmacological blockade of the latter does not attentuate the
increase of BP. Thus, a generally enhanced sympathetic activity during
emotional stress in hypertensives must be assumed.

Stimulus specifity in cardiovascular reactions to noise

Experimental noise induces a reaction pattern different from that of
mental arithmetic. In previous investigations from our laboratory a decrease
of HR to experimental noise was observed (35). Although there is a remarkable interindividual variation (36), in most subjects mean arterial BP,
diastolic BP and TPR increase, while CO and HR decrease (37). Thus, the
hemodynamics of stress reactions may qualitatively differ which contradicts
the concept of a general adaptation syndrom suggested by Selye (38). As
compared to mental arithmetic, noise induces primarily vascular reactions.
The predominance of vascular reactions is underlined by the lack of adaptation of the diastolic BP response to noise (39-41).

The different hemodynamic reactions may indicate a different type of
emotional arousal, for example missing challenge during experimental noise
exposure. In several experiments we further analyzed the role of situational
factors as modifying the hemodynamic response. We found that the increase
of BP to experimental traffic noise could be prevented by the simultaneous
visual perception of the source of the noise on a video screen (42). Furthermore, in contrast to experimental traffic noise, broadband noise did not
induce an increase of BP (43). On the other hand, immediately after mental
arithmetic BP remained markedly elevated during traffic noise exposure (44).
These results demonstrate that the kind of the stimulus, situational
factors, and coping mechanisms influence the cardiovascular reaction to stress.

Despite this complexity, there are also similarities between different
stressors. As with mental arithmetic, antihypertensive drugs did not influence the reactions of BP to noise even though they may change the hemodynamic pattern (45,46). Furthermore subjects with a genetic risk of
hypertension also exhibited enhanced reactions to experimental noise exposure (41,46), suggesting that their hyperreactivity is independent from
the particular stimulus. Repetitive acute stress reactions, for example to
traffic noise (47,48), may be important in the pathogenesis of hypertension.

Sex specifity in blood pressure regulation

As BP in sexually mature women is lower than in men a sex specific regulation of BP also in acute stress reactions was assumed. Therefore several studies were performed to evaluate a possible sex specific cardiovascular reactivity in emotional stress testing. In such an investigation we could find out that under the stress of mental arithmetic the BP of female subjects was lower than that of males (49; fig.3).

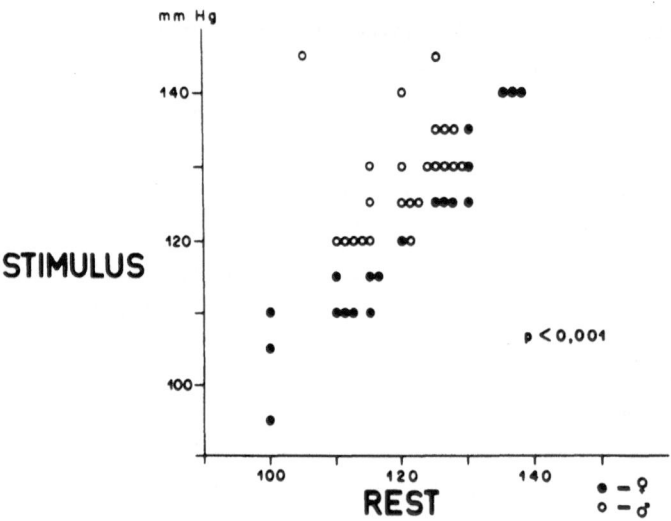

Fig.3. BP under resting conditions and under mental stress in male and female subjects(49).

Besides this quantitative difference in BP reactivity to this unspecific stimulus it was assumed that certain stimuli provoke emotional reactions in male rather than in female subjects, and vice versa. Therefore two films were shown to 24 male and 24 female volunteers (50). On the basis of previous experiments it was expected that the different themes of the films would elicit sex specific emotional reactions. In this way BP reactivity of the male subjects was stronger during the presentation of that film which showed an aggressive discussion between two colleagues. However, to the film causing emotional reactions especially in women, female subjects reacted with other autonomic functions rather than with BP such as HR and respiration.

Thus it is suggestable that women generally exhibit BP reactions less often and to a lower degree than men. Supposing that the female hormones are responsible for this phenomenon the hypothesis of a protective mechanism of estrogen on BP in women was tested in a double-blind study on ovariectomized women. After the administration of a long-acting estrogen the stress reaction of systolic BP was less pronouced than that of the controls (51).

There were further results being in consistence with this concept. Especially in women with a normal menstrual cycle a distinct relationship between the degree of endogenous estrogenic activity and BP could be verified (51). This concerned BP under resting conditions and even more BP response to stress.

These findings are important not only for the understanding of stress reactions but also for some clinical aspects, too. The epidemiological observation that the incidence of a cardiovascular damage is sex-related can be partly explained by this sex specific BP regulation, for our results imply that the heart and the blood vessels of women are strained less often and to a lower extent by increases of BP.

Recently we began to study the mechanism of the sex specific BP regulation. Spontaneously hypertensive rats were used as an experimental model for studying the therapeutic influence of estrogen on female and male rats (52). The findings are suggesting an activation of cerebral presynaptic alpha$_2$-adrenoceptors by estrogen. The effect of treatment with estrogen on BP is depending on a preexisting physiological protection as it is provided in female organisms. (52).

Conclusions

In conclusion emotional stress testing is important for many reasons. This concerns research on pathogenetic mechanisms, clinical diagnostics, and clinical-therapeutic trials. But modifying factors such as age, sex, and defined initial resting values (21) must be considered to enable the comparison of homogeneous groups. Under these aspects future research should concentrate on the long-term effects of acute stress reactions, especially in the manifestation of a cardiovascular disease.

References

1. Levi L. 1972. Stress and distress in response to psychosocial stimuli. Laboratory and real life studies on sympathoadrenomedullary and related reactions. Oxford, Pergamon Press.

2. Levi L. 1975. Emotions. Their parameters and measurement. New York, Raven Press.

3. v. Eiff AW. 1957. Grundumsatz und Psyche. Heidelberg, Springer.

4. v. Eiff AW, Quint A, Kloska G. 1967. Essentielle Hypertonie. ed.: v. Eiff AW. Stuttgart, Thieme, japan. ed. 1971.

5. v. Eiff AW (ed.).1976. Seelische und körperliche Störungen durch Stress. Stuttgart, Fischer.

6. Brod J, Fencl V, Hejl Z, Jirka J. 1959. Circulatory changes underlying blood pressure elevation during acute emotional stress (mental arithmetic) in normotensive and hypertensive subjects. Clin.Sci. 18, 269-272.

7. Richter-Heinrich E, Knust U, Müller W, Schmidt K-H, Sprung H. 1975. Psychophysiological investigations in essential hypertensives. J. Psychosom.Res. 19, 251-258.

8. Sleight P, Fox P, Lopez R, Brooks DE. 1978. The effect of mental arithmetic on blood pressure variability and baroreflex sensitivity in man. Clin.Sci. Mol.Med. 55, 381s-382s.

9. Bonelli J, Hörtnagl H, Brücke Th, Magometschnigg D, Lochs H, Kaik, G. 1979. Effect of calculation stress on hemodynamics and plasma catecholamines before and after ß-blockade with propranolol (Inderal) and mepindolol sulfate. Europ. J.Clin.Pharmacol. 15, 1-8

10. Jern S. 1982. Psychological and hemodynamic factors in borderline hypertension. Acta Med. Scand. Supp. 662.

11. Schulte W, Neus H, v. Eiff AW. 1981. Blutdruckreaktivität unter emotionalem Streß bei unkomplizierten Formen des Hochdrucks. Klin. Wochenschr. 59, 1243-1249.

12. Lorimer AR, Macfarlane PW, Provan G, Duffy T, Lawrie TDV. 1971. Blood pressure and catecholamine responses to "stress" in normotensive and hypertensive subjects. Cardiovasc. Res. 5, 169-173.

13. Nestel PJ. 1969. Blood pressure and catecholamine excretion after mental stress in labile hypertension. Lancet 1, 692-694.

14. Shapiro AP, Moutsos, SE, Krifcher, E. 1963. Patterns of pressor response to noxious stimuli in normal,hypertensive, and diabetic subjects. J. Clin.Invest. 42, 1890-1898.

15. Voudoukis IJ. 1978. Cold pressor test and hypertension. Angiology 29, 429-439.

16. Imhof P, Hürlimann A, Steinmann B. 1957. Über Blutdrucksteigerung bei psychischer Belastung. Cardiologia 31, 272-276.

17. Schulte W, Neus H. 1979. Bedeutung von Streßreaktionen in der Hypertoniediagnostik. Herz/Kreislauf 11, 541-546

18. Schulte W, Neus H, Rüddel H. 1981. Zum Blutdruckverhalten unter emotionalem Streß bei Normotonikern mit familiärer Hypertonieanamnese. Med. Welt 32, 1135-1137.

19. Falkner B, Onesti G, Angelakos ET, Fernandes M, Langman C. 1979. Cardiovascular response to mental stress in normal adolescents with hypertensive parents.   Hypertension 1, 23-30.

20. Light KC, Obrist PA. 1980. Cardiovascular reactivity to behavioral stress in young males with and without marginally elevated casual systolic pressures. Comparison of clinic, home, and laboratory measures. Hypertension 2, 8o2-8o8.

21. v. Eiff AW, Schulte W, Heusch G. 1979. Classification of homogeneous blood pressure groups and the specific hemodynamic pattern of borderline hypertension. In: Prophylactic approach to hypertensive diseases. Perspectives in cardiovascular research. Vol. 4. ed.: Yamori Y, Lorenberg W, Freis ED. New York, Raven Press.

22. Schulte W, Neus H. 1983, in press. Hemodynamics during emotional stress in borderline and mild hypertension. Europ. Heart J.

23. Hallbäck M, Folkow B. 1974. Cardiovascular responses to acute mental "stress" in spontaneously hypertensive rats. Acta Physiol. Scand. 9o, 684-698.

24. Lundin SA, Hallbäck-Nordlander M. 1980. Background of hyperkinetic circulatory state in young spontaneously hypertensive rats. Cardiovasc. Res. 14, 561-567.

25. Neus H, v. Eiff AW, Friedrich G, Heusch G, Schulte W. 1981. Das Problem der Adaptation in der klinisch-therapeutischen Hypertonieforschung. Dtsch. med. Wschr. 1o6, 622-624.

26. Friedrich G, Langewitz W, Neus H, Schirmer G, Thönes M. 1981. Der emotionale Belastungstest in der klinisch-therapeutischen Prüfung von Antihypertensiva. Verh.dtsch.Ges.Inn.Med. 87, 551-554.

27. Folkow B, Hallbäck M, Lundgren Y, Sivertsson R, Weiss L. 1973. Importance of adaptive changes in vascular design for establishment of primary hypertension. Studied in man and spontaneously hypertensive rats. Circ. Res. 32/33, Suppl. 1, 2-16.

28. Friedrich G, Neus H, Rüddel H, Schirmer G, Schulte W. 1981. Zur Problematik der antihypertensiven Therapie des Belastungshochdrucks. Therapiewoche 31, 7766-7768.

29. v. Eiff AW, Czernik A, Zanders H. 1969. Zur medikamentösen Beeinflussung der Sympathikushyperaktivität: Parasympathomimeticum und Beta-Receptoren-Blocker im kurzdauernden pharmakologischen Experiment am Menschen. Klin.Wschr. 47, 701-708.

30. Andrén L, Hansson L. 1980. Circulatory effects of stress in essential hypertension. Acta Med. Scand. Suppl. 640, 69-72.

31. Bonelli J. 1982. Stress, catecholamines and beta blockade. Acta Med. Scand. Suppl. 660, 214-218.

32. Heidbreder E, Ziegler A, Heidland A. 1982. Verhindern sympatholytische Antihypertensiva den Blutdruckanstieg bei mentalem Streß? Herz/Kreislauf 14, 135-141

1002

33. Franz IW. 1980. Differential antihypertensive effect of acebutolol and hydrochlorothiazide/amiloride hydrochloride combination on elevated exercise blood pressures in hypertensive patients. Am.J. Cardiol. 46, 301-305.

34. Lund-Johansen P. 1977. Hemodynamic alterations in hypertension. Spontaneous changes and effects of drug therapy. A review. Acta Med. Scand. Suppl. 603. 1-14.

35. Czernik A. 1976. Körperliche Reaktionen Gesunder auf Lärm. In: Seelische und körperliche Störungen durch Streß. ed.: v. Eiff AW.Stuttgart, Fischer.

36. Neus H, Schirmer G, Rüddel H, Schulte W. 1980. Zur Reaktion der Fingerpulsamplitude auf Belärmung. Int. Arch. Occup. Environ. Health 47, 9-19.

37. Andrén L, Hansson L, Björkman M, Jonsson A, Borg KO. 1979. Hemodynamic and hormonal changes by noise. Acta Med. Scand. 625, 13-18.

38. Selye H. 1978. The stress of life. Revised edition. New York, Mc Graw Hill.

39. v. Eiff AW. 1964. Funktionsspezifische Effekte und Gewöhnungseffekte bei Lärm unterschiedlicher Zeitstruktur. In: Psychologische Fragen der Lärmforschung. Bonn - Bad Godesberg, Deutsche Forschungsgemeinschaft.

40. Mosskow JI, Ettema JH. 1977. Extraauditory effects in long-term exposure to aircraft- and traffic noise. Int. Arch. Occup.Environ. Health 47, 9-19.

41. v. Eiff AW, Friedrich G, Langewitz W, Neus H, Rüddel H, Schirmer G, Schulte W. 1981. Verkehrslärm und Hypertonie-Risiko. Hypothalamus-Theorie der essentiellen Hypertonie. 2. Mitteilung. Münch.med.Wschr. 123, 42o-424.

42. v. Eiff AW, Neus H, Münch K, Schulte W. 1981. Verkehrslärm als Risikofaktor für Hypertonie. Verh. dtsch.Ges.Inn.Med. 87, 549-551.

43. Heusch G, Schulte W, Rüddel H, Neus H. 1981. Lärm und vegetatives Nervensystem. Therapiewoche 31, 33-36.

44. Schulte W, Heusch G, v. Eiff AW. 1977. Der Einfluß von experimentellem Verkehrslärm auf vegetative Funktionen von Normotonikern und Hypertonikern nach Streß. Basic Res. Cardiol. 72, 575-583.

45. Andrén L, Hansson L, Björkman M. 1981. Haemodynamic effects of noise exposure before and after $ß_1$-selective and non-selective ß-adrenoceptor blockade in patients with essential hypertension. Clin. Sci. 61, 89s-91s.

46. Andrén L. 1982. Cardiovascular effects of noise. Acta Med. Scand. Suppl. 657,

47. Knipschild P. 1977. Medical effects of aircraft noise: Community cardiovascular survey. Int. Arch. Occup. Environ. Health 40,185-190.

48. v. Eiff AW, Neus H. 1980. Verkehrslärm und Hypertonie-Risiko. 1. Mitteilung. Münch.Med. Wschr. 122, 894-896.

49. v. Eiff AW. 1970. The role of the autonomic nervous system in the etiology and pathogenesis of essential hypertension.Jpn.Circ. J. 34, 147-153

50. v. Eiff AW, Piekarski C. 1977. Stress reactions of normotensives and hypertensives and the influence of female sex hormones on blood pressure regulation. In: Hypertension and brain mechanism. Progress in brain research. Vol. 47. ed.: de Jong W, Provoost AP, Shapiro AP. Amsterdam, Elsevier.

51. v. Eiff AW, Plotz EJ, Beck KJ, Czernik A. 1971. The effect of estrogens and progestins on blood pressure regulation of normotensive women.Am. J. Obstet. Gynecol. 109, 887-892.

52. v. Eiff AW, Gries J, Kretzschmar R, Lutz HM, Neidhardt RC. 1982. Effect of estradiol on blood pressure, body weight and life span of stroke prone hypertensive rats (SHRSP). Naunyn-Schmiedeberg's Arch. Pharmacol. Suppl. 321, R 18.

NEUROHUMORAL FACTORS INVOLVED IN THE PATHOGENESIS OF HYPERTENSION

by J.L. ELGHOZI, L.C.L. JACOMINI, M.A. DEVYNCK, L.A. KAMAL, J.F. CLOIX, M.G. PERNOLLET, H. DE THE, P. MEYER.

Dpt de Pharmacologie, INSERM U7, CNRS LA 318, Faculté de Médecine Necker-Enfants Malades, 156, rue de Vaugirard, 75015 PARIS - FRANCE.

INTRODUCTION

The sodium, potassium activated adenosine triphosphatase (Na^+, K^+-ATPase) is present in most animal cells. This membrane-embedded enzyme is responsible for the active transport of sodium and potassium across the plasma membrane. The resulting concentration gradients largely determine resting membrane potential and participate in cell volume homeostasis. The cardiac glycoside ouabain and its derivatives such as strophantidin can specifically inhibit the Na^+, K^+-ATPase allowing the sodium and potassium gradients to dissipate. The existence of a specific binding site for these plant alkaloids raises the possibility of the existence of an unknown digitalis-like compound which would regulate the activity of the enzyme through this binding site.

Recent studies have reported the existence of a circulating hormone which inhibits Na^+, K^+,-ATPase activity. Since abnormal sodium metabolism has a critical role in hypertension the question must be asked concerning the role of a circulating Na^+, K^+-ATPase inhibitor in essential hypertension. The inability of the kidneys to cope normally with salt and water load could be compensated by an increase in the hypothetical natriuretic hormone which would facilitate salt and water excretion. The raised arterial pressure would represent a secondary effect of the increased level of the hormone. The membrane sodium-pump inhibition in the arterial smooth muscle could lead to a rise in intracellular sodium and calcium concentration and hence contractility. The site of production of the hormone is not known although it could originate from

neuronal tissue such as the hypothalamus. This attractive neurohumoral hypothesis may be a stimulus for further understanding
the link between hypertension and the activity of the central
nervous system. This mini review summarizes data favouring
this hypothesis.

EVIDENCE FOR A OUABAIN-LIKE HUMORAL AGENT IN HYPERTENSION

Initially Na^+, K^+-ATPase activity was reported to be suppressed in the arteries and veins of dogs with one model of low
renin hypertension (one kidney-one wrapped). The same changes
were observed in other models of low renin hypertension (one
kidney-one clip ; one kidney-DOCA-saline ; reduced renal mass-
saline) but not genetic models of hypertension. The change in
Na^+, K^+-ATPase activity could be induced in normal animals
with rapid volume expansion. Interestingly, this inhibitory
factor could be transferred to the arteries of another animal
via the plasma. Ouabain-like activity was also found in the
plasma of several low renin models of hypertension.

The relationship between the ouabain-like humoral agent
and salt-dependent i.e. low renin hypertension was also investigated in human hypertension. Incubating leukocytes from normotensive subjects in serum obtained from patients with essential hypertension caused an impairment in sodium transport
similar to that found in the leukocytes of hypertensive patients.
Male hypertensive subjects with a high sodium intake had reduced sodium efflux from their red blood cells when these
cells were incubated in the presence of their plasma. A significant correlation between mean arterial pressure and Na^+,
K^+-ATPase inhibition was found. Collectively, these results
suggest that expansion of body fluid secondary either to saline infusion or to a defect in sodium excretion is associated
with a change in the concentration of a substance in the blood
which inhibits Na^+, K^+-ATPase.

BIOCHEMICAL CHARACTERIZATION

The methods used previously for testing an inhibitor of Na^+,

K^+-ATPase in plasma were indirect. Direct tests have now been made that take into account the main actions of endogenous digitalis-like compound.

The inhibitory effect of plasma extracts on tritiated-ouabain binding to erythrocytes was studied on erythrocytes from normotensive subjects devoid of a family history of hypertension. Cells were incubated at 37° C for 5 hours in the presence of tritiated-ouabain concentrations ranging from 2×10^{-9} M to 2.5×10^{-8} M. Parallel incubations were performed in the presence of an excess of unlabelled ouabain (10^{-4} M). Bound and free radioactivity were separated by filtration. Each binding study was performed using five ouabain concentrations. Ouabain binding to erythrocytes reached equilibrium within 5 hours and non-saturable binding accounted for less than 10 % of total binding. Specific binding was saturable and the Scatchard analysis revealed the presence of only one class of sites, characterized by an apparent dissociation constant of 3×10^{-9} M and a mean number of sites of 400 per cell. Since plasma contains potassium ions known to inhibit ouabain binding, experiments were performed to estimate the variations of the affinity constant in the presence of potassium concentrations up to 1mM. Significant changes in affinity was observed only with potassium concentrations higher than 0.25 mM. Since the physiological plasma potassium concentration ranges from 3.5 to 5 mM, plasma extracts were anayzed at a 1 to 20 dilution. At this dilution the final potassium concentration in the incubation medium did not interfere with ouabain binding. The inhibitory effect of the plasma extracts was expressed as the ratio of the apparent affinity constant measured in the presence of the extract to the control affinity constant. A high ratio thus indicated potent inhibition of ouabain binding. The mean values measured for normotensive and hypertensive subjects were 1.4 and 1.6 respectively.

The inhibitory effect of plasma extract on Na^+, K^+-ATPase activity is another test that directly estimates the inhibitory action of the factor on its target enzyme. The

activity of the partially purified canine kidney was determined by hydrolysis of ^{32}P-ATP. Non sodium-pump ATP hydrolysis
was measured in the presence of 0.1 mM ouabain. The reaction
was stopped after 45 min and inorganic phosphate was separated from ATP by absorption on acid-washed charcoal and the
supernatant counted. Na$^+$, K$^+$-ATPase activity was studied in
the linear range corresponding to the hydrolysis of less than
30 % of the total ATP. Another assay is based on the coupling
between regeneration of enzymatically hydrolysed ATP and oxy-
dation of NADH so that the action of various inhibitors can
be monitored by continuously recording the absorbance of
NADH at 340 nm. Plasma is known to contain potent inhibitors
of Na$^+$, K$^+$-ATPase activity such as calcium and vanadate ions.
Inhibition by calcium and vanadate ions was prevented by the
addition of EGTA as calcium chelator and noradrenaline as vanadate inhibitor. Plasma extracts from normotensive and hypertensive subjects were observed to inhibit ATP hydrolysis
by 15 % and 21 % respectively. Interestingly a significant
correlation between inhibition of enzyme activity and ouabain
binding was observed, suggesting that the effect was determined by the same circulating substance.

The uptake of serotonin in blood platelets depends on
the activity of Na$^+$, K$^+$-ATPase. In blood platelets of hypertensive subjects, the uptake of tritiated serotonin is significantly reduced when compared to normotensive subjects. Ouabain inhibits, in a concentration-dependent manner, the uptake of tritiated serotonin in blood platelets from normotensive subjects. The effect of several fractions of plasma
extract was tested on the uptake of tritiated serotonin by
preincubating during 45 min, the platelets with these fractions. The uptake of tritiated serotonin was significantly
inhibited in the presence of particular fractions, with a
maximal inhibition of approximately 40 %. These same fractions
were effective in inhibiting the binding of ouabain to red
blood cells and in inhibiting the activity of Na$^+$, K$^+$-ATPase
from canine kidney.

Finally endogenous sodium-pump inhibitors can cross-react

to a certain extent with digoxin antibody.

PERIPHERAL EFFECTS OF THE OUABAIN-LIKE HUMORAL AGENT

It has been possible to obtain purified extracts of urine which are natriuretic and inhibit Na^+, K^+-ATPase and which are more potent when the urine comes from a volume expanded man or animal. Sodium pump inhibition promotes salt and water excretion by inhibiting sodium reabsorption in the kidney tubules. De Wardener provided evidence for the presence of a natriuretic hormone different from catecholamines and common peptides in plasma and urine of volume expanded animals.

Evidence that a circulating agent can be involved in the genesis of hypertension comes from experiments in which salt-resistant rats received blood from salt-sensitive rats on a high-salt diet. The originally unresponsive rats rapidly developed hypertension. One interpretation for these experiments was that the salt-sensitive rat on a high-salt diet produced a circulating natriuretic hormone which increases the contractility of the smooth muscle of the arterioles. The ouabain-like humoral agent could correspond to the transferred circulating factor.

There is evidence that inhibition of the sodium-pump in smooth muscle, by a reduction in extracellular potassium or by high concentrations of cardiac glycosides can lead to an increase in arteriolar resistance. The increased vascular resting tone and the enhanced reactivity to catecholamines could reflect an increase in intracellular free calcium which is the immediate trigger for vascular smooth muscle contraction. Inhibition of the sodium pump leads to a net accumulation or redistribution of intracellular calcium by two different processes. Firstly, sodium gradient alteration produces electrogenic pump suppression and hence depolarization. Since calcium permeability is sensitive to the membrane potential, this should result in calcium influx. Secondly, an increase in intracellular sodium concentration can activate a reverse sodium-calcium exchange and lead to calcium influx.

Not all the actions of cardiac glycosides can be attributed to Na^+, K^+-pump inhibition. The positive inotropic effect cannot be entirely accounted for by the glycoside-evoked changes in sodium and potassium gradients. Therefore, promotion of positive inotropy may not be a property of endogenous sodium pump inhibitors. However, digitalis-like substance which revitalized the perfused failing heart could be elaborated by the liver.

NEURAL EFFECTS

The Na^+, K^+-ATPase is found in very high activity in neuronal tissue. The enzyme assumes great importance in neuronal homeostasis in the period immediately succeeding membrane depolarization when comparatively large fluxes of sodium and potassium occur, resulting in an action potential. Noradrenaline uptake which takes place at the level of sympathetic nerve terminals may be reduced by ouabain. Moreover, norepinephrine release is increased by ouabain. A carrier-mediated exit of cytoplasmic noradrenaline could be activated by the alteration in sodium gradient induced by ouabain. This voltage independent release is a mechanism of release different from the exocytotic process which occurs after depolarization. The resulting higher level of release of noradrenaline could contribute to the maintenance of elevated vascular smooth muscle tone . The reduced uptake of serotonin reported in blood platelets of hypertensive subjects could result in an increase of serotonin at the level of blood vessels, resulting in increased vascular resistance.

Intracerebroventricular (i.c.v.) administration of ouabain results in neuroexcitatory effects including autonomic nervous system activation and epileptic activity. The degree of inhibition of Na^+, K^+-ATPase is greatest in the hypothalamus and hippocampus after i.c.v. injection of the cardiac glycoside. Depletion of tissue potassium and its replacement by sodium by ouabain   results in neuronal swelling.

In urethane anesthetized rats a constant pattern of cardiovascular responses was observed after i.c.v. administration of ouabain. A primary vagal activation (bradycardia) was followed by a progressive sympathetic overactivity with blood pressure and heart rate increases. Tachyphylaxis was observed. In preliminary experiments, several human plasmatic fractions which inhibit the binding of ouabain to its site on red blood cells and inhibit Na^+, K^+-activated ATPase were tested on anesthetized rats. These fractions injected i.c.v. induced blood pressure increases and interestingly one fraction prevented the cardiovascular response of increased blood pressure by i.c.v. ouabain injected thereafter.

BRAIN SOURCE OF THE OUABAIN-LIKE AGENT

The substance partially isolated from plasma may be central in origin as several investigators have reported the isolation of a substance from whole guinea pig brain or bovine hypothalamus which inhibits active sodium transport and ouabain binding to Na^+, K^+-ATPase. A variety of types of experimental hypertension in rats are prevented by an anteroventral third ventricle (AV3V) lesion. This lesion prevents the appearance of the ouabain-like humoral agent in plasma and prevents the suppression of the vascular sodium pump activity of these rats. Thus the mechanism by which AV3V lesion prevents or ameliorates certain forms of low renin hypertension in rats may be through the interruption of production of this factor. Attempts to purify and identify the biochemical structure of natriuretic hormone are in progress in several laboratories. Conceivably putative endogenous ligands of the digitalis receptor may include both steroids and peptides. Although a minor pathway in mammalian systems, 5-beta reduction affords the metabolic potential for tissue-specific conversion of the mammalian steroid to the cardiac glycoside configuration. ACTH 4-IO has an aminoacid sequence which is common to several recognized hormones (ACTH, βLPH, ɣMSH). This peptide has natriuretic and pressor effects which could result from its ability to inhibit Na^+, K^+-ATPase. Sodium pump activity is also reduced by cyclo (His-Pro), a naturally occuring metabolite of TRH.

CONCLUSION

The sodium transport inhibitors in mammalian plasma, urine and brain have properties similar to those of the plant glycoside, ouabain, in that they inhibit Na^+, K^+-ATPase, displace ouabain bound to cellular membranes and cross-react to a certain extent with digoxin antibody. Many questions must now be asked concerning i) the mechanism of secretion of the endogenous digitalis-like compounds, ii) their sites of production which could be enlarged to tissues other than brain, iii) their nature which is not yet defined although certain steroids and peptide fragments have been shown to inhibit Na^+, K^+-ATPase. Answers to these questions would permit the study of the physiological actions of these endogenous digitalis on the cardiovascular, renal and autonomic nervous systems and allow the measurement of their levels in various types of hypertension leading to a better understanding of the pathogenesis of the disease.

REFERENCES

1. BLAUSTEIN, M.P. 1977. Sodium ions, calcium ions, blood pressure regulation, and hypertension : a reassessment and a hypothesis . Am. J. Physiol. 232, C165-C173.

2. CLOIX, J.F., DEVYNCK, M.A., ELGHOZI, J.L., KAMAL, L., LACERDA-JACOMINI, L.C., MEYER, P., PERNOLLET, M.G., ROSENFELD, J.B., DE THE H. 1983 (in press). : Plasma endogenous pump inhibitor in essential hypertension : measurement and significance. J. Hypertension.

3. DEVYNCK, M.A., PERNOLLET, M.G., BLAUDIN DE THE, H., MEYER, P. 1983 (in press). : Facteur plasmatique interagissant avec la pompe sodium-potassium dans l'hypertension artérielle essentielle. Arch. Mal. Coeur.

4. DE WARDENER, H.E., CLARKSON, E.M. 1982. : The natriuretic hormone : recent developments. Clin. Science, 63,415-420.

5. ELGHOZI, J.L., LE QUAN-BUI, K.H., DEVYNCK, M.A., MEYER, P. 1983 (in press). : Nomifensine antagonizes the ouabain-induced increase in dopamine metabolites in cerebrospinal fluid of the rat. European J. Pharmacol.

6. GLYNN, I.M., RINK, T.J. 1982. : Hypertension and inhibition of the sodium pump : a strong link but in which chain ? Nature. 300, 576-577.

7. GRUBER, K.A., HENNESSY, J.F., BUCKALEW Jr, V.M., LYMANGROVER, J.R. 1982. : Digitalis and natriuretic hormone-like activities reside in a heptapeptide fragment of pro-opiocortin. Clin. Res. 30, 848A.

8. HADDY, F.J. 1982. : Natriuretic hormone. The missing link in low renin hypertension ? Biochem. Pharmacol. 31, 3159-3161.

9. HAMLYN, J.M., RINGEL, R., SCHAEFFER, J., LEVINSON, P.D., HAMILTON, B.P., KOWARSKI, A.A., BLAUSTEIN, M. 1982. : A circulating inhibitor of $(Na^+ + K^+)$ ATPase associated with essential hypertension. Nature. 300, 650-652.

10. LABELLA, F. 1982. : Is there an endogenous digitalis ? Trends Pharmacol. Sci. 3, 334-335.

RESULTS OF EXPERIMENTAL STUDIES FAVORING THE HYPOTHESIS OF THE INFLUENCE OF STRESS ON THE GENESIS OF HYPERTENSION

G.Mancia, A.Ramirez,G.Bertinieri,G.Parati and A.Zanchetti
Istituto di Clinica Medica IV,Università di Milano e Centro di Fisiologia Clinica e Ipertensione ,Ospedale Maggiore, Milano, Italy

A major hypothesis that is advanced upon the origin of essential hypertension is that this condition is initiated by an overactivity of the sympathetic noradrenergic nerves(and an underactivity of the vagal nerves) generated by emotional behaviors definiable as"stress".Although the difficulties of categorizing and quantifying the differentiated forms of "stress" and emotion to which mankind is exposed have made this hypothesis hard to be tested in clinical studies, important progress has been made in this direction by thoughful use of animal research.

We will briefly describe such progress by briefly considering the following topics: 1) animal models in which "stress" has been primarily related to the initiation of a hypertensive state; 2) animal models in which neural, and in particular central, factors have been related to the secondary maintenance of a hypertensive state; 3) regional hemodynamic effects of emotions that may have importance in the production of hypertension.

Animal models of stress-induced hypertension

An animal model that shows the ability of stress to produce a prolonged rise in blood pressure is that based on the behavioral manipulation known as "adversive conditioning" (1,2).By such manipulation an animal can be tought to press a lever in order to avoid an unpleasant or frankly painful stimulus (usually an electrical shock) the delivery of which is preceded by the appearance of food or other pleasant items. The upper panel of Figure 1 shows that during the months over which a

monkey was subjected to adversive conditioning its blood pressure raised
markedly and sustainedly as compared to the lower values measured prior
to the beginning of the conditioning.Because the monkey learnt to press
the lever progressively more frequently the electrical shocks that were
delivered became progressively less during the study (Figure 1,lowest
panel).   Thus the raised blood pressure was not directly due to an increased number of painful stimuli.It was indirectly brought about by the
stressful condition and the anxiety state to which the animal was subjected.

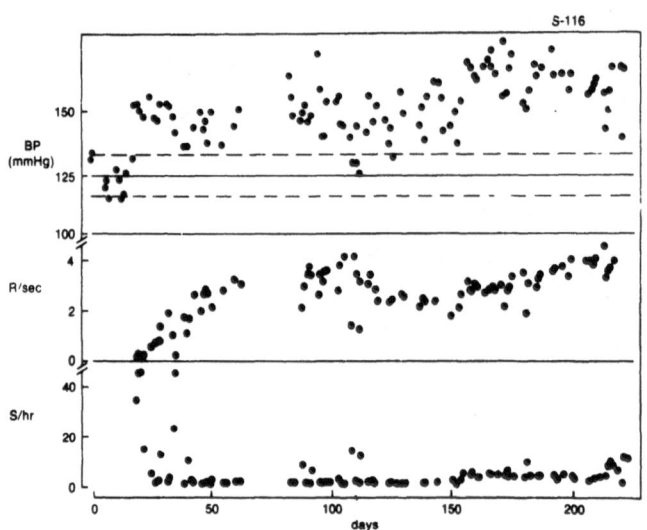

FIGURE 1.   Increases in blood pressure (BP) observed in a monkey subjected to adversive conditioning for more than 200 days.Notice that over
the same period the number of times the monkey pressed the lever(R/sec)
increased which kept the number of electrical shocks (S/hr) at a minimum
(from Herd et al., American J Physiol 217, 24,1969, by permission).

A second animal model of stress-induced hypertension is that ob-

served in mice by what may be labelled "psychosocial factors".Henry et
al. (3) have observed that when mice are raised in colonies males'
behavior rapidly differentiates into three patterns.While some males
develop a sort of "dominant" attitude that lead them to aggressively
defend their food, famales etc.,others behave socially as "subordinate"
or "indifferent".As shown in Figure 2 the "dominant" mice were found to
rapidly develop blood pressure values sustainedly greater than the
"subordinate" or "indifferent" mice. Greater blood pressure values were
found in male mice raised in isolation, after they had been introduced
into the colonies and forced to fight for establishing and defending
their territoriality.Interestingly, the dominant and re-introduced mice
had a higher incidence of blood pressure related complications and of
aortic atherosclerosis.Thus far from representing an innocent condition
their raised blood pressure had all the alarming features known to
characterize a hypertensive state.

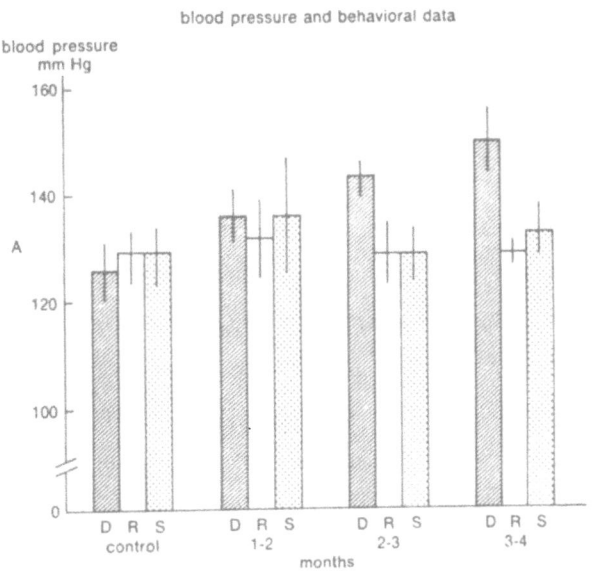

FIGURE 2. Blood pressure values observed in dominant (D),subordinate
(S) and indifferent (R) male mice over a few month period.Explanations
in text (from Henry et al, Circ.Res 36,156,1975, by permission)

A third example of a stress-initiated hypertension is believed to be the spontaneously hypertensive rat(4).Folkow and his group (5) have made important observations which demonstrate that the pressor and tachycardic effects of stressful stimuli are much greater in spontaneously hypertensive rats than in normotensive rats taken as comparison.These differences can be seen also when spontaneously hypertensive rats are compared with renovascular hypertensive rats.Furthermore , greater pressor and tachycardic responses to stress characterize spontaneously hypertensive rats even before the development of hypertnsion. Indeed, this development can be prevented not only by early administration of "antisympathetic" drugs but also by raising these rats in isolation and depriving them of most of their sensory inputs (for example light inputs).All these data suggest that in this strain of rats hypertension is produced through an interaction between normal environmental stimuli and inherently hyperactive central structures integrating the cardiovascular responses to stress (6).This interaction leads to more marked and frequent rises in blood pressure which eventually induce a permanent blood pressure elevation.It will be noticed that this model of hypertension differs from that obtained by adversive conditioning.   The latter is the effect of manipulation of environmental factors only, whereas the former represents the result of both environmental and genetic factors.Because both environmental factors and a genetic background are thought to concur to the production of essential hypertension, this model may be particularly rewarding in understanding this abnormal condition.

The last model we wish to mention is the well known strain of rats studied by Dahl (7) in which hypertension can be produced by the interaction of genetic factors and an increased salt intake.This strain makes a sharp contrast with another strain which shows genetic resistance to the hypertensive effect of increasing salt intake.This contrast is not limited to the salt factor, however.Figure 3 shows the results of a study of Friedman (8) in which the two strains were kept under a similar diet and subjected to a similar prolonged program of food-shock adversive conditioning, aimed to produce a state of stress or anxiety (see above).During the months of this procedure the blood pressure of the salt resistant rats did not show any rise, whereas the blood pressure of the salt sensitive rats increased to sustainedly elevated values.This may suggest that in the sensitive strain the in-herited background is such as to favour non-specifically different environmental hypertensiogenic factors.It may also suggest, however, that this background, though favouring only sodium retention,does so through the multifold mechanisms by which this function is modulated. Some of these mechanisms are neural in nature(see below).

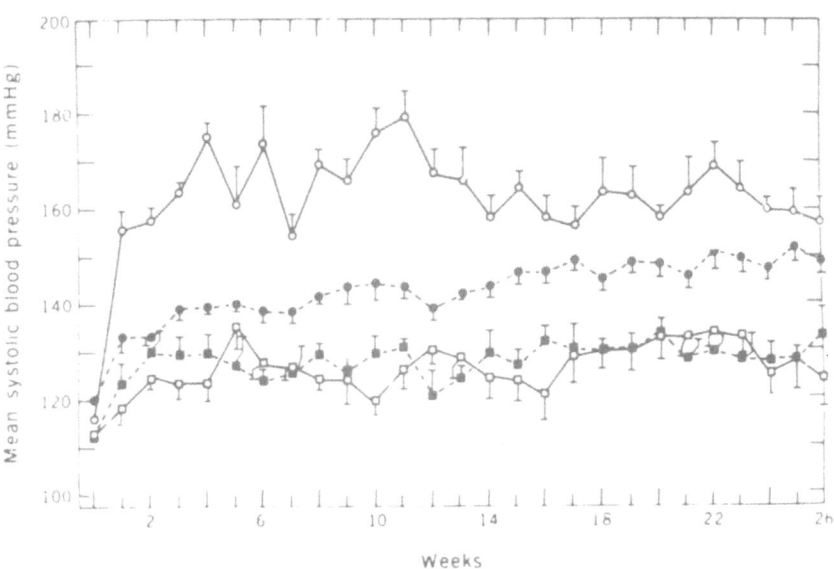

FIGURE 3.   Effects on systolic blood pressure of a food-shock,conflict procedure performed for several weeks on Dahl sensistive (open circles) and resistant (open squares)rats.Closed circles and squares represent sensitive and resistant free feeding controls (from Friedman and Iwai, Science 193,161,1976, by permission)

Description of so many successful models of stress-induced hypertension should not prevent mention of a neurogenic model of hypertension
which has fallen, after years of popularity (9,10), into controversy and
disrepute.We refer to the hypertension that has long been thought to
follow arterial baroreceptor denervation and elimination of reflex inhibitory influence on the vasomotor center. This is relevant to our
issue because baroreceptor denervation enhances and prolongs the blood
pressure rises induced by stress, a condition that might favour hypertension (5). However, recent studies in which blood pressure has been
recorded intra-arterially in unanesthetized animals and the time of the
recording has been prolonged to hours or days (a procedure unfeasible
in earlier days) has shown that denervation of arterial baroreceptors,
although associated with an increased blood pressure variability, does
not cause a sustained hypertension(11). Lack of sustained hypertension
after arterial baroreceptor denervation has been confirmed by us in
unanesthetized cats (Table 1).Furthermore we have shown that blood
pressure does not sustainedly rise even when the baroreceptor
denervation is obtained in conjunction with bilateral cervical vagotomy
(12).This procedure interrupts the aortic baroreceptor fibers that
travel outside the aortic nerves (12).More importantly, it denervates
the cardiopulmonary receptors that inhibit the vasomotor center in
conjunction to the arterial baroreceptors(14).Under these circumstances,
i.e. under deprivation of all reflex vasomotor inhibition, the animals
show a large blood pressure lability.However, average of these labile
values, does not show any blood pressure rise compared to the values
observed with all reflexes intact.

TABLE 1. Effects of sino-aortic denervation(SAD)alone or combined with bilateral cervical vagotomy (vagotomy) on blood pressure mean values and blood pressure variability *

	Intact	SAD	p<	SAD	SAD+Vagotomy	P<
		n = 8			n = 7	
Mean arterial pressure(mmHg)	99+7	93+7	NS	103+7	105+6	NS
Standard deviation (mmHg)	5+1	11+1	0.01	13+1	12+1	NS

* Data represent means (+SE) from recordings performed in unanesthetized, unrestrained cats.In each cat blood pressure was recorded intra-arterially for 8-12 hours.The blood pressure signal was analyzed by a computer every 60msec and the data were averaged for all the recording period. The standard deviation of this average was taken as the measure of blood pressure variability.

Neural factors in non-neurogenically induced hypertensions

An important distinction that must be made in investigation of hypertension is that between primary and secondary pathogenetic factors. While primary factors have an initiating pathogenetic role, secondary factors materialize at a later stage to maintain (or help to maintain) a hypertensive state they have not primarily originated.

A question that should be asked is whether stress, beside playing an initiating role in several hypertension models (see above), can act as a secondary factor in hypertensions that are primarily non-neurogenic. To our knowledge this question is still unanswered.It has been repeatedly shown, however, that in several experimental hypertensions neural mechanisms do represent important secondary factors.For example destruction of sympathetic nerve terminal by 6-hydroxy-dopamine has been shown to abolish the hypertensions induced by DOCA and by renal artery stenosis (15).The latter has been abolished or prevented not only by diffuse neural damage (such as that induced by 6-hydroxy-dopamine) but also by selective renal denervation (16) and by central lesions restricted to tiny areas of the hypothalamus(17).

A clearcut example of the importance of sympathetic activity is sustaining a non-neurally originated hypertension is that provided by

our studies on sleep(18).In the cat this condition is accompained by a
fall in blood pressure particularly during the REM (or deep sleep)phase.
This fall can be enhanced by section of the sino-aortic nerves,after
which the REM sleep mechanisms induce a widespread and drastic reduction
in sympathetic vasoconstrictor tone. Figure 4 shows that in a renovascular hypertensive cat(right panel) this reduction brought blood
pressure to the same low value that had been reached during REM sleep in
a cat without renovascular hypertension(left panel).Thus this hypertension was abolished when the sympathetic activity was turned off by
the depressor influence of sleep.

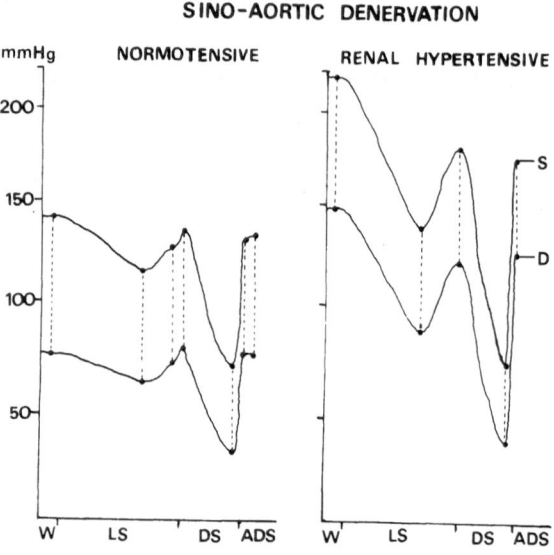

FIGURE 4. Systolic (S) and diastolic (D)blood pressure during the wake-
fulness-sleep cycle in a normotensive and a renovascular hypertensive
cat, both with sino-aortic denervation.Each point represents the average
of 10 different episodes.W:wakefulness;LS:light sleep;DS:REM or deep
sleep;ADS:after REM sleep(from Mancia and Zanchetti,Physiology during
sleep, Academic Press,New York,1980, p.1, by permission).

It remains to be explained why hypertensions originated by non-neural factors eventually stimulate sympathetic cardiovascular control and make it so important in secondary maintenance of the blood pressure elevation.Current hypotheses focus on the relationships between sodium and norepinephrine stores and release in the nerve terminals(15) and on the stimulating properties of angiotensin II at various central and peripheral sympathetic sites(19,20).These hypotheses,however, should include the effects of sympathetic nerves on the major sodium dealing mechanisms of the body.To this topic we will briefly devote the last section of this paper.

Regional hemodynamic effects of emotions

To study cardiovascular effects of emotional behaviors in the cat, we have made use of a large series of emotional stimuli.Some of these stimuli are shown in Figure 5. Briefly, a cat was chronically implanted with an electrode in the hypothalamic defence area to produce, via electrical stimulation, an aggressive behavior of the desired intensity and duration (6).This behavior was employed to evoke a defence reaction in another cat equipped with an arterial catheter and several electro-magnetic flowprobes for recording cardiac output and vasomotor changes in peripheral vascular districts.Detailed responses to natural defence reactions with or without concomitant exercise could thus be observed. To our surprise most emotional behaviors did not change blood pressure markedly.When the behavior consisted of an immobile confrontation of the cat with its aggressor, blood pressure showed little change because the occurrence of peripheral vasoconstriction was offset by a concomitant reduction in heart rate and cardiac output.Blood pressure also showed little change during a more intense emotion induced by fighting between the animals.In this istance the occurrence of tachycardia, increased cardiac output and intense visceral vasoconstriction was offset by a concomitant marked vasodilatation in skeletal muscle areas.

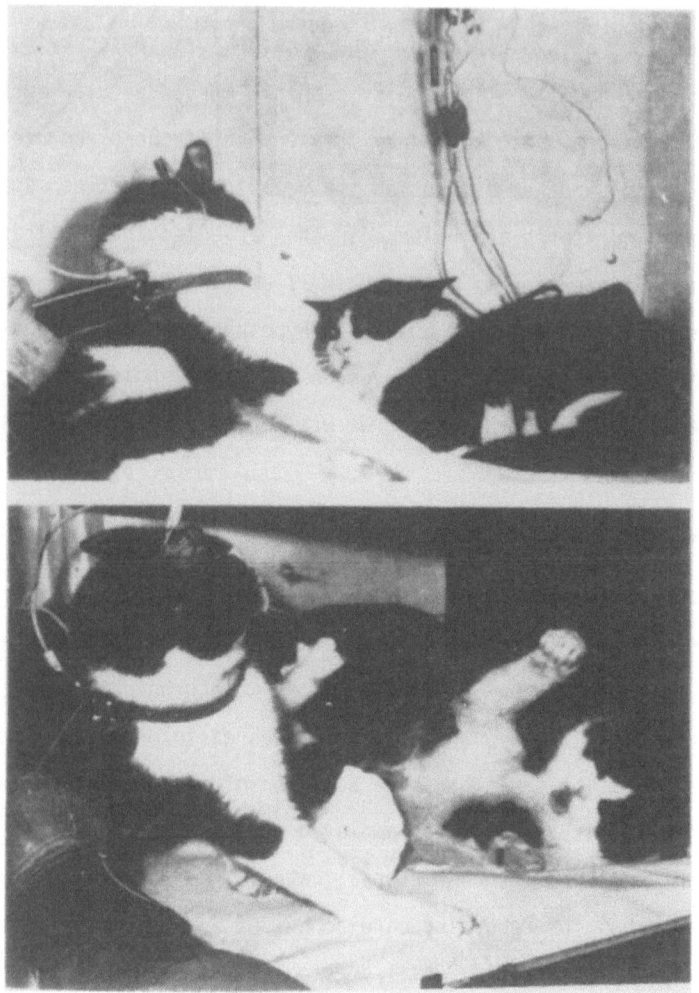

FIGURE 5. The upper panel shows an immobile confrontation of a cat equipped for blood pressure and blood flows measurements with another cat made aggressive by electrical hypothalamic stimulation.The lower panel shows a fighting between the two cats.

These results show that blood pressure changes represent a poor index of stress whose effects may be more visible in some regional circulations.In our experience stress induced particularly marked changes in renal circulation(21).An example of this is given in Figure 6 which shows blood flow recordings from both kidneys one of which(the left)had been previously denervated. Immobile confrontation and fighting again

were accompanied by little change in blood pressure.Both behaviors,
however, caused a marked reduction in renal blood flow whose neurogenic
nature was demonstrated by its occurrence exclusively(immobile
confrontation) or predominantly (fighting) on the innervated side.This
is a clear indication that neural factors may profoundly affect renal
circulation.

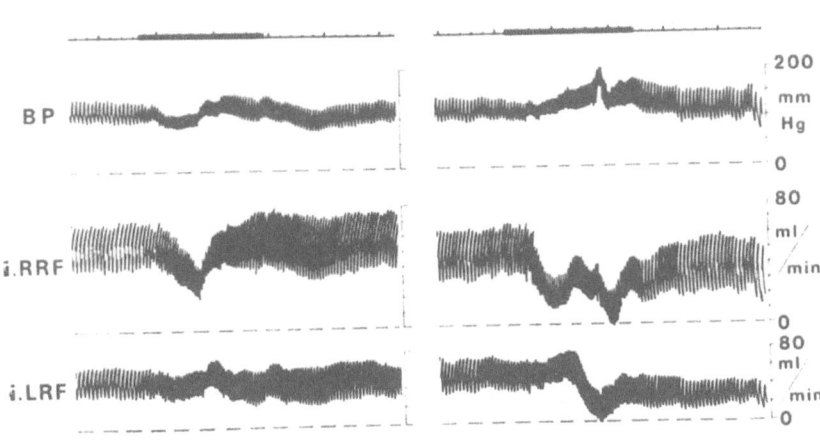

FIGURE 6 . Original tracings showing an episode of immobile confrontation (left) and of supportive fighting with a dog.BP:blood pressure;
i.RRF:instantaneous blood flow to innervated right kidney; i.LRF:
instantaneous blood flow to innervated left kidney(from Mancia et al.
Am.J.Physiol 227,136,1974 by permission).

Recently evidence obtained in several animal species has not only
confirmed that emotions may drastically reduce renal blood flow(22,23)
but also suggested that the reduction may be prolonged and lead to
anatomical damage of renal tissue(24).In addition, evidence has been ob-

tained that emotion and neural influences powerfully modulate kidney functions such as secretion of renin(25),secretion of prostaglandings and tubular reabsorption of sodium (27).

By demonstrating a close link between neural and salt factors these data may help a more comprehensive understanding of the pathogeneses of human hypertensions.

REFERENCES

1. Forsyth RP and Harris RE.Circulatory changes during stressful stimuli in Rhesus monkeys.Circ Res 26,Suppl 1: 13-20,1970
2. Herd JA,Morse WH,Kelleher RT,Jones LG. Arterial hypertension in the squirrel monkey during behavioral experiments. Am J Physiol 217:24-29, 1969
3. Henry JP,Stephens PP, Santisteban GA. A model of psychsocial hypertension showing reversibility and progressions of cardiovascular complications. Circ Res 36:156-164,1975
4. Okamoto K. Spontaneous hypertension in rats. Int Rev Exp Path 7, 227-236, 1969
5. Folkow B. Central neurohumoral mechanisms in spontaneously hypertensive rats compared with human essential hypertension.Cli Sci Mol Med 48, Suppl.2:205s-214s, 1975
6. Mancia G, Zanchetti A. Hypothalamic control of autonomic functions. Handbook of Hypothalamus.Morgane PJ, Panksepp J (Ed).M.Dekker,New York, 1980, Vol.III, Part B, pp. 147-202
7. Dahl LK, Heine M, Tassinari L. Role of genetic factors in subsceptibility to experimental hypertension due to chronic excess salt ingestion.Nature 194:480-482, 1962
8. Friedman R. and Iwai J. Genetic predisposition and stress induced hypertension. Science 193, 161-162, 1976
9. Thomas CB. Experimental hypertension from section of the moderator nerves.John Hopkins Hosp Bull 74, 335-377,1944
10. Heymans C., Bouchaert JJ. Modifications de la pression arterielle après section des quatre nerfs frenateurs chez le chien.CR Soc Biol (Paris) 117,252-255, 1934
11. Cowley AW, Liard LF, Guyton AC. Role of the baroreceptor reflex in daily control of arterial blood pressure and other variables in dogs.CIrc Res 32, 564-576,1973
12. Ramirez A., Bertinieri G, Belli L, Cavallazzi A, Di RIenzo M, Pedotti A, Mancia G. Reflex control of blood pressure and heart rate by arterial baroreceptors and by cardiopulmonary receptors the unanesthetized cat. Submitted for publication
13. Ito CS, Scher DM. Arterial baroreceptor fibers from the aortic region of the dog in the cervical vagus nerve.Circ Res 32,442-446, 1973

14. Mancia G, Donald DE, Shepherd JT. Inhibition of adrenergic outflow to peripheral blood vessels by vagal afferents from the cardiopulmonary region in the dog. Cir Res 33, 713-721, 1973

15. De Champlain J. Experimental aspects of the relationship between the autonomic nervous system and catecholamines in hypertension.In Hypertension. J.Genest, Koiw E, Kuchel O.(Eds). McGraw Hill, New York, 1977, pp.76-92

16. Wintermitz SR, Oparil S. Importance of renal nerves in the pathogenesis of experimental hypertension. Hypertension 4, Suppl.3:108-115, 1982

17. Brody MJ, Fink GD, Buggy J, Haywood JR, Gordon FJ, Johnson AK. The role of the anteroventral third ventricle(AV3V) region in experimental hypertension.Circ Res 43, Suppl 1, 2-13, 1978

18. Mancia G,Zanchetti A. Cardiovascular regulation during sleep in Physiology during sleep.Orem J, Barnes CD(Eds), Academic Press,New York, 1980, pp.1-55

19. Ferrario CM, Barnes KL, Szilagyi SE, Brashihan KB. Physiological and pharmacological characterization of the area postrema pressor pathways in the normal dog. Hypertension 1, 235-245, 1979

20. Zimmerman BG, Rolewicz TF, Dunhanm EW, Gisslen JL. Transmitter release and vascular responses in skin and muscle of hypertensive dogs.J Pharmacol Exp Therap 163, 320-329, 1968

21. Mancia G, Baccelli G, Zanchetti A. Regulation of renal circulation during behavior in the cat. Am J Physiol 227, 536-542, 1974

22. Martin J, Sutherland CJ, Zbrozyna AW. Habituation and conditio ning of the defence reactions and their cardiovascular components in cats and dogs. Pflug.Arch 365,37-47, 1976

23. Zbrozyna AW . Renal vasoconstriction in naturally elicited fear and its habituation in baboons. Cardiov Res 10, 295-300, 1976

24. Von Holst D. Renal failure as a cause of death in Tupaia belangeri exposed to persistent social stress. J Comp Physiol 78, 236-273, 1972

25. Mancia G, Lorenz RR, Shepherd JT. Reflex control of circulation by heart and lungs.Internat.Review of Science Cardiov Physiol II, volume 9 Guyton A.Cowley (Eds), University Press, Baltimore, 1976,pp.111-144

26. Mancia G., Romero JC, Strong CG. Neural influence on canine renal prostaglandin secretion. Acta physiol.Latino am. 1974,24,555-560

27. Bello-Reuss E, Trevino DL, Gottschalk CW. Effect of renal sympathetic nerve stimulation on proximal water and sodium reabsorption. J Clin Invest 57, 1104-1107, 1976

ANIMAL MODELS FOR THE ASSESSMENT OF STRESS ON ARTERIAL BLOOD PRESSURE

D T Greenwood, P W Marshall, and C P Allott.

Bioscience Department, Imperial Chemical Industries Plc, Pharmaceuticals Division, Alderley Park, Macclesfield, Cheshire. United Kingdom.

Introduction

Accumulating evidence suggests that subjects with either borderline or established essential hypertension exhibit abnormal haemodynamic responses to acute stress, including psychoemotional challenge (see reviews by Brod, 1982, and Folkow, 1982). In addition, the importance of supramedullary brain structures, including the hypothalamus and amygdaloid complex both in the control of emotional reactivity and in circulatory homeostasis is increasingly recognised (Folkow and Neil, 1971, Hilton, 1975, Mancia and Zanchetti, 1981). Collectively, these findings may be taken as providing supporting evidence for a possible causal relationship between stress and hypertension. However, whether or not repeated stress can generate neurogenic influences which, either alone or in concert with other environmental or endogenous factors, can subsequently lead to the development and maintenance of hypertension remains a topic of continuing research and considerable debate. Resolution of this issue in an exlusively clinical setting poses formidable if not insurmountable problems and the past two decades have therefore seen an inevitable upsurge of interest in the development of animal models for assessing the effects of stress on arterial blood pressure. The aim of the present article is to briefly survey the various laboratory models which have been described highlighting certain features and developments which appear particularly relevant to the above problem.

Acute and Chronic Studies

A representative selection of animal investigations conducted to evaluate the influence of a variety of different stressful stimuli on arterial blood pressure is surveyed in Tables 1 and 2. The studies quoted include examples of exposure to both acute and chronic stress in rodent and in non rodent species. Obvious practical considerations, coupled with the increasing availability of genetically sensitive, i.e. spontaneously hypertensive, strains have made the rat the most popular experimental subject in studies where the prime objective has been the demonstration and/or characterisation of a possible neurogenic factor in the development and maintenance of hypertension. Larger animal species have featured more prominently in those studies involving more complex analysis of the haemodynamic response to stress.

TABLE 1

RODENT MODELS FOR ASSESSMENT OF STRESS ON ARTERIAL BLOOD PRESSURE

STRESSOR	SPECIES/STRAIN (NT=Normotensive SH=Spontaneously Hypertensive)	REFERENCES
1. Environmental/Sensory:		
Air Blast	Rat,Norway (NT)	Farris et al 1945
Decompression	Rat, Wistar (NT)	Buckley et al 1953
Shaking	Rat, Sprague Dawley (NT)	Bunag et al 1980
Noise/light/vibration	Rat, Wistar (NT)	Hudak and Buckley 1961, Buckley et al 1964
	Rat, Sprague - Dawley (NT)	Rosecrans et al 1966, Perhach et al 1975
	Rat, Wistar (SH)	Hallbach and Folkow 1974
Noise	Rat, Wistar (SH)	Galeno and Brody 1982
Restraint	Rat, Wistar (NT)	Lamprecht et al 1973
Restraint/Cold	Rat, Wistar (SH)	Yamori et al 1969
2. Aversive Conditioning: Classical/Conflict		
Electroshock	Rat, (NT)	Shapiro et al 1957
	Rat, (Dahl-SH)	Dahl et al 1968 Friedman and Dahl, 1975, McCarty et al (1978(a) and (b)
Electroshock/ammonia fumes	Rat, (SHR)	Hoffman and Fitzgerald, 1978
Sidman avoidance	Rat, (NT)	Murray 1978
Avoidance - avoidance	Rat, (NT)	Buckholz et al 1981
	Rat, (F1 offspring SHR x WKY)	Lawler et al 1980 Lawler et al 1981
3. Psychosocial:		
Isolation	Rat, Wistar (NT)	Gardiner and Bennett, 1977
Social Conflict	RAT, (SH)	Hallbach, 1975
	Mouse, (CBA/NT)	Henry, et al (1967, 1975)
	Rat, (Wistar NT)	Alexander, 1974

TABLE 2

MODELS FOR THE ASSESSMENT OF STRESS ON ARTERIAL BLOOD PRESSURE IN NON RODENT SPECIES.

STRESSOR	SPECIES/STRAIN	REFERENCES
Aversive Conditioning:		
Classical avoidance/conflict	Dog	Caraffa-Braga et al 1973
Sidman avoidance	Dog	Anderson and Brady (1972,1973,1976) Lawler et al 1975 Seal and Zbrozyna (1978)
	Rhesus Monkey	Forsyth (1969,1972)
	Squirrel Monkey	Herd et al, 1969, Kelleher et al 1972
	Baboon	Findley et al 1971 Harris et al 1973 Zbrozyna, 1976 Smith et al 1980
Psychosocial:		
Conflict	Cat	Adams et al (1968, 1969) Zanchetti et al 1972, Martin et al 1976

Haemodynamic Response to Stress in Animals

Analysis of the acute haemodynamic response to stress in animals has been the subject of considerable interest for several years and is of direct relevance to earlier comments regarding models of hypertension. Any consideration of this aspect must centre on an appreciation of the so-called "defence" reaction or cardiovascular alerting response (see Mancia and Zanchetti, 1981; Hilton 1982; Folkow, 1982). Stated simply, the cardiovascular components of the defence reaction, comprise a redistribution of blood from the renal and mesenteric vasculature towards cardiac and skeletal muscle in preparation for "fight and flight". This is achieved by marked vasoconstriction of visceral vasculature and dilatation to skeletal muscle; changes which are accompanied by an increased heart rate, cardiac output and blood pressure. Characterisation of the defence reaction evoked by brain stimulation in anaesthetised preparations has yielded valuable information on the various neural pathways involved. However, investigations in anaesthetised preparations are obviously of limited value in the context of chronic psychoemotional stress and subjsequent studies have been extended to awake freely moving animals exposed to a variety of experimental stressors (see Table 2).

The importance of using conscious animals must be emphasised because haemodynamic changes evoked by electrical brain stimulation in anaesthetised animals differ in several respects from those elicited by emotional stimuli in conscious animals. Several different patterns of haemodynamic change have been observed to occur in response to acute stress and these may be considered as comprising possible sequential phases of a stress/cardiovascular response continuum. However, it should be appreciated that not all phases of the cardiovascular response to acute stress are necessarily accompanied by increases in arterial blood pressure, since vasodilation in skeletal vasculature may at times offset vasoconstriction in the visceral vascular beds (see Mancia and Zanchetti 1981). Immobile confrontation may in fact be associated with a bradycardia and fall in cardiac output similar to the orienting response (Adams et al 1968; 1969). Only when locomotor or other active behaviours are considerable, as during actual fight or flight, does the increased visceral vasoconstriction and cardiac output outweigh vasodilation to skeletal muscle and then the blood pressure can rise considerably. Periods preceding operant avoidance sessions have been characterised by a fall in heart rate and cardiac output accompanied by a gradual rise in peripheral resistance and arterial blood pressure (Anderson and Tosheff, 1973, Lawler et al 1975). In contrast active performance of avoidance behaviour is assosicated with a tachycardia and increased cardiac output accompanied by a sustained elevation of arterial blood pressure (Anderson and Tosheff, 1973). Differing neuronal mechanisms subserving these various haemodynamic patterns of response to stress are suggested by their differential sensitivity to adrenoreceptor antagonists (Anderson and Brady, 1976).

The cardiovascular response to acute stress is thus complex and is dependent upon the novelty and severity of the stressor. These responses are also subject to habituation (Martin et al 1976; Seal et al, 1978, Zbrozyna, 1976; Zbrozyna, 1982, Dailey et al 1982) and different components of the cardiovascular alerting response habituate at different rates. The latter may be of special significance where the objective is to establish a persistently elevated blood pressure since as Folkow has pointed out, only those procedures which repeatedly evoke a rise in arterial blood pressure are likely to lead to sustained hypertension.

Models of Hypertension

Since the hypothesis that frequent repetition of acute pressor episodes eventually leads to a sustained rise in blood pressure is central to the clinical debate, it is important to question whether or not the chronic studies referred to offer any corroborative support and, if so, what conclusions can be made concerning the most effective stressor(s).

The variety of experimental designs adopted, together with differences in the nature and duration of stress, animal species, sex and blood pressure monitoring techniques makes it difficult to arrive at any firm conclusions at the present time. Of the various environmental stressors studied, vibration (or shaking)appears to be the most effective acute stressor (Hallbach and Folkow, 1974, Bunag et al 1980) with intense flashing light the least potent. Many chronic investigations in normotensive rat strains have employed combinations of vibration, noise and light with a view to providing a basis for pharmacological intervention (eg. Smookler and Buckley, 1969, Perhach et al 1975).

While recognising that observation of statistically significant increases in blood pressure compared to appropriate controls is a feature of many of the studies quoted, the overall impression gained is that, excepting studies employing animals with a genetic bias, most of the techniques surveyed have resulted in only modest elevations in mean arterial pressure. Generally these have not been greater than most authorities would recognise as mild or borderline. Unfortunately not all investigators have determined whether the observed increase in blood pressure remains elevated after discontinuation of the stressor but where this important aspect has been examined it is rare to find evidence for a prolonged persistence of the effect.

Classical and simple operant conditioning techniques which constitute a more psychoemotional type of stress have also tended to evoke unimpressive elevations of blood pressure in normotensive animals.

Other investigators have adopted alternative approaches using less severe stressors but which perhaps more closely mimic the human situation. The fascinating psychosocial experiments conducted in mice by Henry's group are worthy of particular note. Introduction of socially deprived mice into a competitive environment for just a few days causes significant but reversible increases in blood pressure. After several months of stress the hypertension though more persistent still tends to remit but is associated with arteriosclerotic and myocardial lesions (Henry et al 1967, 1975).

A possible reason for the relative lack of success in inducing persistently high blood pressure with environmental and classical conditioned aversive stimuli, e.g. footshock, is that the animal rapidly habituates to the stressor (see above). Greater success might therefore be expected using operant avoidance conditioning techniques where the animal must make an active response in order to avoid the aversive stimuli, eg. Sidman avoidance. However, such studies while providing further evidence that psychological stress can induce acute elevations of blood pressure, have failed to lead to sustained marked hypertension in genetically non sensitive strains.

Generally, the above findings are in marked contrast to the consistently high levels of persistent hypertension which develop in the SHR; a process which in certain strains is accelerated by stress and retarded by social deprivation (Yamori, 1969 Hallbach, 1975). The latter observations are particularly relevant since the converse is claimed in a normotensive strain where social isolation appears to elevate blood pressure (Gardiner and Bennett, 1977). Clearly a genetic bias towards hypertension is as important a determinent in the outcome of stress studies in animals as it is in humans and this is entirely consistent with the multifactorial nature of the disease. Recent studies by Lawler's Group, in which borderline hypertensive F_1 offspring of a SHR/WKY cross breed were subjected to a novel avoidance conflict paradigm, offer impressive support for this view. Within only 5 weeks these rats developed blood pressures approaching 190 mmHg and the hypertension showed no sign of remission 10 weeks after termination of the stressor (Lawler et al 1980). Coupled with post mortem observations of significant cardiac pathology this would appear to be a more attractive model of stress induced hypertension than those hitherto described. Recent work in dogs by Anderson (1982) provides a further example of the manner in which stress can interact with a second factor, in this instance salt intake, to provoke a more sustained rise in blood pressure than could be elicited by either stimulus alone. Future research might therefore be profitably directed towards a more detailed analysis of the interaction between stress and other predisposing factors.

REFERENCES

Adams, D.B., Baccelli, G., Mancia, G., Zanchetti, A. (1968) Cardiovascular changes during preparation for fighting behaviour in the cat. Nature, 220, 1239

Adams, D.B., Baccelli, G., Mancia, G., Zanchetti, A. (1969) Cardiovascular changes during naturally elicited fighting behaviour in the cat. Am.J. Physiol. 216, 1226

Alexander, N. (1974) Psychosocial hypertension in members of Wistar rat colony. Proc. Soc. Exp. Biol. Med. 146, 162

Anderson, D.E. and Brady, J.V. (1972) Differential preparatory cardiovascular responses to aversive and appetitive behavioural conditioning. Condit. Reflex 7, 82

Anderson, D.E. and Brady, J.V. (1973) Prolonged preavoidance effects upon blood pressure and heart rate in the dog. Psychosom. Med. 35, 4

Anderson, D.E. and Brady, J.V. (1976) Cardiovascular responses to avoidance conditioning: effects of beta adrenergic blockade. Psychosom. Med. 38, 181

Anderson, D.E. (1982) Behavioural hypertension mediated by salt intake. In: Circulation Neurobiology and Behavior. Smith Galosy and Weiss (Eds) Elsevier Science Publishing Co. Inc. P247-257

Anderson, D.E. and Tosheff (1973). Cardiac output and total peripheral resistance changes during pre-avoidance periods in the dog. J. Appl. Physiol. 35, 650

Brod, J., Fencl, V. Hejl, Z and Jorka, J. (1959) Circulatory changes underlying blood pressure elevation during acute emotional stress (mental arithmetic) in normotensive and hypertensive subjects. Clin. Sci 18, 269

Brod, J. (1982) Environmental stress and hypertension - Introduction Contr. Nephrol, 30, 1

Buckholz, R.A., Lawler, J.E. and Barker, G.F. (1981) The effects of avoidance and conflict schedules on the blood pressure and heart rate of rats. Physiol. Behav. 26, 853

Buckley, J.P., Edwards, L.D. and Hiestand, W.A. (1953) Effect of discontinuous decompression on blood pressure in the rat. Am. J. Physiol. 175, 93

Buckley, J.P., Kato, H., Kinnard, W.J., Aceto, M.D.G. and Esterbz, J.M. (1964) Effects of reserpine and chlorpromazine on rats subjected to experimental stress. Psychopharmacologia 6, 87

Buckley, J.P., Vogin, E.E. and Kinnard, W.J. (1966) Effects of pentobarbital, acetylsalicylic acid and reserpine on blood pressure and survival of rats subjected to experimental stress. J. Pharmac. Sci. 55, 572

Buckley, J.P., Parham, C. and Smookler, H.H. (1968) Effects of reserpine on rats subjected to prolonged experimental stress. Arch. int. Pharmacodyn 172, 292

Bunag, R.D., and Riley, E. (1979) Chronic hypothalmic stimulation in awake rats fails to induce hypertension. Hypertension 1 (5), 498

Bunag, R.D., Takeda, K, and Riley, E. (1980) Spontaneous remission of hypertension in awake rats chronically exposed to shaker stress. Hypertension 2, 311

Caraffa-Braga, E., Granata, L. and Pinotti, O. (1973) Changes in blood-flow distribution during acute emotional stress in dogs. Pflugers Arch 339, 203

Dailey, W., Valtair, J. and Amsel, A. (1982) Bidirectional heart-rate change to photic stimulation in infant rats: implication for orienting/defensive reflex distinction. Behavioral and Neural Biology 35, 96

Dahl, L.K., Knudsen, K.D., Heine, M. and Leitl. G. (1968) Hypertension and stress. Nature 219, 735

Farris, E.J., Yeakel, E.H. and Medoff, H.S. (1945) Development of hypertension in emotional grey Norway rats after air blasting. Am.J. Physiol. 144, 331

Findley, J., Brady, J.V., Robinson, W., Gilliam, W., (1971) Continuous cardiovascular monitoring in the baboon during long-term behavioural performances. Comm. Behav. Biol. 6, 49

Folkow, B and Neil, E. (1971)
Circulation (Oxford University Press, London)

Folkow, B (1982) Physiological aspects of primary hypertension
Physiological Reveiws 62 (2), 347

Forsyth, R.P. (1969) Blood pressure responses to long-term avoidance schedules in the restrained rhesus monkey. Psychosom. Med. 31, 300

Forsyth, R.P. (1972) Sympathetic nervous system control of distribution of cardiac output in unanaesthetised monkeys.
Federation Proceedings 31, 1240

Friedman, R. and Dahl, L.K. (1975) The effect of chronic conflict on the blood pressure of rats with a genetic susceptibility to experimental hypertension. Psychosom. Med. 37, 402

Galeno, T.M. and Brody, M.J. (1982) Hemodynamic responses to noise stress in the spontaneously hypertensive rat (SHR). Federation Proceedings 41,(4), 1093

Gardiner, S.M. and Bennett, T. (1977) The effects of short-term isolation on systolic blood pressure and heart rate in rats. Med. Biol. 55, 325-329

Hallbach, M. and Folkow, B. (1974) Cardiovascular responses to acute mental 'stress' in spontaneously hypertensive rats. Acta Physiol. Scand. 90, 684

Hallbach, M. (1975) Consequence of social isolation on blood pressure, cardiovascular reactivity and design in spontaneously hypertensive rats. Acta Physiol. Scand. 93, 455

Harris, A.H., Gillians, W.J., Findley, J.D. and Brady, J.V. (1973) Instrumental conditioning of large magnitude, daily 12-hour blood pressure elevations in the baboon. Science 182, 175

Henry, J.P., Stephens and Santisteban, G.A. (1975) A model of psychosocial hypertension showing reversibility and progression of cardiovascular complications. Circ. Res. 36, 156

Henry, J.P. Meehan, J.P. and Stephens, P.M. (1967) The use of psychosocial stimuli to induce prolonged systolic hypertension in mice. Psychosom. Med. 29, 408

Herd, J.A., Morse, W., Kellerher, R.T. and Jones, J.G. (1969) Arterial hypertension in the squirrel monkey during behavioral experiments. Am. J. Physiol. 217, 24

Hilton, S.M. (1975) Ways of viewing the central nervous control of the circulation - old and new. Brain Res. 87, 213

Hoffman, J.W. and Fitzgerald, R.D. (1978) Classically-conditioned heart rate and blood pressure in rats based on either electric shock or ammonia fumes reinforcements. Physiol. Behav. 21, 735

Hudak, W.J., and Buckley, J.P. (1961) Production of hypertensive rats by experimental stress. J. Pharm. Sci. 50, 263

Kelleher, R., Morse, W. and Herd, J.A. (1972) Effects of propranolol, phentolamine, and methyl atropine on cardiovascular function in the squirrel monkey during behavioural experiments. J. Pharmacol. Exp. Ther. 182, 204

Lamprecht, F., Williams, R.B. and Kopin, I.J. (1973) Serum dopamine-beta-hydroxylase during development of immobilization-induced hypertension. Endocrinology, 92, 953

Lawler, J.E., Obrist, P.A. and Lawler, K.A., (1975) Cardiovascular function during pre-avoidance, avoidance and post-avoidance in dogs. Psychophysiol. 12, 4

Lawler, J.E., Barker, G.F., Hubbard, J.W. and Allen, M.T. (1980) The effects of conflict on tonic levels of blood pressure in the genetically borderline hypertensive rats. Psychophysiol. 17,363

Lawler, J.E., Barker, G.F., Hubbard, B.S. and Schaub, R.G. (1981) Effects of stress on blood pressure and cardiac pathology in rats with borderline hypertension. Hypertension, 3, (4), 496

Mancia, G. and Zanchetti, A (1981) "Hypothalamic Control of Autonomic Functions" In: Handbook of the Hypothalamus. Morgane, P.J. and Panksepp, J. (eds) Marcel Dekker. 147-202

Martin, J., Sutherland, C.J. and Zbrozyna, A.W. (1976) Habituation and conditioning of the defence reaction and their cardiovascular components in cats and dogs. Pflugers. Arch. 365, 37

McCarty, R., Chuieh, C.C. and Kopin, I.J. (1978a) Spontaneously hypertensive rats: adrenergic hyperresponsivity to anticipation of electric shock. Behav. Biol. 23, 180

McCarty, R., Chuieh, C.C. and Kopin, I.J., (1978b) Behavioural and cardiovascular responses of spontaneously hypertensive and normotensive rats to inescapable foot shock. Behav. Biol. 22, 405

McCarty, R. and Kopin, I.J. (1978) Sympatho-adrenal medullary activity and behaviour during exposure to foot shock stress; a comparison of seven rat strains. Physiol. Behav. 21, 567

Murray, D.M. (1978)
Unpublished doctoral dissertation, University of Tennessee, Knoxville.

Okamoto K. (1969) Spontaneous hypertension in rats. Int. Rev. Exp. Path. 7, 227

Perhach, J.L. Ferguson, H.C. and McKinney, G.R. (1975) Evaluation of antihypertensive agents in the stress-induced hypertensive rat. Life Sci. 16, 1731

Rosecrans, J.A. Watzman, N. and Buckley, J.P. (1966) The production of hypertension in male albino rats subjected to experimental stress. Biochemical Pharmac. 15, 1707

Seal, J.B., and Zbrozyna, A.W. (1978) Renal vasoconstriction and its habituation in the course of repeated auditory stimulation and naturally elicited defence reactions in the dog. J.Physiol., Lond. 280, 56-57p

Shapiro, A.P. and Melhado, J. (1957) Factors affecting development of hypertensive vascular disease after renal injury in rats. Proc. Soc. Exp. Biol. Med 96, 619

Smith, O.A. Astley, C.A., DeVito, J.L. Stein, J.M. and Walsh, K.E. (1980) Functional analysis of hypothalamic control of the cardiovascular responses accompanying emotional behaviour. Federation Proc. 39, 2487

Smookler, H.H. and Buckley, J.P. (1969) Relationships between brain catecholamine synthesis, pituitary adrenal function and the production of hypertension during prolonged exposure to environmental stress. Int. J. Neuropharmac. 8, 33

Yamori, Y., Matsumoto, M., Yamabe, M. and Okamoto, K. (1969) Augmentation of spontaneous hypertension by chronic stress in rats. Jap. Circulat. J. 33, 399

Zanchetti, A., Baccelli, E., Mancia, G., Ellison, G.D. (1972) Emotion and the Cardiovascular System in the Cat. In: Physiology, Emotion and Psychosomatic Illness (Ciba Foundation Symp. 8 New Ser.) Ass. Scientific Publ. Amsterdam. 201-219

Zbrozyna, A.W. (1976) Renal vasoconstriction in naturally elicited fear and its habituation in baboons. Cardiovasc, Res. 10, 295

Zbrozyna, A.W. (1982) Habituation of cardiovascular response to aversive stimulation and its significance for the development of essential hypertension. Contr. Nephrol. 30, 87p

VALIDATION AND QUANTIFICATION OF MENTAL STRESS TESTS, AND THEIR APPLICATION TO ACUTE CARDIOVASCULAR PATIENTS

ANDREW STEPTOE

St George's Hospital Medical School

University of London

U.K.

A wide range of mental stress tests are used to investigate the acute cardiovascular reactions of healthy subjects and patients with essential hypertension and ischaemic heart disease. When considering the value of different procedures, a number of factors may be relevant:

a) Unobtrusive administration of the test, so that it does not interfere with cardiovascular monitoring.

b) The cardiovascular responses should be reproducible. This is essential when patients or healthy volunteers are being studied under varying conditions, and when the procedure is being used to assess the effects of pharmacological or behavioural interventions. One of the major problems in acute testing is that rapid habituation takes place, so that smaller responses may occur after repeated presentation. The issue will be considered in more detail below.

c) Tests should be independent of intellectual status. If a procedure depends on intelligence, then some people will find it much easier than others, and may therefore produce smaller cardiovascular reactions. Comparisons between patients will then be compromised.

d) The test should be relevant to the individual, so that his or her mental and emotional resources are actively engaged. If this does not happen, cardiovascular reactions may be modest.

.

e) Probably the most important characteristic, as far as validation and quantification of tests is concerned, is that the motor response components of the procedure are minimal and standardised. If the test involves extensive motor activity (such as large movements), these actions may themselves provoke cardiovascular adjustments irrespective of the psychological challenge. If the task is not externally paced, subjects may apparently produce different cardiovascular reactions purely on the basis of their work or response rate. The similarities between the cardiovascular adjustments produced during exercise and psychological challenge have been discussed by Cohen and Obrist (1). Evidently, even small variations in motor activity will have cardiovascular consequences that may confound mental stress testing. Friedmann et al (2) have recently shown that blood presure (BP) and heart rate (HR) rise when people read aloud, and that the increase depends on the rate of reading. In addition, it appears that BP reactions during speech differ as a function of cardiovascular status, with hypertensives showing larger changes than normotensives (3).

It is difficult to satisfy all these requirements of test procedures in practice. Unless they are considered however, differences in cardiovascular responses between people, or between occasions in the same individual, may be due to many factors apart from sensitivity to mental stress. The main purpose of stress testing - the elicitation of the circulatory concomitants of psychological and emotional challenge - may then be confounded.

A broad distinction can be made between three different types of test procedure. Each has advantages and some drawbacks, and may be appropriate under particular circumstances.

PERSONALLY RELEVANT TESTS

Personally relevant tests include stress interviews, in which people are questioned about personal or social aspects of their lives while cardiovascular parameters are recorded. Such procedures were used in early psychosomatic studies of BP reactions in essential hypertension, and were also employed to document the coronary blood flow changes associated with emotionally-charged interactions (4,5). Theorell et al (6) administered stress interviews while patients with ischaemic heart disease were being monitored with a ballistocardiograph. The different emotions evoked during the interview were associated with varying degrees of cardiac contractility. More recently, interviews about personal life, key relationships and areas of private conflict have been assessed during the cardiovascular monitoring of patients with essential hypertension using sophisticated methods of determining regional blood flow (7,8).

The advantage of this procedure is that material can be tailored to the patient, thereby increasing the probability of generating large reactions. It is possible to adjust the interview so as to elicit specific emotions (eg, anxiety or grief) in a way which cannot be done so conveniently with other tests. Instead of interviews, some investigators have used role playing in order to study the impact of particular emotions such as hostility on BP reactions (9, 10). Personally relevant tests are useful diagnostically, since areas of life experience associated with cardiovascular disorders may be identified. This can facilitate the behavioural analysis of cardiovascular disorders (11), while demonstrating dramatically to patients that their emotions and experience have cardiovascular correlates.

Unfortunately, these procedures have several disadvantages. Almost by definition, such methods cannot be standardised or else their personally relevant qualities would be lost. Reproducibility is therefore poor, while motor response components are also variable. The procedures are psychologically 'reactive', making repeat assessments unreliable. Because of these problems, comparisons between groups treated differently (eg, patients and controls) are difficult to perform. Moreover, despite personal probing, subjects may refuse to become actively involved. This is illustrated in the study of reactions to the Thematic Apperception Test by Weiner et al (12). Hypertensives produced smaller BP and HR reactions than normotensives, suggesting that they were hypo-responsive. However, this was probably due to a lack of emotional engagement in these patients.

Recent efforts have been made to circumvent these limitations, while maintaining the element of personal relevance. Several investigators have monitored physiological responses during the administration of the structured interview used to assess Type A coronary-prone behaviour (eg, 13). Since the interview is designed to elicit Type A behaviours for the purpose of assessment, it is expected to be particularly relevant to coronary-prone individuals. The data however are inconsistent; Type A people have been shown to produce larger systolic BP and HR reactions during the interview in some studies but not others (14). Nevertheless, the use of semi-structured interviews may improve the reproducibility of personally relevant mental stress tests.

BEHAVIOURAL AND INFORMATION-PROCESSING TASKS

These are tests in which people have to solve problems or respond to stimuli in a systematic fashion. They are commonly used in acute cardiovascular settings, and the traditional mental arithmetic technique (serial subtraction) is a test of this type. Serial subtraction as generally presented is not a good procedure, since subjects are usually asked to respond out loud as rapidly as possible. This violates the requirement for standardised motor responses outlined earlier. However, this problem can be resolved by presenting mental arithmetic tasks at a fixed rate. Steptoe and Ross (15) eliminated the variable somatomotor adjustments associated with speaking by presenting subjects with complete arithmetic problems of the type '31 + 27 = 60' at a fixed rate; subjects were required to press one of two buttons depending on whether the answer given was correct or incorrect.

Several other problem solving tasks have been used during the assessment of cardiovascular reactions. They include visual concept formation tasks (16), digit-symbol substitution (15), and general knowledge quizzes (13). Raven's progressive matrices have been performed during the monitoring of systolic time intervals, plasma catecholamines and renal arteriography (17, 18). In all these cases, behavioural data in the form of performance measures may be collected, so that cardiovascular reactions can be related to ability and achievement.

The cardiovascular reactions to tasks of this type are not constant, but tend to diminish on repeated presentations. This may be due to increased familiarity with task demands, practice effects or adaptation to physiological measurement procedures. Using taxing mental arithmetic and digit-substitution tasks, we have found that HR reactions fall to half their initial value after five sessions (15). But even though the absolute magnitude of reactions may decline, there is evidence that individual differences in reactivity are consistent. Initial reactions have been shown to correlate with subsequent responses to the same tasks (15). Manuck and Schaefer (16) divided normotensive volunteers into groups of reactors and non-reactors on the basis of initial responses. The BP and HR reactions of these groups were still distinct on re-test after one week, and differences in systolic BP and HR were maintained after more than a year (19). On the other hand, the observation by Neus et al (20) that normotensives and hypertensives adapt to psychological challenges at different rates poses more severe difficulties in the analysis of repeated stress experiments.

A second serious difficulty is that many tasks of this type are dependent on intellectual status. Some, such as digit-symbol substitution and Raven's matrices, are in fact components of intelligence tests. Many investigators have therefore focussed their efforts on developing behavioural tasks that are less reliant on intellect. Obrist (21) has made substantial advances in understanding the role of psychological stress in the aetiology of essential hypertension through studying haemodynamic changes during a shock avoidance reaction time task. Considerable use has also been made of tasks based on the Stroop colour/word interference phenomenon. Melville and Raftery (22) reported mean increases of 26.1/25.6mm Hg from a group of newly-diagnosed essential hypertensives monitored with intra-arterial cannulae, and marked changes in catecholamine secretion and systolic time intervals have also been reported (23, 24). Another procedure which may become increasingly useful is to monitor physiological activity during performance of video games. Glass et al (25) showed that Type A and non-Type A men respond differently in cardiovascular and neuroendocrine parameters during such games, while hypertensives may be distinguished from normotensives under similar conditions (26). There is some evidence that reactions to video games are maintained on repeated presentation. Unpublished data from our laboratory on a group of untreated male hypertensives (resting BP 154.2/98.6mm) indicates that diastolic BP and HR reactions were maintained after three months, while systolic BP responses were slightly reduced. However, video games may have variable somatomotor correlates, and these have not yet been explored in detail (27).

PASSIVE STRESS CONDITIONS

The methods described in the last section are all active tasks, in the sense that they require some behavioural output from the subject. It is also possible to impose stressors that do not demand any behavioural response at all. These passive stressors include disturbing films (22), the cold pressor and other painful stimuli such as venipuncture (23) and noise (28). While such procedures may produce reliable haemodynamic and neuroendocrine responses, it should be emphasised that they are not equivalent to active behavioural tasks. There is increasing evidence that cardiovascular reactions depend on several aspects of the psychological demand, including the nature of behavioural coping demands, the degree of behavioural control and engagement/involvement (29). For example, catecholamine secretion may increase during passive stress experiences in the absence of concomitant tachycardia (30). Passive stressors such as the cold pressor produce substantial diastolic BP reactions, while active tasks may be characterised by heightened modifications of systolic BP and HR (21). This suggests that active behavioural tasks are associated with stimulation of cardiac sympathetic (β-adrenergic) pathways, while peripheral vasomotor responses may be more important under other conditions. Haemodynamic reaction patterns may also vary with clinical status, since Brod (31) found that changes in regional blood flow better discriminated normotensives from hypertensives than overall BP responses. These data indicate that careful consideration must be given to the nature of the system and patients under study, when selecting appropriate stress tests.

METHODS OF ASSESSMENT AND ANALYSIS

Once appropriate tests have been chosen, several aspects of analysis must be considered. First, it is possible that three distinct components of the cardio-vascular reaction pattern may be relevant to pathology: the anticipatory response, reactions to the test itself, and the recovery period. Research in animals suggests that the anticipation of aversive stimulation is associated with a different pattern of cardiovascular adjustment from that

present during the stimulation itself (32). The interval
following the termination of the stressor may also be
relevant, since breakdown in adaptation may be manifest
through delays in returning to initial levels. For
example, Brod (31) found that it took nearly 30 minutes for
BP to return to pre-test levels following a four minute
arithmetic challenge in hypertensives, compared with an
average of less than 10 minutes in normotensives. Similar
patterns have been observed in neuroendocrine parameters
(32), although Anderson et al (33) recently failed to
detect any differences in recovery rate.

The second issue which must be resolved in analysis is
the problem of baseline variation. It is well-established
that haemodynamic and neuroendocrine parameters are not
stable, but are sensitive to circadian, physical and
situational factors. The act of monitoring autonomic
variables or taking blood for the measurement of endocrine
parameters may itself provoke responses (29, Chapter 4).
Some authorities have recommended the measurement of basal
haemodynamic function only after overnight stay in hospital
under controlled conditions, but this is seldom
practicable. The issue is important in mental stress
testing, since preparation for the test procedure may
itself provoke cardiovascular reactions in some
individuals; reactions to the stressors themselves may then
be modified. It would seem that valid baselines from which
to assess reactions to psychological stressors cannot be
determined if subjects are assessed only on one occasion;
repeated measurement during sessions in which no tasks are
administered is also necessary.

A third controversial aspect of analysis is the method
of assessing responses. A variety of techniques are used,
including analysis of absolute levels, change scores in raw
or percentage forms, covariance analyses and residualised
scores. Groups cannot be compared using absolute scores
unless basal levels are identical. To say for example that
a group of hypertensives showed a mean BP of 170mm during

mental stress compared with 145mm in normotensives conveys no information about their reactivity, since baseline differences may have been present. The so-called 'tension' scores used by early psychophysiological researchers were compromised for this reason (see 35). Assessment of raw change scores may have similar problems. For example, a study by Shapiro et al (36) demonstrated greater mean BP changes amongst hypertensives in response to a number of psychological stimuli. However, re-calculation of these reactions as percentage changes from basal levels reveal few striking effects, due to resting differences. The absolute reactions of 30.1 and 19.6mm Hg to the cold pressor are equivalent to increases of 23.0% and 23.4% in hypertensives and normotensives respectively. The impression that hypertensive reactions are disproportionate is thus not confirmed. This issue has yet to be resolved satisfactorily by statisticians. Nevertheless, methods of analysis must take baseline differences into account, rather than rely on arbitary numerical advantage.

CONCLUSIONS

A wide range of psychological stress procedures are available to the clinician and researcher, and many of these can be administered in a standardised fashion so that responses may be quantified and compared. There is however no such thing as an archetypal mental stress test, since procedures differ on many significant dimensions, including degree of personal emotional involvement, behavioural reponse demands, underlying haemodynamic adjustments and neuroendocrine concomitants. Selection of appropriate tests depends on the nature of the subject population and the constraints imposed by physiological monitoring techniques.

Several aspects of mental stress testing are in urgent need of elaboration in future investigations. It is important to define more precisely the haemodynamic adjustments underlying cardiovascular reactions. Few advances have been made since Brod's classic investigations of changes in regional blood flow, although modern techniques of cardiovascular monitoring are beginning to be applied in this area (eg, 17, 18). Our understanding of the significance of these reactions will continue to be limited if investigators confined themselves to measurement of BP, HR and other general parameters.

A second important issue is the relevance of acute stress testing to the experiences of everyday life. Are the demands imposed on subjects, and the resultant cardiovascular reactions, typical of settings outside the laboratory? The increasing sophistication of ambulatory monitoring procedures is beginning to demonstrate the validity of acute stress tests as models of real life demands. But until the parallel is established, the significance of reactions to acute stressors for the aetiology of cardiovascular disorders must remain tentative.

REFERENCES
1. Cohen, D. H. and Obrist, P. A. 1975. Interactions between behavior and the cardiovascular system. Circulation Res. 37, 693-706
2. Friedmann, E., Thomas, S. A., Kulick-Ciuffo, D. et al. 1982. The effects of normal and rapid speech on blood pressure. Psychosom. Med. 44, 545-553
3. Lynch, J. J., Long, J. W., Thomas, S. A. et al. 1981. The effects of talking on the blood pressure of hypertensive and normotensive individuals. Psychosom. Med. 43, 25-33
4. Wolf, S., Cardon, P. V., Shepard, E. M. et al. 1955. Life Stress and Essential Hypertension. Baltimore, Williams and Wilkins
5. Adsett, C. A., Schottstaedt, W. W. and Wolf, S. G. 1962. Changes in coronary blood flow and other haemodynamic indicators induced by stressful interviews. Psychosom. Med. 24, 331-336
6. Theorell, T., Blunk, D. and Wolf, S. 1974. Emotions and cardiac contractility as reflected in ballistocardiographic recordings. Pavl. J. Biol. Sci. 9, 65-75
7. Groen, J. J., Hansen, B., Herrman, J. N. et al. 1982. Effects of experimental emotional stress and physical exercise on the circulation in hypertensive patients and control subjects. J. Psychosom. Res. 26, 141-154

8. Svensson, J. C. and Theorell, T. 1982. Cardiovascular effects of anxiety induced by interviewing young hypertensive male subjects. J. Psychosom. Res. 26, 359-170

9. Keane, T. M., Martin, J. E., Berler, E. S. et al. 1982. Are hypertensives less assertive? A controlled evaluation. J. Consult. Clin. Psychol. 50, 499-508

10. Holroyd, K. A. and Gorkin, L. 1983. Young adults at risk for hypertension: effects of family history and anger management in determining responses to interpersonal conflict. J. Psychosom. Res. 27, 131-138

11. Kallinke, D., Kulick, B. and Heim, P. 1982. Behaviour analysis and treatment of essential hypertensives. J. Psychosom. Res. 26, 541-550

12. Weiner, H., Singer, M. T. and Reiser, M. F. 1962. Cardiovascular responses and their psychological correlates. Psychosom. Med. 24, 477-498

13. Dembroski, T. M., MacDougall, J. M. and Lushene, R. 1979. Interpersonal interaction and cardiovascular response in Type A subjects and coronary patients. J. Human Stress. 5(4), 28-36

14. Krantz, D. S., Schaeffer, M. A., Davia, J. E. et al. 1981. Extent of coronary atherosclerosis, Type A behavior and cardiovascular response to social interaction. Psychophysiology. 18, 654-664

15. Steptoe, A. and Ross, A. 1982. Voluntary control of cardiovascular reactions to demanding tasks. Biofeedback Self-Regul. 7, 149-166

16. Manuck, S. B. and Schaefer, D. C. 1978. Stability of individual differences in cardiovascular reactivity. Physiol. Behav. 21, 675-678

17. McCubbin, J. A., Richardson, J. E., Langer, A. W. et al. 1983. Sympathetic neuronal function and left ventricular performance during behavioral stress in humans: the relationship between plasma catecholamines and systolic time intervals. Psychophysiology. 20, 102-110

18. Hollenberg, M. K., Williams, G. H. and Adams, D. F. 1981. Essential hypertension: abnormal renal, vascular and endocrine responses to a mild psychological stimulus. Hypertension. 3, 11-17

19. Manuck, S. B. and Garland, F. N. 1980. Stability of individual differences in cardiovascular reactivity: a thirteen month follow up. Physiol. Behav. 24, 621-624

20. Neus, H., Von Eiff, A. W., Friedrich, G. et al. 1981. Das Problem der Adaptation in der Klinisch-therapeutischen Hypertonieforschung. Dtsch. Med. Wschr. 106, 622-624

21. Obrist, P. A. 1981. Cardiovascular Psychophysiology. New York, Plenum Press

22. Melville, D. I. and Raftery, E. B. 1981. Blood pressure changes during acute mental stress in hypertensive subjects using the Oxford Intra-Arterial system. J. Psychosom. Res. 25, 487-498

23. Frankenhauser, M., Dunne, E. and Lundberg, U. 1976. Sex differences in sympathetic-adrenal medullary responses induced by different stressors. Psychopharmacologia. 47, 1-5

24. Newlin, D. B. and Levenson, R. W. 1982. Cardiovascular responses of individuals with Type A behavior pattern and parental coronary heart disease. J. Psychosom. Res. 26, 393-402

25. Glass, D. C., Krakoff, L. R., Contrada, R. et al. 1980. Effect of harassment and competition upon cardiovascular plasma catecholamine resonses in Type A and Type B individuals. Psychophysiology. 17, 453-463

26. Steptoe, A. In press. Stress, helplessness and control: the implications of laboratory studies. J. Psychosom. Res.

27. Turner, J. R., Carroll, D. and Courtenay, H. In press. Cardiac and metabolic responses to space invaders: an instance of metabolically-exaggerated cardiac adjustment? Psychophysiology.

28. Von Eiff, A. W., Freidrich, G. and Neus, H. 1982. Traffic noise, a factor in the pathogenesis of essential hypertension. Contr. Nephrol. 30, 82-86

29. Steptoe, A. 1981. Psychological Factors in Cardiovascular Disorders. London, Academic Press

30. Carruthers, M. and Taggart, P. 1973. Vagotonicity of violence: biochemical and cardiac responses to violent films and television programmes. Br. Med. J. III, 384-389

31. Brod, J. 1960. Essential hypertension: haemodynamic observations with a bearing on its pathogenesis. Lancet. II, 773-778

32. Baumann, R., Ziprian, H., Godicke, W. et al. 1973. The influence of acute psychic stress situations on biochemical and vegetative parameters of essential hypertensives at the early stages of the disease. Psychother. Psychosom. 22, 131-140

33. Anderson, D. E. and Tosheff, J. G. 1973. Cardiac output and total peripheral resistance changes during preavoidance periods in dogs. J. appl. Physiol. 34, 650-654

34. Anderson, C. D., Stoyva, J. M. and Vaughn, L. J. 1982. A test of delayed recovery following stressful stimulation in four psychosomatic disorders. J. Psychosom. Res. 26, 571-580

35. Gannon, L. 1981. The psychophysiology of psychosomatic disorders. In: Psychosomatic Disorders. Ed. Haynes, S. N. and Gannon, L. New York, Praeger

36. Shapiro, A. P., Moutsos, S. E., Krifcher, E. 1963. Patterns of pressor response to noxious stimuli in normal, hypertensive and diabetic subjects. J. Clin. Invest. 42, 1890-1898

METHODS AND LIMITS FOR THE DETECTION OF THE RESPONSE OF CORONARY CIRCULATION TO ACUTE STRESS.

A. BIAGINI, A. L'ABBATE

C.N.R. Institute of Clinical Physiology and Institute Patologia Medica, University of Pisa, Pisa, Italy.

1. INTRODUCTION

In order to design a clinical protocol to assess the possible role of stress in the genesis of acute myocardial ischemia, it is probably useful to define, in addition to the appropriate stressor tests, the applicability, significance and limits of the available techniques for the detection of ischemia. In fact the kind of stressor tests and protocol to be adopted will largely depend on the technique employed and viceversa.

Aim of this report is to review the different techniques that can be employed to assess the myocardial function during acute stressfull situations, the specificity and sensitivity of the measurements, the possibility of an active involvment of the patient during the procedure.

2. NON INVASIVE METHODS

2.1. Symptomatology

At the present time it is not clear yet which mechanisms are responsible for pain sensation during myocardial ischemia. We don't know either when the symptoms experienced by the patient are a sicure consequence of myocardial ischemia. It is well known that pain is a late phenomenon (1-2) and that only 25-30% of the ischemic episodes are painful (3-4). The specificity and the sensitivity of this "marker" of ischemia is therefore low. In presence of an objective documentation of ischemia, cardiac pain can be considered only a marker of the severity of the ischemia being the painfull episodes usually longer than the painless (5).

2.2. Electrocardiogram

Myocardial ischemia causes changes in the electrical properties of the myocardium which are detected in the surface electrocardiogram. Using a 12 leads standard apparatus it is possible to grossly identify the myocardial region undergoing ischemia and its transmural distribution. During the examination the patient has to lie or sit down quietly in order to avoid artifacts due to muscolar activity. In the recent years a system that allows 24-hour monitoring of 2 electrocardiographic leads in ambulant patient has became available (6). The validity of this method is well established for the detection of arrhythmias (7) while its usefulness in detecting myocardial ischemia is still controversial (8). The sensitivity of the electrocardiogram is probably not very high as not infrequently the surface electrocardiogram does not show any change in spite of well evident myocardial ischemia as detected by other techniques.

The specificity of what are considered to be the "major" signs of ischemia, like ST segment elevation or depression is certainly high, while the meaning of the "minor" changes, like T and U wave alterations or QRS amplitude variations, is more questionable especially in normal population as compared to patients with ascertained ischemic heart disease (9).

In spite of these limits, the electrocardiogram is still

to be considered the most usefull, easy, available and applicable technique for the detection of myocardial ischemia.

2.3. Echocardiography

This technique allows to appreciate changes in the contractility and in the dimension of the ventricular cavities that occur as consequence of myocardial ischemia (10). During the examination the patient has to lie down quietly in his bed breathing slowly while the ultrasonic transducer is firmly applied on his chest wall in different standardized positions. With this technique is possible to monitorize the cardiac function only for short periods of times (some minutes) however it is possible to repeat the measure whenever required.

The sensitivity and the specificity of this technique are very high, as an immediate loss of contractility of the ischemic regions occurs even before than metabolical or electrical changes become clinically apparent.

In same patient especially in those with lung emphysema, the procedure is inadequate due to the poor quality of the echoacustic window. In addition some region of the heart, particularly the apex, often are not well visualized even by a bidimensional apparatus. At the present time systems allowing a rapid quantification of the recorded parameters are not yet available.

2.4. Blood pool gating

This technique allows to appreciate total and regional variations in the volume of the ventricles for long periods of time (hours). In contrast to echocardiography this technique requires the use of a computer which makes also possible a rapid quantification of the parameters (11). Furthermore in the recent years more sophisticated algorhythms have been developed that have greatly enhanced the sensitivity of this technique (12).

During the examination the patient has to lie down quietly and it is necessary that he remains motionless. The activity in the myocardium is recorded by a gamma camera or by a scintillation counter (nuclear stethoscope). Using the latter equipment only information on changes in ventricular dimension but not of regional changes of contractility can be obtained (13).

Considering that with blood pool gating it is possible to appreciate regional changes of contractility its sensitivity and specificity is high similarly to echocardiography. However differently than echocardiography, only some myocardial regions, depending on the view used can be visualized during each acquisition (usually the septum, the apex and the lateral wall). Due to the time required for each acquisition only one projection can be used for detection of transient phenomena.

2.5. Scintigraphic studies with potassium analogs

In the recent years many different tracers has been used to assess regional blood flow distribution by external detection with a gamma camera. At the present time the most used tracer for clinical purpose is the potassium analog Thallium-201. Following the intravenous injection, the tracer accumulates into the myocardium according to the flow and the Na-K pump-rate (14). When injected during ischemia the early scintigraphic images reflect the myocardial blood flow distribution at the time of the injection (15). Using early sequential multiple views, it is possible to spatially define the underperfused myocardial region.

During the acquisition time (about 2-3 minutes per view) the patient has to lie down motionless on his bed. The images obtained can be read directly or after computerized process that minimally enhance the sensitivity of this technique. A tomographic approach allows a more precise definition of the myocardium (16).

The procedure can be repeated in the same patient but not before 6-7 days due to the relatively long half-life of the tracer (approximately 4 days).

2.6. Regional myocardial metabolic studies

In addition to changes in regional perfusion it is possible to assess regional changes in the metabolism of glucose (17) or of fatty acids (18) during ischemia by nuclear techniques and the intravenous injection of special radiopharmaceuticals. Using tomographic techniques the spatial resolution can be greatly enhanced, however these methods are still far for being clinically applicable and are mainly confined to research studies.

3. INVASIVE METHODS
3.1. Angiography

A direct anatomical evaluation of the coronary vessels is obtained by injection of contrast medium in the coronary arteries. Multiple injections of the vessel under study can be performed so that it is possible to visualize the changes in vascular lumen during an ischemic episodes (19). During the procedure the patient has to lie down but he can be asked to actively participate to a test even with his arms or legs. Coronary angiography usually takes less than half an hour to be performed but if necessary it can last much longer. It is not a repeatable techniques unless required by clinical evaluation.

It appears feasible to use coronary angiography to assess the effect of stressor tests on the epicardial vessels. It could be used in conjunction with coronary sinus flow detection and echocardiography in order to obtain information both on myocardial blood flow and on regional contractility.

3.2. Left ventricular pressure monitoring

Continuous recording of left ventricular pressure allows to obtain information on changes in systolic, diastolic and dP/dt (an index of the overall contractility of the left ventricle) during ischemia.

Changes in these parameters are related to the entity of left ventricular impairment. By this technique no information on regional ventricular function can be obtained (20). The sensitivity of the pressure monitoring depends on the amount of myocardium that became functionally inpaired, as this technique explores the global function of the left ventricular pump.

The limits are the invasivity of the method, the necessity to continuously perfuse the catheter, the risk of arrhythmias or tromboembolic accidents. For this reasons usually the monitoring lasts not longer than some hours. In the recent years catheters that do not need to be perfused have been developed. By these catheters relative changes of cardiac output can be also assessed (21).

During the procedure the patient has to lie down in its bed near a pressure transducer connected to an amplifier The pressure signals usually are stored on magnetic tapes

or directly on a computer; they can be processed analogic-
ally or by computerized programs (22). Usually during left
ventricular pressure monitoring also the right pressure are
recorded (see below). In especially equipped centers it is
also possible to perform at the same time echocardiographic
studies or perfusion studies by nuclear techniques (see
below).

3.3. Right ventricular pressure monitoring

Usually a Swan-Ganz catheter is introduced via an ante-
cubital vein and advanced into the pulmonary artery. The
procedure can be performed at the bed- side. Pressure from
the right atrium and the pulmonary artery can be recorded
simultaneously, while inflating a balloon pulmonary wedge
pressure is obtained. In addition the cardiac output can be
measured via thermodilution method. While the information
obtained by this procedure are clinically relevant in
seriously ill patients, they are less appropriate for the
study of patients with transient ischemia unless associated
with left ventricular pressure monitoring.

3.4. Coronary sinus catheterization

Via a vein of the arm it is possible to catheterize the
coronary sinus or the great cardiac vein or both. The
catheter can be left in situ for long periods of time (24
hours). Spoiling of the arm vein due to frequent using the
percutaneous technique, prevents the repetition of the
procedure. It is possible to withdraw blood by the
catheter during an ischemic episode and to measure the
production of lactate and discharge of potassium which are
specific indicators of myocardial ischemia (25).

At the present time catheters are available for coronary venous blood oxygen saturation monitoring an indirect parameter of myocardial blood flow (24). Moreover by thermodilution catheter it is possible to perform serial measurement of coronary sinus or great cardiac vein flow thus obtaining information on regional (anterior vs posterior wall) or total myocardial blood flow.

This technique can easily be used in combination with other methods capable to appreciate changes in contractility (echocardiography, blood pool gating) or in regional myocardial blood flow (nuclear techniques).

3.5. Myocardial scintigraphy with radiolabeled microspheres

Radiolabeled serum human albumin microspheres injected in the left ventricle distribute to each organ according to its flow. A certain amount of microspheres will then enter the coronary bed and, due to their size, they will stop at the arteriolar level. Imaging by the gamma camera of microspheres distribution will give information on regional myocardial blood flow at the time of the injection. Using computerized ECG-gated acquisition it is also possible to get information on myocardial contractility during all the time the heart wall remains labeled by the microspheres (several hours) (25). It is possible in the same study to perform two injections of microspheres in two different conditions (i.e. basally and during ischemia) using microspheres labeled with two different isotopes, although for technical reasons the results are not so satisfactory. The major limit of this method, which has many advantages on Thallium-201 scintigraphy, is its invasivity.

1056

Table 1. Methods for measuring coronary blood flow in man.

Tracer	Method	Total Flow or Flow/g	Regional Flow absolute value	Regional Flow distribution	Repetitive Measurements	Catheterization Cor. sinus	Catheterization left
Microspheres	Fractionation	No	No	Yes	Yes (limit)	No	Yes
K, Rb, Tl...	Fractionation	No	No	Yes	No	No	No
Heat	Dilution curves	Yes	Yes	No	Yes	Yes	No

3.6. Invasive assessment of myocardial perfusion by short life isotopes

The introduction of miniciclotrons for medical purposes has made possible the use of isotopes with very short half life (seconds or minutes). Diffusible tracers, like Kripton 81m, are continuously delivered in the ascending aorta via a special shaped catheter. The tracer distributes to the myocardium according to its blood flow (26). It is therefore possible to monitorize changes in myocardial blood flow for all the time the tracer is injected. During the procedure the patient has to lie down in the bed. The activity recorded on the precordium is usually stored on computer disk and processed by computerized programs. Due to the short half-life of the tracer, myocardial blood flow can be monitorized for quite long periods of time (hours), but the procedure due to its invasivity cannot be repeated. This is the only available method which allows a relatively long lasting monitoring of myocardial regional blood flow, although the problems arising from the need of a constant delivery of tracer to each coronary artery have not been entirely solved.

4. CONCLUSION

In conclusion we have tried to briefly review the available methods for the direct or indirect detection of myocardial ischemia. In practice for the great majority of them the patient is limited in movement and stressor tests which require the use of the hands appear hardly applicable. In these cases information processing tasks, such as solving arithmetic problems or answer to knowledge quittes or passive tests can easily be performed. By contrast under electrocardiographic and even hemodynamic monitoring less constraints are imposed to the patient and a larger number of tests are applicable. Apart from the procedure constraints, the choice of the technique to be employed and the appropriate test will be largely dependent on the parameters one wants to measure. This will be relevant especially in the design of protocols designed to understand the mechanism underlying an acute pathological cardiovascular response to certain stressor. Measurements of other parameters, such as coronary blood flow, large coronary artery diameter, or regional myocardial contractility in addition to the traditional measurement of heart-rate and blood pressure will certainly enhance our understanding of the pathogenetic mechanisms and significance of cardiovascular response to acute stressors.

ACKNOWLEDGMENTS

We are deeply grateful to all the members of the Coronary Group of the Clinical Physiology Institute, who have helped us in collecting the material for this report.

Moreover the authors gratefully acknowledge the co-operation of Miss Emanuela Campani, Miss Daniela Banti e Mrs Hilda Biagini de Ruyter, for assistance in manuscript preparation.

REFERENCES
1. Chierchia S, Lazzari M, Simonetti I, Maseri A. 1980. Hemodynamic monitoring in angina at rest. Herz, 5, 188.
2. Distante A, Rovai D, Palombo C, Maseri A, L'Abbate A. 1981. Diagnosi di ischemia miocardica transitoria nell'uomo mediante ecocardiografia M-mode. Giorn. It. Cardiol., 22, 1252.
3. Maseri A, Severi S, De Nes M, L'Abbate A, Chierchia S, Marzilli M, Ballestra AM, Parodi O, Biagini A, Distante

1059

A. 1978. "Variant" angina: one aspect of a continuous spectrum of vasospastic myocardial ischemia. Pathogenetic mechanisms, estimated incidence and clinical and coronary arteriographic findings in 138 patients. Am. J. Cardiol., 42, 1019.

4. Biagini A, Testa R, Carpeggiani C, Mazzei MG, Emdin M, Michelassi C, L'Abbate A. 1982. Contributo dell'elettrocardiografia dinamica alla conoscenza della cardiopatia ischemica silente. In: "La Cardiopatia Ischemica Silente", Prati PL, ed, p 81.

5. Chierchia s, Lazzari M, Freedman B, Brunelli C, Maseri A. 1983. Impairment of myocardial perfusion and function during painless myocardial ischemia. J. Am. Coll. Cardiol., 30, 924.

6. Holter NJ. 1961. New method for heart studies. Continuous electrocardiography of a life subject over long periods is now practical. Science, 184, 1214.

7. Harrison DC, Fitzgerald JW, Winkle RA. 1976. Ambulatory electrocardiography for diagnosis and treatment of cardiac arrhythmias. N. Engl. J. Med. 294, 373.

8. Bragg-Remschel DA, Anderson CH, Winkle RA. 1982. Frequency response characteristics of ambulatory ECG monitoring system and their implications for ST segment analysis. Am. Heart J., 103, 20.

9. Biagini A, Carpeggiani C, Mazzei MG, Testa R, L'Abbate A. 1982 (in press). Significance of electrocardiographic changes in acute transient myocardial ischemia. In Ambulatory monitoring. Cardiovascular system and allied applications, Marchesi C, ed., Martinus Nijhoff BV Publisher.

10. Distante A, Rovai D, Picano E, Moscarelli E, Palombo C, Morales MA, Michelassi C, L'Abbate A. (in press). Transient changes in left ventricular mechanics during attacks of Prinzmetal angina: an M-mode echocardiographic study. Am. Heart J.

11. Strauss HW, Pitt B. 1977. Gated cardiac blood pool scan: use in patients with coronary heart disease. Prog. Cardiovasc. Dis., 20, 207.

12. Ratib O, Eberhard H, Schon H, et al. 1982. Phase analysis of radionuclide ventriculograms for the detection of coronary artery disease. Am. Heart J., 104, 1.

13. Wagner HN, Jr, Wake R, Nickoloff E, Natarajan TK. 1976. The nuclear stethoscope: a simple device for generation of left ventricular volume curves. Am. J. Cardiol., 38, 747.

14. L'Abbate A, Biagini A, Michelassi C, Maseri A. 1979.

Myocardial kinetics of Thallium and Potassium in man. Circulation, 60, 776.

15. Maseri A, Parodi O, Severi S, Pesola A. 1976. Transient transmural reduction of myocardial blood flow demonstrated by Thallium-201 scintigraphy as a cause of variant angina. Circulation, 56, 280.

16. Schelbert HR, Phelps ME, Hoffman E. 1980. Regional myocardial blood flow, metabolism and function assessed non-invasively with positron emission tomography. Am. J. Cardiol., 46, 1269.

17. L'Abbate A, Camici P, Trivella MG, Pelosi G. 1981. Regional myocardial glucose utilization assessed by (^{14}C) deoxiglucose. Basic Res. Cardiol., 76, 394.

18. TerPogossian MM, Klein MS, Markham J, Roberts R, Sobel BE. 1980. Regional assessment of myocardial metabolic integrity in vivo by positron-emission tomography with 11C-labeled palmitate. Circulation, 61, 242.

19. Maseri A, L'Abbate A, Pesola A, Ballestra AM, Marzilli M, Severi S, Maltinti G, De Nes M, Parodi O, Biagini A. 1977. Coronary vasospasm in angina pectoris. Lancet, 1, 713.

20. Maseri A, Mimmo R, Chierchia S, Marchesi C, Pesola A, L'Abbate A. 1975. Coronary artery spasm as a cause of acute myocardial ischemia in man. Chest, 68, 625.

21. Millar HD, Baker LE. 1973. A stable ultraminiature catheter-tip pressure transducer. Med. Biol. Eng. 2, 86.

22. Marchesi C, Chierchia S, Maseri A. 1977. Left and right ventricular pressure monitoring in CCU. Method and significance. IEEE Proc. Computers in Cardiology. Rotterdam, 579.

23. Gorlin R. 1972. Assessment of hypoxia in the human heart. Cardiology, 57, 24.

24. Chierchia S, Brunelli C, Simonetti I, Lazzari M, Maseri A. 1980. Sequence of events in angina at rest: primary reduction in coronary flow. Circulation, 61, 759.

25. Parodi O, Bencivelli W, Camici P, Marzullo P, Davies JG, Maseri A, L'Abbate A. 1981. A new technique for the simultaneous assessment of myocardial perfusion and contractility in man. Am. J. Cardiol., 47, 394.

26. Turner JH, Selwyn AP, Jones T, Evans TR, Raphael MJ, Lavender JP. 1976. Continuous imaging of regional myocardial blood flow in dogs using Krypton-81m. Cardiovasc. Res., 10, 398.